Society of Earth Scientists Series

Series Editor

Satish C. Tripathi, Lucknow, India

The Society of Earth Scientists Series aims to publish selected conference proceedings, monographs, edited topical books/text books by leading scientists and experts in the field of geophysics, geology, atmospheric and environmental science, meteorology and oceanography as Special Publications of The Society of Earth Scientists. The objective is to highlight recent multidisciplinary scientific research and to strengthen the scientific literature related to Earth Sciences. Quality scientific contributions from all across the Globe are invited for publication under this series. Series Editor: Dr. Satish C. Tripathi

Bandana Samant · Deepali Thakre
Editors

Applications of Palynology in Stratigraphy and Climate Studies

Springer

Editors
Bandana Samant
Present Address:
Department of Geology
School of Environment and Earth Sciences
Central University of Punjab
Bathinda, India

Postgraduate Department of Geology
RTM Nagpur University
Nagpur, India

Deepali Thakre
Postgraduate Department of Geology
RTM Nagpur University
Nagpur, India

ISSN 2194-9204 ISSN 2194-9212 (electronic)
Society of Earth Scientists Series
ISBN 978-3-031-51876-8 ISBN 978-3-031-51877-5 (eBook)
https://doi.org/10.1007/978-3-031-51877-5

This Springer imprint is published by the registered company Springer Nature Switzerland AG
The registered company address is: Gewerbestrasse 11, 6330 Cham, Switzerland

Paper in this product is recyclable.

Dedicated to my esteemed Ph.D. supervisor Dr. N. R. Phadtare, Scientist (retd.), Wadia Institute of Himalayan Geology, Dehradun. I always admire him as a great teacher and a mentor for his unwavering inspiration and guidance all through my academic career.

With immense gratitude
Bandana Samant

Preface

Palynology, which deals with the study of organic walled microfossils has now emerged as a time-tested tool in stratigraphy, reconstruction of past vegetation, understanding depositional environments and climate. This book deliberates an overview of the role palynology plays in solving various geological problems, right from the Precambrian to Recent. Precambrian palynology is significant in understanding the origin of life on the earth, biostratigraphy, and the early atmosphere. The fluvio-lacustrine Gondwana sediments have rich coal deposits. Palynological studies of Gondwana sediments help in age determination and coal seam correlation. Palynology also plays a key role in understanding the changing climate and depositional environment of the Gondwana Supergroup. The intertrappean sediments of the Early Cretaceous Rajmahal Traps are significant in understanding *Ptillophyllum* flora and the evolution of the early angiosperm on the Indian subcontinent. Late Cretaceous-Early Palaeocene Deccan volcanic activity is an important geological event on the Indian subcontinent. This massive volcanic event is considered as one of the causes of the mass extinction of biota across the globe. The Deccan volcanic associated palynoflora rich infra and intertrappean sediments are useful in understanding floral diversity, age, depositional environments, and paleoclimate before the onset of volcanic activity as well as during volcanic activity. The palynoflora of this time is also useful in understanding the origin and evolution of many angiosperm families and tracking palynofloral changes across the Cretaceous-Palaeocene boundary. Post-Deccan Paleogene palynology has played a pivotal role in deciphering the floral change associated with the equatorial latitudinal position of India and understanding their age and depositional environments. Significantly, the palynoflora of the Palaeocene-Eocene deposits of the Indian subcontinent show similarity with the present day tropical and subtropical flora of Africa and Southeast Asia.

The role of palynology in comprehending climate change in the pre-industrial Common Era is enormous. This period has seen global climate events that have influenced socio-economic events and the ancient Indian economy. Alike, the significance of palynology in deciphering the impact of anthropogenic activity on the environment is quite remarkable. These topics have been thoroughly analysed and covered in this book. The study of modern pollen rain and its taxonomic characterization helps in knowing the climatic conditions of any region. This methodology can be effectively used to understand past climatic conditions. Another emerging aspect of palynology is its utilization in the field of archaeology. These days, archaeobotany and palynology are actively used in decoding floral biodiversity, paleoclimate, and human-plant relationships. Organic geochemistry and its use as a hydrocarbon biomarker are emerging fields in petroleum exploration. Organic matter in plants acts as a biomarker for elucidating the source of organic matter and past depositional environment interpretation. Details of organic geochemistry techniques and their role in understanding the paleo-vegetation of the Cambay Basin of western India are outlined in this book. Last but not the least, coal and petroleum resources are the backbone of energy resources and the economic development of any country and the paper highlights the role of palynology in the exploration of coal and petroleum.

Additionally, this book brings into focus the contribution of Indian palynologists in understanding the past floral diversity, climate, and depositional environments of the Indian subcontinent.

Nagpur, India

Bandana Samant
Deepali Thakre

Acknowledgements

First and foremost, my deepest gratitude to my Ph.D. supervisor, Dr. N. R. Phadtare, Scientist (retd.), Wadia Institute of Himalayan Geology, Dehra Dun for introducing me to the fascinating world of palynology. His mentorship and expertise helped in my academic journey and passion for palynology.

My deepest gratitude to Dr. D. M. Mohabey, Director General and HOD (retd.), Geological Survey of India for inspiring and encouraging me to work on Deccan palynostratigraphy. His consistent encouragement has been a driving force behind my research.

I extend my sincere appreciation to all my research students including Dr. Hemant Sonkusare, Anup Dhobale, Neha Thakre, Abhilash Sen, Akash Karande, and Dr. Apoorva Sitholey for their invaluable assistance during the compilation of this book.

I am thankful to the Head Department of Geology, RTM Nagpur University, for providing working facilities which have been important for conducting the research work.

I am deeply indebted to all the reviewers for meticulously going through the manuscript, offering valuable feedback and suggestions. Their expertise and careful examination have been invaluable in refining the papers.

I thank each and every author and contributor for submitting chapters for the book. Their valuable contributions helped significantly to the diversity of the contents presented.

Finally, my thanks are due to all my family members, for their patience and continuous encouragement during the compilation of this book.

This book would not have been possible without the support and contribution from each one of you.

Bandana Samant

Contents

Editors and Contributors

About the Editors

Prof. Bandana Samant obtained her M.Sc. degree in geology from H. N. Bahuguna Garhwal University, Uttarakhand, India. She did her Ph.D. in palynology of lignite deposits of Gujarat from the Wadia Institute of Himalayan Geology, Dehra Dun, Uttarakhand. Dr. Samant carried out postdoctoral work as a Research Associate, Senior Research Associate, and Research Scientist under the projects awarded by the Department of Science and Technology, and the Council of Scientific and Industrial Research. She has worked as a faculty at the Banaras Hindu University, Varanasi and RTM Nagpur University, Nagpur. Presently working at the Department of Geology, School of Environment and Earth Sciences, Central University of Punjab, Bathinda, Punjab, India. So far, she has completed seven research projects awarded by national funding agencies. She has published about 63 research papers, including one in 'Nature Communication'. She also worked as a collaborator in National Science Foundation, USA funded research projects. She was a member of the National Working Group of the prestigious International Geoscience programme, IGCP-609. Since last 20 years, she has been working on the palynoflora of Deccan volcanic associated sediments to understand the response of the flora to Deccan volcanic eruptions. Dr. Samant has successfully guided three research students for Ph.D.

Dr. Deepali Thakre completed her Postgraduate degree in geology from the Postgraduate Department of RTM Nagpur University. She did Ph.D. work on the palynological study of Deccan volcanic associated sediments of Chhindwara district, MP, for assessing the effect of volcanism on contemporary flora towards the K-Pg boundary, from RTM Nagpur University. She worked as a Junior Geologist in the Directorate of Geology and Mining (DGM). Prior to her appointment in DGM, she was a Junior Research fellow at the Space Application Centre-ISRO, Ahmedabad, Gujarat and in SERB-DST sponsored project at RTM Nagpur University, Nagpur. Her research work is focused on understanding the environmental and climate perturbation brought by the Deccan volcanic eruption, its effect on flora, and the dynamics of palynofloral turnover close to the Cretaceous-Paleogene boundary, especially in terrestrial environments.

Contributors

Subir Bera Centre of Advanced Study, Palaeobotany-Palynology Laboratory, Department of Botany, University of Calcutta, Kolkata, India

Oindrila Biswas Centre of Advanced Study, Palaeobotany-Palynology Laboratory, Department of Botany, University of Calcutta, Kolkata, India;
Department of Botany, School of Basic Sciences, SRM University Sikkim, Gangtok, India

Neelam Das Birbal Sahni Institute of Palaeosciences, Lucknow, India

Suryendu Dutta Department of Earth Sciences, Indian Institute of Technology Bombay, Mumbai, India

Suyash Gupta Birbal Sahni Institute of Palaeosciences, Lucknow, India

Ratan Kar Birbal Sahni Institute of Palaeosciences, Lucknow, Uttar Pradesh, India

Suraj Kumar Birbal Sahni Institute of Palaeosciences, Lucknow, India;
Academy of Scientific and Innovative Research (AcSIR), Ghaziabad, India

Krishna Kali Mandal Dakshin Chobbish Pargana Protna Itihas Gobeshona Kendra, Dhopagachi, Baruipur, West Bengal, India

Nivedita Mehrotra Birbal Sahni Institute of Palaeosciences, Lucknow, Uttar Pradesh, India

D. M. Mohabey Postgraduate Department of Geology, RTM Nagpur University, Nagpur, Maharashtra, India

Srikanta Murthy Birbal Sahni Institute of Palaeosciences, Lucknow, India

Deepti Niyolia Department of Earth Sciences, Indian Institute of Technology Bombay, Mumbai, India

Dipak Kumar Paruya Centre of Advanced Study, Palaeobotany-Palynology Laboratory, Department of Botany, University of Calcutta, Kolkata, India

Sankar Suresh Kumar Pillai Birbal Sahni Institute of Palaeosciences, Lucknow, India;
Academy of Scientific and Innovative Research (AcSIR), Ghaziabad, India

Mohammad Firoze Quamar Birbal Sahni Institute of Palaeosciences, Lucknow, Uttar Pradesh, India

Rajeev Ranjan Birbal Sahni Institute of Palaeosciences (BSIP), Uttar Pradesh, Lucknow, India;
Babasaheb Bhimrao Ambedkar University (A Central University) (BBAU), Uttar Pradesh, Lucknow, India

Mrutyunjaya Sahoo Department of Geology, Ravenshaw University, Cuttack, Odisha, India

Sumit Kumar Sahu Department of Earth Sciences, Indian Institute of Technology Bombay, Mumbai, India

Bandana Samant *Present Address:* Department of Geology, School of Environment and Earth Sciences, Central University of Punjab, Bathinda, Punjab, India;
Postgraduate Department of Geology, RTM Nagpur University, Nagpur, Maharashtra, India

Binod Saradar Centre of Advanced Study, Palaeobotany-Palynology Laboratory, Department of Botany, University of Calcutta, Kolkata, India

Anju Saxena Birbal Sahni Institute of Palaeosciences, Lucknow, India

Abhilash Sen Postgraduate Department of Geology, RTM Nagpur University, Nagpur, Maharashtra, India

Santosh K. Shah Birbal Sahni Institute of Palaeosciences, Lucknow, Uttar Pradesh, India

Yogesh Pal Singh Birbal Sahni Institute of Palaeosciences, Lucknow, India;
Centre of Advanced Studies in Geology, University of Lucknow, Lucknow, India

Deepali Thakre Postgraduate Department of Geology, RTM Nagpur University, Nagpur, Maharashtra, India

Neha Thakre Postgraduate Department of Geology, RTM Nagpur University, Nagpur, Maharashtra, India

Swati Tripathi Birbal Sahni Institute of Palaeosciences (BSIP), Uttar Pradesh, Lucknow, India

Poonam Verma Birbal Sahni Institute of Palaeosciences (BSIP), Lucknow, India

Precambrian Microfossils: Indicators of Early Life and Environments on the Earth

Bandana Samant

Abstract Precambrian biota, which primarily includes bacteria, prokaryotes, protists, algal-stromatolites, organic- walled microfossils (micro and macrofossils) and trace fossils, plays a significant role in understanding early life as well as their diversification on the Earth. In addition, they are also used in biostratigraphy, to understand the depositional environment, early biosphere, and atmosphere. A review of Precambrian biota, recovered from the Archean, and Proterozoic sedimentary rocks of India, reveals the presence of diverse assemblages of organic remains. These remains play a significant role in understanding Indian Precambrian biostratigraphy and early life.

Keywords Precambrian · Microfossils · Organic walled microfossils · Evolution · Biostratigraphy

1 Introduction

The Precambrian Era, which is broadly divided into the Archean (4000–2500 Ma) and Proterozoic (2500-541 Ma), is the period when simple life forms evolved on the earth. Over billions of years of the evolutionary cycle, simple life evolved into complex forms (Woese et al. 1990). Previously, records of microscopic life from the Precambrian sediments were viewed with doubt and scepticism. However, current studies and records of numerous biotic remains of the Precambrian viz, bacteria, protists, stromatolites, organic walled microfossils, and biomarkers, are now considered an important tool in biostratigraphy and understanding the early evolution of life and depositional environments (Sharma et al. 2021).

Present Address:
B. Samant (✉)
Department of Geology, School of Environment and Earth Sciences, Central University of Punjab, Bathinda 151401, Punjab, India
e-mail: bandanabhu@gmail.com

Postgraduate Department of Geology, RTM Nagpur University, Nagpur 440010, Maharashtra, India

B. Samant and D. Thakre (eds.), *Applications of Palynology in Stratigraphy and Climate Studies*, Society of Earth Scientists Series, https://doi.org/10.1007/978-3-031-51877-5_1

Stromatolites are the oldest Precambrian organo-sedimentary structures and they are thought to be an indicator of life in Precambrian sediments. Stromatolites found in the North Port Chert Member of the Dresser Formation are the best preserved, and probably the oldest biogenic structures (Djokic et al. 2017; Baumgartner et al. 2019). The oldest accepted fossilized microorganisms have been reported from the 3770/4280 Ma Nuvvuagittuq Belt of Quebec, Canada (Dodd et al. 2017). Another significant microbial sedimentary structures from the Precambrian rocks are Microbially Induced Sedimentary Structures (MISS), recorded by Noffke (2000; Noffke et al. 2001a, b). So far, 17 different types of MISS have been reported. These forms are widely known from Australia (Noffke et al. 2013), Africa (Noffke et al. 2006; 2013) and Greenland (Retallack and Noffke 2018). Holland (2006) proposed the significance of these microfossils in assessing the oxygen level of the primitive atmosphere. Overall, the Precambrian biota plays an important role in understanding evolution and diversification of life in the present-day earth's biosphere. This chapter discusses the Indian Archeozoic and Proterozoic micro- and macrofossil records, as well as their biostratigraphic significance and the role they play in deciphering the depositional environments.

2 Indian Archeozoic Microfossils

The record of microfossils from the Archeozoic cratonic rocks is scarce. Archean microbial mats, filamentous microfossils, and stromatolites are recorded from the Dharwar and Singhbhum Cratons (summarised in Sharma et al. 2021). Recently, stromatolite like structures called "Egg Craton" were recorded from the Dharwar Craton (Shukla and Sharma 2020a), it is the second occurrence of such structures in the world (Allwood et al. 2006).

3 Indian Proterozoic Microfossils

Proterozoic sedimentary basins of the Indian shield are mainly the Bijawar, Sonrai, Gwalior, Abujhmar, Bhima, Chhattisgarh, Cuddapah, Khariar, Ampani, Indravati, Sukma, Pranhita-Godavari, Kaladgi, Bhima, and Vindhyan (Ramakrishnan and Vaidyanathan 2008). A variety of MISS have been recorded from the Proterozoic rocks of northern India (Banerjee and Jeevankumar 2005; Banerjee et al. 2006, 2010; Raghav et al. 2005; Kumar and Pandey 2009; Kumar and Ahmad 2014 etc.). These structures have also been recorded from the Cuddapah Basin, suggesting a shallow marine, reducing depositional environment (Goswami et al. 2017). The oldest record of Palaeoproterozoic stromatolites is from the carbonate sedimentary rocks of the Cuddapah Supergroup (King 1872). Later on, Loon and Mazumder (2013) recorded Palaeoproterozoic stromatolites from the Iron Ore Series rocks of Singhbhum craton.

Many workers have reported various forms of stromatolite from the Cuddapah Supergroup (summarised by Sharma et al. 2021). According to Khelen et al. (2017), the various shapes of stromatolites are related to the interaction between mat growth, current structures, and sediment supply. They (op. cit.) also used them to reconstruct marine depositional environments, such as shallow shelf (marginal, inner and outer), open ocean, and anoxic conditions. The carbon and oxygen isotope studies of stromatolites have been used for understanding the paleotemperature of the Vampalli Formation of the Cuddapah Group (Chakrabarti et al. 2011). It is noteworthy that the Indian Proterozoic sedimentary rocks show a large amount of phosphatic deposition, and this phosphatic concentration is attributed to the scavenging of the microbial carpets from the sea water (Banerjee et al. 1986). Salujha et al. (1972a, b) reported the various species of *Leiosphaeridia* and *Kildinosphaera.* Shukla (2011) described Neoproterozoic biofils from Bhima Formation. Shukla and Sharma 2020b proposed that the biofilms facilitate precipitation of phosphorous and iron rich mineral forming bands. Carbonaceous shales associated with stromatolites have yielded a good number of organic walled microfossils (Jankauskas et al. 1989) that have helped in understanding the evolution and diversification of Precambrian life (Knoll 1996; Sergeev et al. 2008, 2012) (Fig. 1).

Organic walled microfossils (OWM) include cyanobacterial and eukaryotic fossil remains. The Great Oxidation Event (GOE) of 2.45–2.32 GA is an important period in the earth's history as it changed the pace of life on the earth (Schirrmeister et al. 2013) and caused the diversification of cyanobacteria. On the Indian subcontinent, diversification of Organic walled microfossils is also observed in the Mesoroterozoic and Neoproterozoic sediments. A good number of OWM have been recorded from the Vindhyan, Chhattisgarh, Kurnool, Bhima, Kaladgi and Krol Belt of Lesser Himalaya (Kumar and Srivastava 1995; Moitra 1999; Prasad et al. 2005; Sharma 2006; Sharma et al. 2009; 2016; Singh and Sharma 2014, 2016, 2019; Singh et al. 2011; Sharma and Shukla 2012a, b; Pandey and Kumar 2013; Shukla and Tiwari 2014).

Acritarchs, an important tool for the biostratigraphy of Proterozoic rocks, especially *Tappania* and *Valeria,* from Australia, (Javaux et al. 2001; Lamb et al. 2009; Knoll 2014; Adam et al. 2017) and China (Yin 1997), are also recorded from India. Various species of *Tappania* have been documented from the Bahraich Group in Ganga Basin (Prasad and Asher 2001), the Semri Group (Lower Vindhyan) from Vindhyan Basin (Prasad et al. 2005, 2007) and the Saraipali Formation (Singhora Group), Chhattisgarh Basin (Singh et al. 2019c). Acritarchs are also recorded from the Chhattisgarh Supergroup (Singh et al. 2019a, b, c), Vindhyan Supergroup (Maithy 1992; Venkatachala et al. 1996; Prasad et al. 2005, 2007; Prasad 2007), and subsurface sediments of the Marwar Supergroup, Bikaner-Nagaur Basin (Prasad et al. 2010) and Ganga Basin (Prasad and Asher 2001). *Tappania planta,* a eukaryotic fossil, and the botuliform microfossil *Jacutianema solubila* have been recovered from the Chhattisgarh Basin (Singh et al. 2016, 2019c), Ganga Basin (Prasad and Asher 2001) and the Vindhyan Basin (Prasad et al. 2005, 2007; Singh and Sharma 2016; Asher et al. 2017; Singh et al. 2019a). Tonian age marker acritarchs, *Trachyhystrichopshaera* have also been recorded from the subsurface sediments of the Ganga Basin (Prasad and Asher 2001; Asher et al. 2017; Tang et al. 2017).

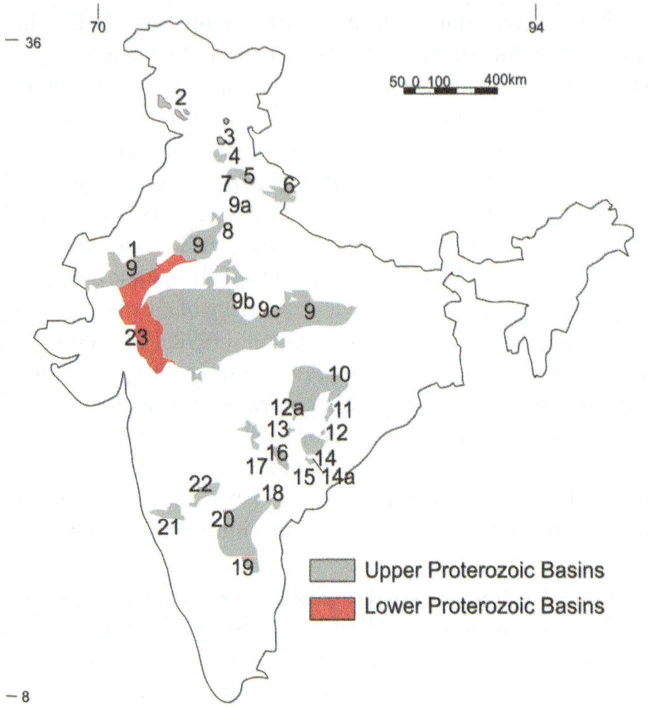

Fig. 1 Map showing Proterozoic basins of India, 1. Marwar Supergroup, 2. Jammu Lst, 3. Larji Fm., 4. Shali Fm., 5. Deoban Fm., 6. Pittoragarh Fm., 7. Sataun Lst., 8. Delhi Supergroup, 9. Vindhyan Supergroup, 9a. Gwalior Basin, 9b. Bijawar Basin, 9c. Sonrai Basin, 10. Chhattisgarh Basin, 11. Khariar Basin, 12. Ampani Basin, 12a. Keskal and Singanpur outliers, 13. Abujmar Basin, 14. Indravati Basin, 14a. Chedrapal outlier, 15. Sabari/Sukma Basin, 16. Pranhita Godavari Basin, 17. Pakhal Basin, 18. Palnad Basin, 19. Cuddapah Basin, 20. Kurnool Basin, 21. Kaladgi Basin, 22. Bhima Basin, 23. Aravalli Supergroup (modified after Sharma et al. 2020; Kale 1991)

Cyanobacterial microfossils are recorded from the Aravalli Group (Banerjee 1974), Cuddapah, and Karnool basins (Salujha et al. 1972a, b; Schopf and Prasad 1978; Gururaja et al. 1979; Gururaja and Rao 1980; Sharma and Shukla 2012a, 2012b, 2016; Shukla and Sharma 2020a). A very large size coccoidal microfossils have also been recorded from the Vindhyan Supergroup (Shukla and Sharma 2016; Sharma and Shukla 2019a). Heterocysts and akinetes are known from many Paleoproterozoic and Mesoproterozoic rocks of India (Sergeev et al. 2012; Sharma and Shukla 2019a). The oldest recorded eukaryotic fossil from India, *Shuiyousphaeridium echinulatum*, is from the Semri Group in Vindhyan (Kumar and Srivastava 1991; Anbarasu 2001; Singh and Sharma 2014). Rhodophytes are recorded from the Son valley of Chitrakoot (Bengtson et al. 2017; Sallstedt et al. 2018). Remains of *Grypania,* an oldest eucaryotes which evolved around 1.9–1.8 Ga (Han and Runnegar 1992) have also been recorded from the Rohtas Group of Vindhyan Supergroup (Sharma and Shukla 2009).

In addition, carbonaceous microfossils are also recorded from the Indian Proterozoic rocks. Algal carbonaceous fossils are recorded from the Vindhyan, Chhattisgarh, Kurnool, and Bhima formations (Sharma and Singh 2019b). Late Proterozoic Chuarids (*Chuaria* and *Tawuid*) and other coccoid microfossils are recorded from the Kurnool Group (Das Sharma et al. 1992; Ashok Kumar et al. 1993; Sharma and Shukla 2012a, b, 2016). These microfossils are important age marker fossils and are significant in giving information about the transition of organisms from unicellularity to multicellularity. The largest carbonaceous fossil of *Tawuia dalensis* was recorded from the Early Neoproterozoic rocks of the Lesser Himalaya (Sharma et al. 2016). Well preserved benthic multicellular algae of the Phaeophyceae are also known from the Chhattisgarh Basin (Babu and Singh 2011; 2013) (Fig. 2).

The Ediacaran period represents the end of the Precambrian. It is represented by an assemblage of sessile soft bodied organisms that are preserved as impressions. A lot of burrowing and bioturbations are reported from the Cuddapah and Kurnool basins (Kale et al. 1997; Gururajan et al. 2000; Sharma 2008; Vijayam 1968; Arya and Rao 1979; Sharma 2008). Ediacaran disc like structures and some strange fossils are also known from Ediacaran successions (Shukla and Sharma 2020b; Kumar and Ahmad 2016; Pandey and Sharma 2017; Pandey et al. 2019a, b). A diverse Ediacaran Complex Acanthomorphs palynoflora is recovered from the Vindhyan Supergroup (Prasad and Asher 2016). *Obruchevell*, a Vendian marker, is recorded from the Owk Shale of the Kurnool Group (Sharma and Shukla 2012a).

The Krol belt in the Lesser Himalaya represents a thick, uninterrupted sequence of the Ediacaran-Lower Cambrian interval. These sequences have diverse acanthomorphic acritarch sequences. Shukla and Tiwari (2014) reported large sized acanthomorphic acritarchs from the Krol Formation in the Ediacaran horizon. Acritarchs are also recorded from the Infra Krol sequences. (Tiwari and Knoll 1994; Tiwari 1996; Tiwari and Pant 2004). Ichnofossils, which are important in understanding metazoan activity on the earth and benthic activity of organisms in marine life (Jensen et al. 2000) have been recorded from the Krol-Tal formations (Singh et al. 2019a, b, c). From the Mussoorie and Garhwal areas, trace fossils and shelly fauna were recorded by Banerjee and Narain 1976; Singh and Rai 1983; Azmi 1983; Azmi and Joshi 1983; Azmi and Pancholi 1983). Ichnofossils were also recorded from Marwar Supergroup; Kurnool and Bhima groups.

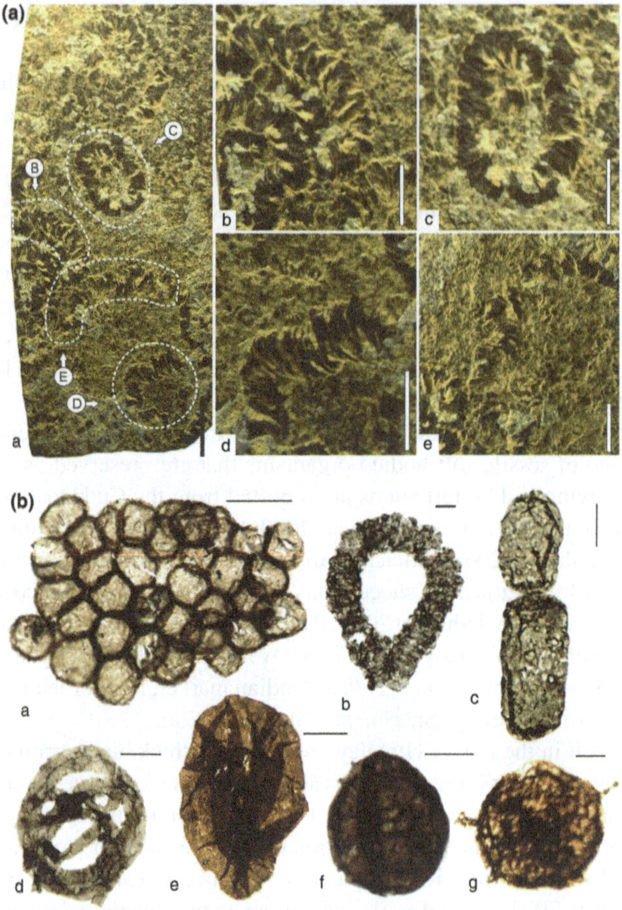

Fig. 2 **a** Organic walled tubular fossil *Shaanxilithes ningqiangensis* recorded from the very top of the Krol Group and in the basalmost Tal Group in Nigalidhar syncline of India preserved as organic compression on the bedding surface. (A) assemblage of *Shaanxilithes ningqiangensis* showing the annulations and high sinuosity; (B–E) magnified views of *S. ningqiangensis* demarcated by boxes at different levels. (A) showing the diagnostic properties; B, D–E) curved ribbon like feature showing annulations; (C) discoidal shaped *S. ningqiangensis* with poorly preserved annulations (Museum specimen no. BSIP 41,863; Scale bar A = 1 cm, for B–E = 0.5 cm). **b** Microfossils assemblage of the Owk Shale, Kurnool Basin, South India. (A) Spheroidal small vesicles tightly arranged in planar colonies *Ostiana microcystis*; (B) spirally coiled cyanobacterium *Obruchevella delicata*; (C) multicellular botuliform microfossil *Jacutianema solubila*; (D) loosely coiled filament of *Obruchevella parva*; (E) compressed, thin-walled spheroidal vesicles of *Leiosphaeridia tenuissima*; (F and G) Ediacaran complex acanthomorph, (F) *Cavaspina* aff. *C. acuminate*; (G) *Variomargosphaeridium* aff. *V. litoschum* (A: Slide no. BSIP-16351; B: Slide no. BSIP-16643; C: Slide no. BSIP-14401; D: Slide no. BSIP-16643; E: Slide no. BSIP-14405; F: Slide no. BSIP- 16,348; G: Slide no. BSIP-16346; Scale bar for B = 10 μm; for rest of the specimens = 20 μm) (A, C, E, F, G after Shukla et al. 2020a, b), (Reprinted from Sharma et al. 2021, copyright (2021), published with permission from, Springer; License Number: 5544181150056)

4 Summary

The preceding review of Indian Archean and Proterozoic rocks show the presence of well preserved and quantitatively and qualitatively diverse fossil remains, that not only played a crucial role in Precambrian biostratigraphy but also aided in comprehending the evolution and diversification of early life and the early biosphere of the earth.

Acknowledgements I am grateful to Dr. Bijai Prasad, Scientist (Retired), Oil and Natural Gas Commission Limited, Dehra Dun for going through the manuscript meticulously and providing valuable suggestions that helped in improvising the manuscript. My thanks are also to the Head, Department of Geology, RTMNU, Nagpur for providing working facility.

References

Adam ZR, Skidmore ML, Mogk DW et al (2017) A Laurentian record of the earliest fossil eukaryotes. Geology 45(5):387–390

Allwood AC, Walter MR, Kamber BS, Marshall CP, Burch IW (2006) Stromatolite reef from the Early Archaean era of Australia. Nature 441:714–718

Anbarasu K (2001) Acritarchs from Mesoproterozoic Chitrakoot Formation, Semri Group, Chitrakoot area Central India. J Geol Soc India 57:179–183

Arya BC, Rao CN (1979) Bioturbation structures from the middle Proterozoic Narji Formation, Kurnool Group Andhra Pradesh India. Sed Geol 22:127–134

Asher R, Kapoor PN, Prasad B (2017) Ediacaran-early Cambrian Acritarchs and associated organic-walled microfossils from the subsurface Vindhyan succession of Chambal valley Rajasthan India. ONGC Bull 52(1):145–158

Ashok Kumar P, Anantaram S, Patil RS et al (1993) Palaeontological and biostratigraphic studies of late PROTEROZOIC and early Paleozoic basins of Andhra Pradesh and Karnataka. Rec Geol Surv India 126(5):296

Azmi RJ (1983) Microfauna and age of the Lower Tal Phosphorite of Mussoorie Syncline, Garhwal Lesser Himalaya India. Himalayan Geol 11:198–223

Azmi RJ, Pancholi VP (1983) Early Cambrian (Tommotian conodonts and other shelly microfauna from the upper Krol of Mussoorie syncline, Garhwal Lesser Himalaya with Remarks on the Precambrian- Cambrian Boundary. Himalayan Geol 11:361–372

Azmi RJ, Joshi MN (1983) Conodonts and other biostratigraphic evidences on the age and evolution of the Krol Belt. Himalayan Geol 11:373–409

Babu R, Singh VK (2011) Record of aquatic carbonaceous metaphytic remains from the Proterozoic Singhora Group of Chhattisgarh Supergroup, India and their significance. J Evol Biol Res 3:47–66

Babu R, Singh K (2013) An evaluation of carbonaceous metaphytic remains from the Proterozoic Singhora Group of Chhattisgarh Supergroup, India. Spec Publ Geol Soc India 1:325–338

Banerjee DM (1974) Microfossils from the Late Precambrian Aravalli stromatolites of Udaipur, Rajasthan, India. Recent Res Geol 1:263–268

Banerjee DM, Narain MJ (1976) Trace fossils in the Lower Tal Formation of Mussoorie and their environmental significance. J Sediment Petrol 46:235–239

Banerjee DM, Schidlowski M, Arneth JD (1986) Genesis of upper Proterozoic-Cambrian phophorites deposits of India: isotopic inferences from carbonate flourapaptite, carbonate and organic carbon. Precambr Res 33:239–253

Banerjee S, Jeevankumar S (2005) Microbially originated wrinkle structures on sandstone and their stratigraphic context: Palaeoproterozoic Koldaha Shale, central India. Sed Geol 176:211–224

Banerjee N, Furnes H, Muehlenbachs K et al (2006) Preservation of ca. 3.4-3.5 Ga microbial biomarkers in pillow lavas and hyaloclastites from the Barberton Greenstone Belt, South Africa. Earth Planet Sci Lett 241:707–722

Banerjee S et al (2010) Microbially related structures in siliciclastic sediment resembling Ediacaran fossils: examples from India. Microbial mats: modern and ancient microorganisms in stratified systems. Springer, Dordrecht, pp 109–112

Baumgartner RJ, Van Kranendonk MJ, Wacey D et al (2019) Nano-porous pyrite and organic matter in 3.5-billion-year-old stromatolites record primordial life. Geology 47(11):1039–1043

Bengtson S, Sallstedt T, Belivanova V, et al (2017) Three-dimensional preservation of cellular and subcellular structures suggests 1.6 billion-year-old crown-group red algae. Paleos Biol 15:e2000735

Chakrabarti G, Shome D, Kumar S et al (2011) Carbon and oxygen isotopic variations in stromatolitic dolomites of Palaeoproterozoic Vempalle formation, Cuddapah Basin, India. Carbonate Evaporites 26:181–191

Das Sharma DC, Raha PK, Moitra AK et al (1992) Discovery of Precambrian-Cambrian transition fossil *Sabellitids* from India. Curr Sci 63(3):140–142

Dodd MS, Papineau D, Grenne T et al (2017) Evidence for early life in Earth's oldest hydrothermal vent precipitates. Nature 543(7643):60–64

Djokic T, van Kranendonk MJ, Campbell KA et al (2017) Earliest signs of life on land preserved in ca 3.5 Ga hot spring deposits. Nat Commun 8(1):1–9

Goswami S, Bhagat S, Zakaulla S et al (2017) Role of organic matter in uranium mineralisation in Vempalle dolostone; Cuddapah Basin, India. J Geol Soc India 89:145–154

Gururaja MN, Jagannatha Rao BR, Bhaskara Rao B (1979) Stromatolitic microbiota from black chert of Cumbum Formation, Upper Cuddapah. Andhra Pradesh J Geol Soc India 20:138–140

Gururaja MN, Ashok Kumar P, Rao CV et al (2000) Precambrian- Cambrian boundary strata in Cuddapah Basin, Andhra Pradesh. Geol Surv India Spec Publ 55:155–162

Han TM, Runnegar B (1992) Megascopic eukaryotic algae from the 2.1 billion year old Negaunne iron formation, Michigan. Science 257(5067):232–235

Holland HD (2006) The oxygenation of the atmosphere and oceans. Philos Trans R Soc Biol Sci 361:903–915. https://doi.org/10.1098/rstb.2006.1838

Jankauskas TV, Mikhailova N, Hermann TN (1989) Mikrofossilii dokembriya SSSR (Precambrian microfossils of the USSR): Leningrad, 188. Trudy Instituta GeologiiGeochronologii Dokembria SSSR Akademii Nauk. Nauka (In Russian)

Javaux EJ, Knoll AH, Walter MR (2001) Morphological and ecological complexity in earlyeukaryotic ecosystems. Nature 412(6842):66–69

Jensen SB, Saylor Z, Gehling JG et al (2000) Complex trace fossils from the terminal Proterozoic of Namibia. Geology 28:143–146

Kale VS (1991) Constraints on the evolution of Purana basins of Peninsuar India. J Geol Soc India 38:231–252

Kale VS, Patil SS, Satoskar V et al (1997) Occurrence of Planolites from the Nagarjuna Sagar Area, Northwestern Cuddapah Basin. J Geol Soc India 49:589–596

Khelen AC, Manikyamba C, Ganguly S et al (2017) Geochemical and stable isotope signatures of Proterozoic stromatolitic carbonates from the Vempalle and Tadpatri formations, Cuddapah Supergroup, India: Implications on paleoenvironment and depositional conditions. Precambr Res 298:365–384

King W (1872) On the Kadapah and Karnul formations in the Madras Presidency. Mem Geol Surv India 8:1–313

Knoll AH (1996) Archean and Proterozoic paleontology. In: Jansonius J, McGregor DC (eds) Palynology: principles and applications, vo. 1. American Association of Stratigraphic Palynologists Foundation, pp 51–80

Knoll AH (2014) Paleobiological perspectives on early eukaryotic evolution. Cold Spring Harb Perspect Biol 6(1):1–14

Kumar S, Srivastava P (1991) Microfossils from the non-stromatolitic middle Proterozoic Vindhyan Chert, Chitrakut area, Uttar Pradesh. J Geol Soc India 38:511–515

Kumar S, Srivastava P (1995) Microfossils from the Kheinjua Formation, Mesoproterozoic Semri Group, Newari area, central India. Precambr Res 74:91–117

Kumar S, Pandey SK (2009) A note on the occurrence of *Arumberiabanksi* from the Jodhpur Sandstone, Marwar Supergroup, Jodhpurarea, Rajasthan. J Palaeontol Soc India 54(2):171–178

Kumar S, Ahmad S (2014) Microbially Induced Sedimentary Structures (MISS) from the Ediacaran Jodhpur Sandstone, Marwar Supergroup, western Rajasthan. J Asian Earth Sci 91:352–361

Kumar S, Ahmad S (2016) Problematic structures from the Ediacaran Jodhpur Sandstone, Rajasthan, India and their possible affinity. J Palaeontol Soc India 60(2):21–26

Lamb DM, Awramik SM, Chapman DJ et al (2009) Evidence for eukaryotic diversification in the ~1800 million-year-old Changzhougou Formation. North China. Precambr Res 173(1–4):93–104

Maithy PK (1992) Palaeobiology of Vindhyan. Palaeobotanist 40:52–72

Moitra AK (1999) Biostratigraphic study of stromatolites and microbiota of the Chhattisgarh Basin, M.P., India. Palaeontol Indica 51:1–95

Noffke N (2000) Extensive microbial mats and their influences on the erosional and depositional dynamics of a siliciclastic cold water environment (Lower Arenigian, Montagne Noire, France). Sediment Geol 136(3–4):207–215

Noffke N, Gerdes G, Klenke T et al (2001a) Microbially induced sedimentary structures indicating climatological, hydrological and depositional conditions within recent and Pleistocene coastal facies zones (Southern Tunisia). Facies 44(1):23–30

Noffke N, Gerdes G, Klenke T et al (2001b) Microbially induced sedimentary structures: a new category within the classification of primary sedimentary structures. J Sediment Res 71(5):649–656

Noffke N, Eriksson KA, Hazen RM, et al (2006) A new window into Early Archean life: Microbial mats in Earth's oldest siliciclastic tidal deposits (3.2 Ga) Moodies Group, South Africa. Geology 34(4):253–256

Noffke N, Christian D, Wacey D, et al (2013) Microbially induced sedimentary structures recording an ancient ecosystem in the ca. 3.48 billion-year-old dresser Formation, Pilbara, Western Australia. Astrobiology 13(12):1103–1124

Pandey SK, Kumar S (2013) Organic walled microbiota from the silicified Algal clasts, Bhander Limestone, Satna Area, Madhya Pradesh. J Geol Soc India 82:499–508

Pandey SK, Sharma M (2017) Enigmatic Ediacaran Megascopic bedding plane structures on the Sonia Sandstone, the Jodhpur Group, Marwar Supergroup, India: seaweed or problematica? Geol J 52:784–807

Pandey SK, Jonathan C, Nema P et al (2019a) Ladakh: diverse, high-altitude extreme environments for off earth analogue and astrobiology research. Int J Astrobiol 19(1):78–98

Pandey SK, Bykova N, Sharma M (2019b) Current status of the Ediacaran-Cambrian Bilara Group, Marwar Supergroup. India. Estudios Geol 75(2):29–30

Prasad B (2007) *Obruchevella* and other Terminal Proterozoic (Vendian) organic-walled microfossils from the Bhander Group (Vindhyan Supergroup), Madhya Pradesh. J Geol Soc India 69:295–310

Prasad B, Asher R (2001) Acritarch biostratigraphy and lithostratigraphic classification of Proterozoic and lower Paleozoic sediments (Pre-Unconformity sequence) of Ganga Basin, India. Paleontographica Indica 5:1–151

Prasad B, Asher R (2016) Record of Ediacaran complex acanthomorphic acritarchs from the lower Vindhyan succession of the Chambal Valley (East Rajasthan), India and their biostratigraphic significance. J Palaeontol Soc India 61:29–62

Prasad B, Uniyal SN, Asher R (2005) Organic-walled microfossils from the Proterozoic Vindhyan Supergroup of Son Valley, Madhya Pradesh, India. Palaeobotanist 54(1–3):13–60

Prasad B, Uniyal SN, Asher R (2007) Organic-walled microfossils from the Proterozoic Vindhyan Supergroup of Son Valley, Madhya Pradesh, India. In: Proc. XXII Ind. Colloq. Mircopaleontol. Stratgraph, vol 1. Norosa Publishing House, New Delhi, pp 7–16

Prasad B, Asher R, Borgohai B (2010) Late Neoproterozoic (Ediacaran)– Early Paleozoic (Cambrian) acritarchs from the Marwar Supergroup, Bikaner-Nagaur Basin, Rajasthan. J Geol Soc India 75:415–431

Raghav KS, De C, Jain RL (2005) The first record of Vendian medusoids and trace fossils-bearing algal mat grounds from the basal part of the Marwar Supergroup of Rajasthan, India. Indian Miner 59:23–30

Ramakrishna M, Vaidyanadhan R (2008) Geology of India. Geological Society of India

Retallack G, Noffke N (2018) Are there ancient soils in the 3.7 Ga Isua Greenstone Belt, Greenland? Palaeoecol Palaeooceanogr Palaeoclimatol 514:18–30

Sallstedt T, Bengtson S, Broman C (2018) Evidence of oxygenic phototrophy in ancient phosphatic stromatolites from the Paleoproterozoic Vindhyan and Aravalli Supergroups. India. Geobiology 16(2):139–159

Salujha SK, Rehman K, Arora CM (1972a) Early Palaeozoic microplankton from the Kurnools, Andhra Pradesh. J Palynol 8:123–131

Salujha SK, Rehman K, Arora CM (1972b) Microplanktons from the Bhimas. J Paleontol Soc India 15:10–16

Schirrmeister BE, de Vos JM, Antonelli A et al (2013) Evolution of multicellularity coincided with increased diversification of cyanobacteria and the great oxidation event. Proc Natl Acad Sci 110:1791–1796

Schopf JW, Prasad KN (1978) Microfossils in *Collenia*-like stromatolites from the Proterozoic Vempalle Formation of the Cuddapah Basin, India. Precambr Res 6(3–4):347–366

Sergeev VN, Sharma M, Shukla Y (2008) Mesoproterozoic silicified microbiotas of Russia and India characteristics and contrasts. Palaeobotanist 57:323–358

Sergeev VN, Sharma M, Shukla Y (2012) Proterozoic fossil cyanobacteria. Palaeobotanist 61:189–358

Sharma M (2006) Small-sized akinetes from the Mesoproterozoic Salkhan Limestone, SemriGroup, Bihar, India. J Palaeontol Soc India 51:109–118

Sharma M (2008) Neoproterozoic biotic signatures in the peninsular Indian basins—an overview. Mem Geol Soc India 74:119–131

Sharma M, Mishra S, Dutta S et al (2009) On the affinity of *Chuaria-Tawuia* Complex; a multidisciplinary study. Precambr Res 173:123–136

Sharma, M Shulka Y (2009) Mesoproterozoic coiled megascopic fossil Grypania spiralis from the Rohtas Formation, Semri Group, Bihar, India. Curr Sci 96(12):1636–1639

Sharma M, Shukla Y (2012a) Occurrence of helically coiled microfossil *Obruchevella* in the Owk shale of the Kurnool Group and its significance. J Earth Syst Sci 121(3):755–768

Sharma M, Shukla Y (2012b) Megascopic carbonaceous compression fossils from the Neoproterozoic Bhima Basin, Karnataka, South India. Geol Soc London 360:277–293

Sharma M, Shukla Y (2016) The palaeobiological remains of the Owk Shale, Kurnool Basin: a discussion on the age of the basin. J Palaeontol Soc India 61(2):175–187

Sharma M, Tiwari M, Ahmad S et al (2016) Palaeobiology of Indian Proterozoic and early Cambrian successions- recent developments. Proc Indian Natl Sci Acad 82:559–579

Sharma M, Shukla B (2019a) Akinetes from Late Palaeoproterozoic Salkhan Limestone (ca.1600 Ma) of India: a proxy for understanding life in extreme conditions. Front Microbiol 10(397)

Sharma M, Singh VK (2019b) Megascopic carbonaceous remains from Proterozoic basins of India. In: Mondal M (ed) Geological evolution of the Precambrian Indian Shield. Society of Earth Scientists Series, Springer, pp 725–749

Sharma M. Pande SK, Kumar S (2020) Vindhyan Basin, Son Valley area, Central India (Field Guide). In: 36th International Geological Congress, New Delhi. pp 1–176

Sharma M, Singh VK, Pandey SK et al (2021) Precambrian and early Cambrian paleobiology of India: *Quo Vadis*, vol 87. In: Proceedings of the Indian National Science Academy., New Delhi, pp 199–233. https://doi.org/10.1007/s43538-021-00029-2

Shukla Y (2011) Palaeobiology of terminal Proterozoic Bhima Basin, Karnataka: University of Lucknow, India. Unpublished thesis

Shukla R, Tiwari M (2014) Ediacaran acanthomorphic acritarchs from the Outer Krol Belt, Lesser Himalaya, India: Their significance for global correlation. Palaeoworld 23:209–224

Shukla B, Sharma M (2016) A new assemblage of large-sized microfossils from the Salkhan Limestone (ca 1600 Ma), Semri Group, Vindhyan Supergroup, India. J Palaeontol Soc India 61:287–299

Shukla Y, Sharma M (2020a) 'Egg-carton' shaped plausible organo-sedimentary structures from the Archaean Dharwar Craton. India Int J Earth Sci 109(3):931–932

Shukla Y, Sharma M (2020b) Ediacaran discs from the Bhima Group, Karnataka, South India. J Geol Soc India 95:483–490

Shukla,Y, Sharma M, Sergeev VN (2020a) Organic walled microfossils from the Neoproterozoic Owk Shale, Kurnool Group, South India. Palaeoworld 29:490–511

Shukla Y, Sharma M, Noffke N, Callefo F (2020b) Biofilm microfacies in phosphatic units of the Neoproterozoic Halkal Shale, Bhima Basin, South India. Precambrian Res 349:105501

Singh IB, Rai V (1983) Fauna and biogenic structures in Krol-Tal succession (Vendian–Early Cambrian), Lesser Himalaya: their biostratigraphic and palaeontological significance. J Palaeontol Soc India 28:69–80

Singh VK, Babu R, Shukla M (2011) Heterolithic prokaryotes from the coated grains bearing carbonate facies of Bhander Group, Madhya Pradesh, India. J Appl Biosci 37:80–90

Singh VK, Sharma M (2014) Morphologically complex organic-walled microfossils (OWM) from the Late Palaeoproterozoic-Early Mesoproterozoic Chitrakut Formation, Vindhyan Supergroup, Central India and their implications on the antiquity of eukaryotes. J Palaeontol Soc India 59:89–102

Singh VK, Sharma M (2016) Mesoproterozoic organic-walled microfossils from the Chaporadih-Formation, Chandarpur Group, Chhattisgarh Supergroup, Odisha, India. J Palaeontol Soc India 61:75–84

Singh BP, Virmani N, Bhargava ON et al (2016) Trilobite fauna of basal Cambrian Series 3 (Stage 5) from the Parahio Valley (Spiti), Northwest Himalaya, India and its biostratigraphic significance. Ann Paléontol 102(1):59–67

Singh BP, Bhargava ON, Mikul R et al (2019a) Integrated sedimentological, ichnological and sequence stratigraphical studies of the Koti Dhaman Formation (Tal Group), Nigali Dhar Syncline, Lesser Himalaya, India: paleoenvironmental, paleoecological, paleogeographic significance. Ichnos 27(2):1–34

Singh VK, Sharma M (2019) Morphologically complex organic-walled microfossils (OWM) from the Late Palaeoproterozoic-Early Mesoproterozoic Chitrakut Formation, Vindhyan Supergroup, Central India and their implications on the antiquity of eukaryotes. J Palaeontol Soc India 59:89–102

Singh VK, Sharma M, Sergeev VN (2019a) Helically coiled cyanobacterial microfossil *Obruchevella* from the Saradih Limestone, Raipur Group, Chhattisgarh Supergroup, India. J Palaeontol Soc India 64:304–313

Singh VK, Sharma M, Sergeev VN (2019b) A new record of acanthomorphic acritarch *Tappania* Yin from the early mesoproterozoic Saraipali Formation, Singhora Group, Chhattisgarh Supergroup, India and its biostratigraphic significance. J Geol Soc India 94(5):471–479

Tang Q, Hughes NC, McKenzie NR et al (2017) Late Mesoproterozoic–early Neoproterozoic organic-walled microfossils from the Madhubani Group of the Ganga Valley, northern India. Palaeontology 60(6):869–891

Tiwari M, Knoll AH (1994) Large acanthomorphic acritarchs from the Infra Krol Formation of the Lesser Himalaya and their stratigraphic significance. Him Geol 5:193–201

Tiwari M (1996) Precambrian Cambrian boundary microbiota from the Chert Phosphorite Member of Tal Formation in the Korgai Syncline, Lesser Himalaya, India. Curr Sci 71:559–566
Tiwari M, Pant C (2004) Neoproterozoic silicified microfossils in Infra Krol Formation, Lesser Himalaya India. Him Geol 25:1–21
van Loon AJ, Mazumder R (2013) First find of biogenic activity in the Palaeoproterozoic of the Singhbhum craton (E India). Geologos 19(3):185–192
Venkatachala BS, Sharma M, Shukla M (1996) Age and life of the Vindhyans—facts and conjectures. Mem Geol Soc India 36:137–165
Vijayam BE (1968) Worm burrows in Narji Limestone, near Govindinne, Kurnool district, Andhra Pradesh. Curr Sci 37(5):142
Woese Carl R, Kandler O, Wheelis M (1990) Towards a natural system of organisms: proposal for the domains Archaea, Bacteria, and Eucarya. In: Proc Natl Acad Sci USA 87(12):4576–4579
Yin L (1997) Acanthomorphic acritarchs from Middle-Upper Proterozoic shales of the Ruyang Group in Shanxing, China. Rev Paleobotany Palynol 98:15–25

Significance of Palynology in Understanding Age, Palaeoclimate and Correlation of Indian Gondwana Sediments

Mrutyunjaya Sahoo, Srikanta Murthy, Anju Saxena, Sankar Suresh Kumar Pillai, and Sumit Kumar Sahu

Abstract The Gondwana sedimentary deposits are spread across the Indian Peninsula. The Gondwana sequences in Peninsular India are mostly non-marine. Based on morphological features of palynoflora and megaflora, palynomorphs from different biostratigraphic sediments (Talchir to Supra Panchet) of Gondwana show a high degree of variation. Based on the microstructures of exine among spores, the primitive and younger states of the sediments are confirmed with time. The morphological characteristics of pollen spores help in their evolutionary sequence through different time scales, hence playing a vital role in biohorizon stratigraphy. Apart from this, palynomorphs also help in delineating the biostratigraphic age, First Appearance Datum (FAD), LAD (Last Appearance Datum), DOD (Dominance of Datum), assemblage zones, and palaeoclimate. The current paper provides an overview of the role of palynology in determining the biostratigraphic ages, palaeoclimatic conditions, and correlation of Indian Gondwana sediments. The data presented here is the basic framework for further refinement of chronology based on palynology. It is a compilation of published data from peninsular India.

Keywords Palynology · Indian Gondwana · Correlation · Palynostratigraphy · Palaeoclimate

---

M. Sahoo
Department of Geology, Ravenshaw University, Cuttack 753003, Odisha, India

S. Murthy · A. Saxena · S. S. K. Pillai (✉) · S. K. Sahu
Birbal Sahni Institute of Palaeosciences, 53 University Road, Lucknow 226007, India
e-mail: sureshk_pillai@bsip.res.in

S. S. K. Pillai · S. K. Sahu
Academy of Scientific and Innovative Research (AcSIR), Ghaziabad 201002, India

B. Samant and D. Thakre (eds.), *Applications of Palynology in Stratigraphy and Climate Studies*, Society of Earth Scientists Series, https://doi.org/10.1007/978-3-031-51877-5_2

1 Introduction

The sedimentary rock strata ranging in age from the Permo-Carboniferous to the Early Cretaceous (Asselian to Albian; 299 to 100 Ma), found in patches in different parts of India, South Africa, South America, Madagascar, Australia, and Antarctica, are referred to as Gondwana. These landmasses were joined together in the geological past, constituting the supercontinent called Gondwanaland. Some characteristic features are the presence of coal, fossilized remains of land plants, and vertebrates. The social importance of the Gondwana Supergroup is conferred by the fact that more than 99% of the total coal resource of the Indian subcontinent is hosted in the Gondwana basins. The palaeontological and stratigraphical studies of the Indian Gondwana have revealed that the accumulation of sediments occurred in several discrete inter-cratonic basins in the peninsula. The major Gondwana basins in India are in the peninsular region (Fig. 1) and are dispersed along modern-day river valleys, viz., the Godavari, Son-Mahanadi, Narmada, and Pranihta-Godavari. The Indian Gondwana basins are usually named after these rivers. The Gondwana sediments show distinct characteristics as most of its land area occupied high latitudes in the southern hemisphere (Gondwanaland), providing a unique environment (Veevers 1993; Tiwari and Kumar 2002). Under the influence of Gondwana's climate, ecology, land-sea distribution, and geotectonics, the typical flora and fauna of the Gondwana region maintained their identities during the Permo-Carboniferous to Early Cretaceous (mainly in the Permian). The palaeogeographic configuration mentioned above is supported by the strong association between the floras of Gondwanaland during the Permian (the *Glossopteris* floral province) and most of the Mesozoic.

Palaeontology plays the most important role in understanding the then geological events of the past, and among the studies of palaeontology, palynology is a high-precision tool for the process. The term was proposed by Hyde and Williams (1944). Palynology studies acid resistant organic microfossils (5–500 μm in diameter) such as spores, pollen, dinoflagellate cysts, microscopic algae, megaspores, etc. Throughout much of geologic time, these microscopic plant particles have accumulated in sediments without losing their morphological characteristics. Their recovery from residues enables the palynologist to establish a correlation based on time equivalence. The palynologist analyses the particulate samples collected from the air, from water, or deposits, including sediments of any age. The condition and identification of those particles, organic and inorganic, give the palynologist clues to the life, environment, and energetic conditions that produced them. Flora and fauna during the Permian and Mesozoic were influenced by different climates, ecology, and land-sea distribution, but they maintained their identities (Tiwari and Vijaya 1995). Hence, it is an essential tool for studying different geological and biological events.

Palynology serves as an essential tool, since it has been used worldwide for dating, correlation, palaeovegetational, and palaeoclimatic interpretation of the Gondwana sediments (Tiwari and Tripathi 1992; McLoughlin 2001; Backhouse et al. 2002; Souza and Marques-Toigo 2005; Jha 2006; Quattrocchio et al. 2006; Shi et al. 2010; Tripathi et al. 2012; Jha et al. 2012; Jha et al. 2014; Souza et al. 2015; Sahoo

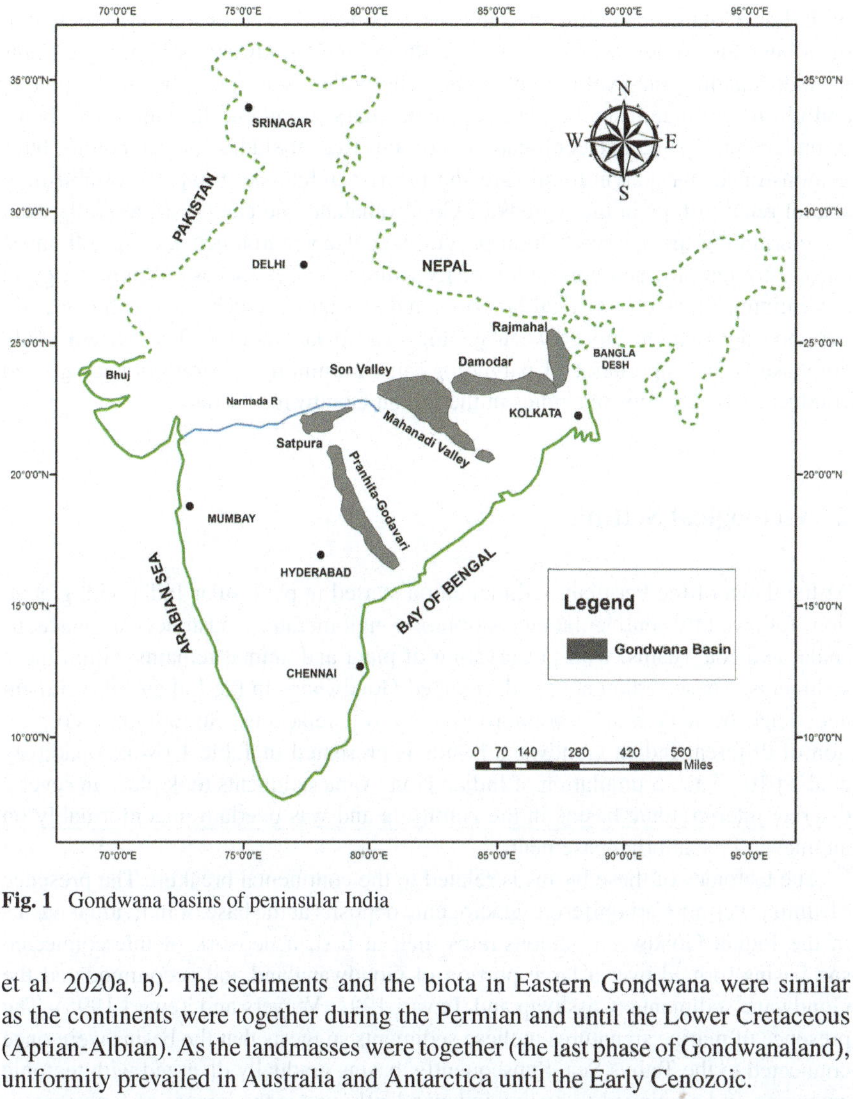

Fig. 1 Gondwana basins of peninsular India

et al. 2020a, b). The sediments and the biota in Eastern Gondwana were similar as the continents were together during the Permian and until the Lower Cretaceous (Aptian-Albian). As the landmasses were together (the last phase of Gondwanaland), uniformity prevailed in Australia and Antarctica until the Early Cenozoic.

There is always a debate between geologists and palaeontologists concerning the age of sediments. But the bioentities (megaflora or palynoflora) are imperative for determining the age of the deposits. The biological entities of the Permian are essential tools for correlating their relationships throughout the continents of Gondwanaland. It helps in delineating the relationship between the regions and calibrating the chronostratigraphy. Palaeontologists are considering the palynological tool for filling the geological gaps and prevailing problems based on the chronology of the Indian Gondwana sequences (Tiwari and Tripathi 1992 and references therein).

The palynological study of the Indian Gondwana sequence dates back to the nineteen-thirties. Since then, data on the morpho-taxonomy and palyno-stratigraphy of Indian Gondwana sediments have been collected. The preservation potential of pollens and spores was most prolific during the Gondwana period, which aids in understanding the evolutionary trend. The pollen-type sporopollenin is unique and ultimately manifests the plant's genetic character and at the same time, helps in understanding the palaeoclimatic condition. Recent studies of palynology have demonstrated that palynofossils are the primary reference group for establishing spatial relationships in the regions of Gondwanaland and can be successfully used in determining the age, past floral provinces, palaeoenvironments, palaeoclimates, chronostratigraphy, and correlation of its sediments. The success of palynology in determining the age of critical horizons and stratigraphic correlations has greatly enriched the relationship between geology and palaeontology. The current study emphasises the importance of palynology in determining biostratigraphic age and conducting palaeoclimate studies in the Indian Gondwana basins.

2 Geological Setting

At the dawn of the Permian, sedimentation started in peninsular India with glaciation, followed by continental deposition of conglomerate, sandstone, carbonaceous shale, and coal seams, with preservation of plant and animal remains within these sediments. These sediments are designated Gondwanas in the Indian subcontinent and range from Permo-Carboniferous to Early Cretaceous. Stratigraphic correlation of different Indian Gondwana basins is presented in Table 1 (Mukhopadhyay et al. 2010). The accumulation of Indian Gondwana sediments took place in several discrete inter-cratonic basins in the peninsula and was overlain unconformably on an uneven Precambrian basement.

The tectonics of these basins is related to the continental breakup. The presence of similar Permo-Carboniferous glaciogenic deposits at the base of individual basins in the Indian Gondwana demonstrates that, at first, a network of interconnected sag basins formed over a large portion of Gondwanaland and accommodated the Gondwana sediments (Casshyap and Tewari 1991; Veevers and Tewari 1995). The presence of marine signatures in these sediments suggests that the basins were once connected to the Tethys Sea. Subsequently, basins gradually diverged with tectonic extension and continued to evolve independently under the impact of fault (extensional and strike-slip faults along Precambrian lineaments) controlled subsidence. All the Gondwana basins of the Indian peninsula formed as a result of a single regional tectonic event. The bulk kinematics was an E-W directed lateral extension (Chakraborty et al. 2003). The basins are distributed along the ENE-WSW trending valleys of the Narmada, Son, and Damodar rivers, the NNW-SSE trending valley of the Pranhita–Godavari River, and the NW-SE trending valley of the Mahanadi River. The Pranhita–Godavari basin was an elongated half-graben, whereas the rhombic and rectangular basins like Satpura, Karanpura, Bokaro, Jharia, Raniganj, Talcher, and Daltonganj are strike-slip basins (Ghosh and Sengupta 2020).

Table 1 Stratigraphic correlation of different Indian Gondwana basins (after Mukhopadhyay et al. 2010)

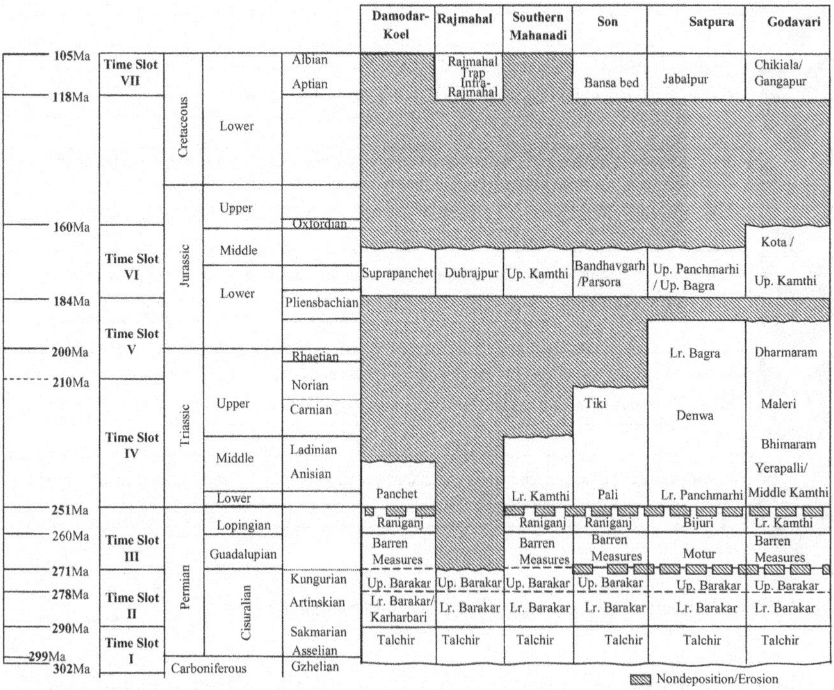

3 Methodology

The standard maceration technique is used to recover spores and pollen grains (Bharadwaj 1962; Bharadwaj and Saluja 1964). The dried samples are crushed into small pieces (2–3 mm in size) and treated with 40% hydrofluoric acid (HF) to dissolve the siliceous matter. Then the samples are washed with water to make them acid-free. Samples are then treated for 3–4 days with commercial nitric acid (HNO_3; 63.01%), with frequent addition and stirring of fresh HNO_3 from time to time for organic matter digestion. Each sample is observed under the microscope at every step of the maceration before further treatment. The samples are finally treated with 10–20% potassium hydroxide (KOH) after thorough washing with water to remove the hummus and obtain clear palynomorphs. Finally, the samples are washed and sieved with a sieve of 400 μm size. After a series of centrifuging steps and gently shaking the watch glasses to separate the spores on the surface from the sand and other heavier particles at the bottom, a good concentration of clear miospores is obtained. These residues are mixed with polyvinyl alcohol (PVA), smeared over a covered glass, and dried at room temperature. Finally, the coverslip is mounted with Canada balsam (mounting medium) to prepare permanent slides. The palynomorphs are counted and photographed in transmitted light under a high-power microscope.

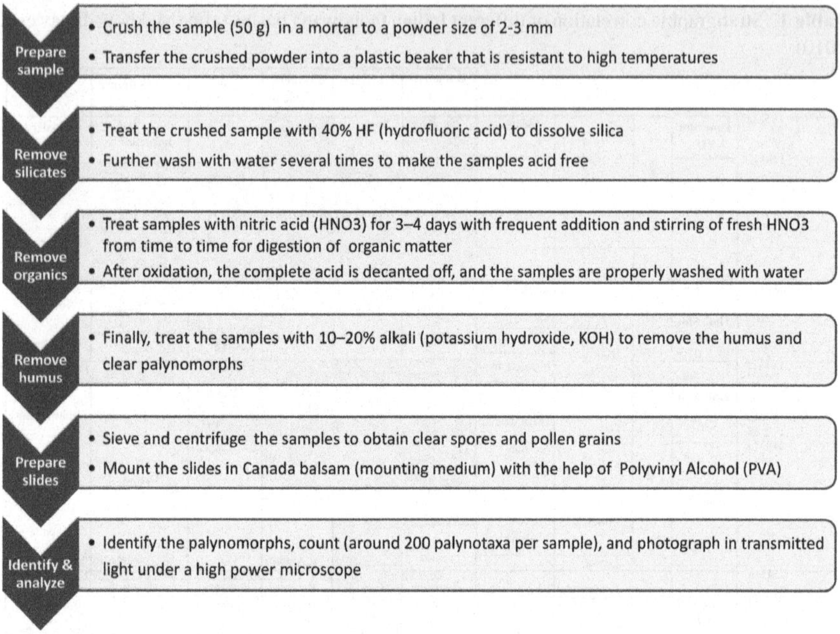

Fig. 2 Pollen extraction methodology (Bharadwaj and Saluja 1964)

A summary of the above methodology is presented in Fig. 2. Pollen grains are microscopic, so their identification and quantification depend on human observers using light microscopes. Modern improvements in computing and imaging hardware and software now bring the automation of pollen analyses within reach (Holt and Bennett 2014).

4 Biostratigraphic Age Assessment of Indian Gondwana Sequences Using Palynomorphs

The stratigraphic units in the Indian Gondwana were initially acknowledged based on distinct lithology. Later, biostratigraphic zones (megaflora assemblages) were identified and tagged with the lithostratigraphic units. In the last four to five decades, palynology (palynoassemblages) was used as a precise parameter for biostratigraphy (Tiwari and Tripathi 1992). Megafloral biostratigraphy has limitations in the preservation and abundance of specimens. Thus, palynology is suitable for refining biostratigraphy and establishing the biochronological arrangement of the Indian Gondwana.

The study of stratigraphy employs fossils to determine rocks' relative ages and establish the sedimentary rock successions within and between depositional basins.

Three terms are often used concerning the bio entities for describing the biostratigraphic age: First Appearance Datum (FAD), Dominance of Datum (DOD) and Last Appearance Datum (LAD). FAD is the first appearance of a species in the geological record. It denotes the geologically oldest fossil discovered to date of a particular species. DOD is used for the specific species dominating during that certain geological period. LAD is used for a species that appeared for the last time in a geological period. The term "biozone" refers to a division of geological strata based on the fossil taxa that are unique to each interval. Figure 3 (after Tiwari and Tripathi 1987) shows the palyno-compositions of each formation (from Talchir to Panchet) and palynofloral changes in successive boundaries, as well as the percentage frequency distribution of significant palynotaxa across Indian Gondwana sequences.

4.1 Talchir Formation (Early Permian, Asselian–Sakmarian)

The deposition of sediments from peninsular India during the Permian starts from the Talchir Formation. The formation represents the lowermost and oldest sequence of Gondwana. It overlies unconformably on an uneven Precambrian granitic basement. These sediments are found in almost all Indian Gondwana basins and are glacial in origin, with lacustrine and shallow marine environments in most of the formation (Casshyap and Tewari 1987). There are various reports of marine signatures from the Talchir Formation (Rao et al. 1977; Venkatachala and Tiwari 1987; Tiwari et al. 1994; Shanker et al. 1996). The thickness of this formation usually ranges from 100 to 120 m, but it reaches a maximum thickness of 300 m in the Damodar Basin. The basal most unit of the Talchir Formation is designated as 'Talchir Boulder Bed' or Tillite. However, hitherto no palynological entities have been reported from these tillite beds.

(a) **Lithology**: Talchir sedimentation in India commenced with a glacial episode in the basal Permian. Glaciogenic deposits viz., tillite, sandstone, shale, conglomerate, boulder bed, varve deposits and the presence of coal seams characterize the Talchir Formation (Dutta 2002). The abundance of chlorite as a primary clay mineral in the shales imparts a greenish or olive-green colour. The shale breaks up into thin pencil-like or prismatic fragments, for which they are often called 'needle shales'.

(b) **Assemblage zones**: Tiwari and Tripathi (1992) have categorized 20 palynoassemblage zones and 30 acme zones based on palynocomosition, LADs and FADs of selected marker species. The three lower assemblage zones are recognized in the Talchir Formation of India viz., (i) *Potonieisporites neglectus* (oldest) (ii) *Plicatipollenites gondwanensis* and (iii) *Parasaccites korbaensis* (youngest). They were assigned to Early Permian based on the general dominance of radial and bilateral monosaccate pollen and trilete spore *Jayantisporites*, respectively.

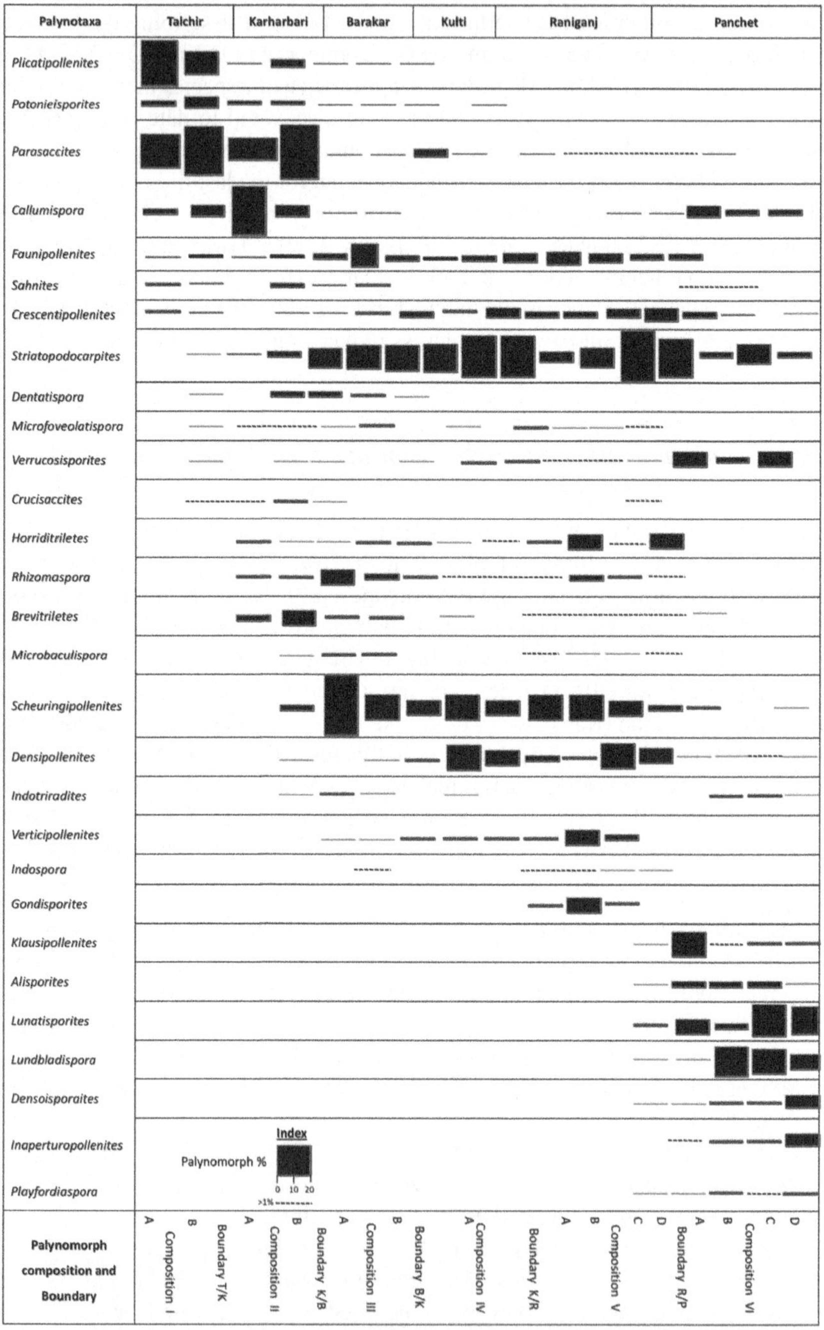

Fig. 3 Distribution of significant palynotaxa in terms of percentage frequency across Indian Gondwana sequences (Tiwari and Tripathi 1987)

(c) **FAD**: *Faunipollenites perexiguus, Tuberisaccites tuberculatus, Sahnites thomasi, Jayantisporites, Parasaccites bilateralis. Microbaculispora tentula, Microfoveolatispora foveolata, Divarisaccus lelei, Tiwariasporis gondwanensis, Marsupipotlenites triradiatus, Welwitscbiapites magnus* and *Stellapollenites talchirensis.*
(d) **DOD**: *Parasaccites, Plicatipollenites* and *Potonieisporites.*
(e) **LAD**: *Crucisaccites monoletus* makes top of the Talchir Formation.
(f) **Reports from different basins of India**: Palynomorphs from the Talchir Formation have been reported from Son-Mahanadi (Meena et al. 2013), Talcher (Tripathi 1993), Ib River (Meena and Murthy 2011), Singrauli (Vijaya et al. 2012a), Satpura (Bharadwaj et al. 1978a; Srivastava and Sarate 1988), Wardha (Sarate and Murthy 2017), Damodar (Vijaya et al. 2012b; Lele 1975), Sout Rewa (Chandra and Lele 1979) and Godavari (Jha and Aggarwal 2012; Srivastava and Jha 1988) basins.

It is worthwhile to mention that more recently Murthy et al. (2020a) have tried to reassess the age and discussed in detail about the lithostratigraphic versus biostratigraphic palynodating limitations in peninsular Gondwana sequences, and described a Late Pennsylvanian palynoassemblage from the Talchir Formation from the bore core of the Rajmahal Basin. They have identified two palynoassemblages. Palynoassemblage-I is marked by the dominance of monosaccate pollen grains having the characteristic occurrence of *Potonieisporites, Plicatipollenites* and *Cannanoropollis* (*Parasaccites*) and is corroborated to the *Potonieisporites neglectus* Assemblage Zone of Tiwari and Tripathi (1992) of the Lower Talchir Formation and assigned a Late Pennsylvanian age. Palynoassemblage-II dominates monosaccate pollen grains with a lesser amount of bisaccate pollen grains and trilete spores and has been equated with *Plicatipollenites gondwanensis* Assemblage Zone (Tiwari and Tripathi 1992) of the Upper Talchir Formation. For that, an Early Cisuralian age is assigned.

4.2 *Karharbari Formation (Middle Early Permian, Sakmarian–Artinskian)*

This formation overlies the Talcher Formation. Its occurrence is local as it has been identified only in a few basins though it has distinct characteristic plant fossils. It has higher-grade coal seams as compared to the underlain Talchir Formation.

(a) **Lithology**: Medium to coarse grained whitish arkosic sandstones, carbonaceous shale, grey shale and coal seams characterise this formation.
(b) **Assemblage zones**: One palynozone was identified in this formation: *Crucisaccites monoletus* Assemblage Zone (Tiwari and Tripathi 1992). The older Karharbari Formation shows higher diversity in radial monosaccate pollen as a

continuation of Upper Talchir palynoflora, but it has distinctions having first appearances of several striates, non-striates, and apiculate trilete spore species that increase in the population. The younger horizon of Karharbari shows a recurrence of dominance in radial monosaccate pollen (*Parasaccites*) and a decline in triletes (*Callumispora*) (Tiwari and Tripathi 1987).

(c) **FAD**: *Tiwariasporis gondwanensis*, *Mansopipollenites triadites* and *Stellapollenites talchirensis*.

(d) **DOD**: *Parasaccites*, *Microbaculispora*, *Brevitriletes*, *Crucisaccites monoletus*, *Callumispora gretensis*, *C. barakarensis*, *Parasaccites obscures*, and *Crescentipollenites rhombicus*.

(e) **LAD**: *Rhizomaspora indica*, *Indotriradites korbaensis and Dentatispora gondwanensis*.

(f) **Reports from different basins of India**: Karaharbari palynoflora have been reported from Godavari (Srivastava and Jha 1988; Jha and Sarate 2001; Jha and Aggarwal 2012), Talcher (Tripathi and Bhattacharya 2001; Meena 2003), Singrauli (Vijaya et al. 2012a), Wardha (Aggarwal et al. 2019), Damodar (Srivastava 1973; Bharadwaj 1974) and Satpura (Srivastava and Sarate 1989) basins.

4.3 Barakar Formation (Late Early Permian, Artinskian–Kungurian)

This formation overlay the Talchir Formation and has the most commercially exploitable coal seams. The thickness of this formation ranges from 600 to 1100 m.

(a) **Lithology**: Fine to coarse-grained feldspathic whitish sandstones, siltstone, black to grey shale, sandy shale, fireclay and coal seams with polymictic conglomerate at the base make up this formation.

(b) **Assemblage zones**: The Barakar Formation, being the primary coal-bearing formation, contains diversified palynofossils with a wide range of characters and various pathways of evolutionary trends (Tiwari and Tripathi 1987). Two palynological zones have been identified from the Barakar Formation: (i) *Scheuringipollenites barakarensis* Assemblage Zone (Lower Barakar), and (ii) *Faunipollenites varius* Assemblage Zone (Upper Barakar) (Tiwari and Tripathi 1992). In the Lower Barakar, striate pollens are dominant, and in the Upper Barakar, non-striate disaccate pollens are predominant, replacing the striate pollens. Trilete spores are more abundant in the bottom and middle portions of the Barakar and less productive in the upper part. The successive younger palynofloras of the Upper Permian (Barren Measures and Raniganj formations) are more similar to the Barakar palynoflora.

(c) **FAD**: *Rhizomaspora indica*, *Indotriradites sparsus*, *Striatites communis*, and *Vatieripolenites gibbosus*.

(d) **DOD**: *Scheuringipollenites*, *Faunipollenites* and *Striatopodocarpites*.

(e) **LAD**: *Corisaccites alutas, Horriditriletes curvibaculosus, Barakarites indicus, Densipollenites densus, Striatopodocarpites ovatus, Cyclobaculisporites minimus* and *Gondisporites raniganjensis.*
(f) **Reports from different basins of India**: Barakar palynoflora have been reported from Son-Mahanadi (Agnihotri et al. 2018; Saxena et al. 2019), Talcher (Das 1958; Tripathi 1993; Tripathi and Bhattacharya 2001; Meena 2003; Ram-Awatar 1987; Mishra et al. 2022), Singrauli (Vijaya et al. 2012a), Godavari (Jha 2014; Jha and Aggarwal 2012; Srivastava and Jha 1988), Rajmahal (Pillai et al. 2020; Murthy et al. 2021), Wardha (Murthy et al. 2017; Aggarwal et al. 2019), Damodar (Vijay et al. 2012; Murthy et al. 2020b), South Karanpura (Murthy et al. 2014), Taptapani-Ramkola (Tripathi et al. 2012), Mand-Raigarh (Murthy et al. 2014), Raniganj (Gautam et al. 2022) and Satpura (Murthy et al. 2013) basins.

4.4 Barren Measures Formation (Middle Permian, Guadalupian)

This formation is mostly barren of coal seams, hence the name. However, in some basins, only a few coal seams have been documented (Dutta 2002). It lies conformably over the Barakar Formation. Local names for the unit include the Ironstone Shale or Kulti Formation in the Raniganj Basin and the Motur Formation in Satpura. The thickness of this formation is highly variable and usually ranges from 100 to 500 m.

(a) **Lithology**: Coarse to medium grained greenish-grey feldspathic sandstones with shreds and lenses of chocolate-coloured clay, siltstone, dark grey micaceous shale, carbonaceous shale and purple-brown shale and clay ironstone characterise this formation.
(b) **Assemblage zones**: This assemblage continues to be dominated by striate disaccate pollen, primarily *Striatopodocarpites*, comparable to the Barakar Assemblage Zone. The trilete spores reduce significantly in this assemblage. One palynozone was identified in this formation: *Densipollenites indicus* Assemblage Zone (Tiwari and Tripathi 1992).
(c) **FAD**: *Densipollenites densus.*
(d) **DOD**: *Densipollenites, Striatopodocarpites, Gondisporites* and *Verticipollenites.*
(e) **LAD**: *Verticipollenites oblongus, Distriatites bilateris* and *Didecitriletes ericianus.*
(f) **Reports from different basins of India**: Report of palynomorphs from Barren Measures Formation is scarce. Palynoflora from this formation have been reported from Godavari (Srivastava and Jha 1988), Damodar (Vijaya et al. 2012a, b), Singrauli (Vijaya et al. 2012b) and Talcher (Tripathi and Bhattacharya 2001) basins.

4.5 Raniganj Formation (Late Permian–Lopingian)

Raniganj Formation overlay the Barren Measures Formation. It is known as Bijori Formation in the Satpura Basin and Lower Kamthi in Mahanadi and Godavari basins.

(a) **Lithology**: Fine to coarse-grained, pebbly cross-bedded ferruginous sandstones, carbonaceous and grey shales, pink clays and siltstone make up this formation. The coal seams in Raniganj Formation are thinner and less frequent than the Barakar Formation. The thickness of the formation ranges widely from 200 to 700 m and sometimes even more.
(b) **Assemblage zones**: Similar to the Barren Measures, the striate bisaccate pollens (mostly *Striatopodocarpites*) continue to dominate this formation, whereas, in the middle of the formation, the frequency of trilete spores (*Microbaculispora, Microfoveolatispora, Cyclogranisporites* and *Horriditriletes*) increases and gradually decreases towards the upper portion. The final stage of this formation observes a rise in radial monosaccate pollen. In order of oldest to youngest, Tiwari and Tripathi (1987) assigned four palynozones (generic) to the Raniganj Formation: *Striatopodocarpites, Gondisporites, Densipollenites*, and *Crescentipollenites*, while later Tiwari and Tripathi (1992) assigned two palynoassemblage zones to the Raniganj Formation, namely the *Gondisporites raniganjensis* zone and the *Densipollenites magnicorpus* zone, which represent the Late Permian and Latest Permian, respectively.
(c) **FAD**: *Distriomonosaccites ovalis, Distriatites bilateris* and *Verticipollenites oblongus*.
(d) **DOD**: *Striatopodocarpites, Faunipollenites* and *Crescentipollenites*.
(e) **LAD**: *Lundblandispora brevierla, Lundblandispora microconata, Lunatisporites diffusus, Playfordiaspora cancellosa* and *Klausipollenites schaubergeri*.
(f) **Reports from different basins of India**: Raniganj palynoflora have been reported from Damodar (Murthy 2017; Vijaya et al. 2012b; Bhattacharya et al. 2021), Godavari (Jha et al. 2014; Jha and Aggarwal 2012), Talcher (Tiwari et al. 1991; Tripathi and Bhattacharya 2001; Sahoo et al. 2020a, b), Ib-River (Patel et al. 2021), Singrauli (Vijaya et al. 2012a), and South Rewa (Gautam et al. 2016) basins.

4.6 Panchet Formation (Early Triassic, Induan–Olenekian)

This formation overlays the Raniganj Formation and is the youngest in the Lower Gondwana. The Panchet Formation is known as Upper Kamthi in Mahanadi Basin, Middle-Upper Kamthi in the Pranihta Godavari Basin, Pali in the Son Basin, and Pachmarhi in the Satpura Basin.

(a) **Lithology**: Sandstone, red shale, greenish yellow shale, and coal-bearing sediments characterise this formation. In the lower portion of this formation, sandstones are arkosic, while in the upper portion, they are sub-arkosic. Intraformational fragments of predominantly greenish-yellow shale in sandstone are distinguishing features of this formation. The yellowish hue of the sandstone is easily distinguishable from the greenish sandstones of the underlying Raniganj Formation.

(b) **Assemblage zones**: Tiwari and Tripathi (1992) defined two palynozones in the Panchet Formation: (i) *Klausipollenites schaubergeri* Assemblage Zone (Lower Panchet) and (ii) *Playfordiaspora cancellosa* Assemblage Zone (Upper Panchet). From oldest to youngest, Tiwari and Tripathi (1987) described four palynozones (generic) that represent the Panchet Formation: *Klausipollenites–Striatopodocarpites*, *Callumispora-Verrucosisporites*, *Lunatisporites* and *Lundbladispora-Densoisporites* assemblage zones.

    Notably, the genus *Callumispora*, one of the distinctive compositions of the Talchir and Karaharbari formations, appears in the early stage of this formation.

(c) **FAD**: *Lundbladispora*, *Lunatisporites pellucidus*, *Playfordiaspora*, *Densoisporites*, *Verrucosisporites triassicus* and *Cavalisingulate*-zonate spore.

(d) **DOD**: *Verrucosisporites*, *Lunatisporites*, *Lundbladispora*, *Densoisporites*.

(e) **LAD**: *Lundbladispora baculata*, *Lundbladispora densicorpus*, *Indotriradites mamillatus*, *Verrucosispora densus*, *Gondirospora morondavensis* and *G. indica*.

(f) **Reports from different basins of India**: Palynoflora from Panchet Formation have been reported from Godavari (Jha and Aggarwal 2012), Talcher (Tripathi 1996; Tripathi and Bhattacharya 2001; Meena 2003; Bhattacharya et al. 2001), Ib River (Pillai et al. 2016), South Rewa (Gautam et al. 2016; Pillai et al. 2018), Singrauli (Vijaya et al. 2012b), Purnea (Prasad and Pundir 2017), Damodar (Bhattacharya et al. 2021) and Satpura (Bharadwaj et al. 1978b; Murthy and Sarate 2015) basins.

4.7 Supra Panchet (Mahadeva/Parsora) Formation (Late Triassic/Jurassic, Carnian–Norian/Toarcian–Bajocian)

This formation overlay the Panchet Formation with a pronounced unconformity. The Supra Panchet Formation is known as Tiki-Parsora in Son Basin, Dubrajpur in Rajmahal and Maleri-Dharmaram in Pranihta Godavari Basin.

(a) **Lithology**: Coarse to medium-grained sandstone, pebbly sandstone, enormous conglomerate bands and reddish to purplish siltstone make up the Supra Panchet Formation. Pebbles, cobbles and boulders are found throughout this unit and are made of vein quartz, jasper, chert, and other chemically stable elements.

(b) **Assemblage zones**: Four palynozones represent the Supra Panchet Formation (Tiwari and Tripathi 1992): (i) *Goubinispora morandavensis*, (ii) *Rajmahalispora rugulata*, (iii) *Brachysaccus ovalis* and (iv) *Dubrajisporites triassicus* assemblage zones.
(c) **Reports from different basins of India**: Palynoflora from Supra Panchet Formation have been reported from the Talcher (Tripathi 2009), Singrauli (Vijaya et al. 2012a) and South Rewa (Gautam et al. 2016) basins.

4.8 Jabalpur Formation (Early Cretaceous, Berriasian–Barremian)

The Jabalpur Formation represents the youngest formation of Gondwana in India. It is named after the Jabalpur City, Madhya Pradesh and forms the highest strata in Mahadeva hills. This is otherwise known as the Bansa Bed in the Son Basin, the Rajmahal Trap in the Rajmahal Basin and Gangapur/Chikiala sequences in the Pranhita–Godavari Basin. The Jabalpur Formation rests unconformably on Archaean rocks and is overlain by the Lameta Formation or Deccan traps (infra–intertrappeans). As a result, the period of non-deposition between the two formations spans a significant portion of the Cretaceous (Albian-Campanian).

(a) **Lithology**: It consists of thick, soft, white to brown massive sandstone overlain by fire clays, coal and sandy shale. The shale and clay horizons are rich in macro- and palynofloras. A 2.5-m-thick coal seam has also been observed in a dug well (Dogra et al. 1994; Rajanikanth and Chinnappa 2016).
(b) **Assemblage zones**: One palynozone was identified in this formation: *Narbadapollis indicus* Assemblage Zone. The significant palynotaxa of this assemblage are *Narbadapollis indicus*, *Platysaccus bharadwajii*, *Araucariacites limbatus*, *Cycadopites couperi*, *Podocarpites elipticus*, *Araucariacites limbatus*, *Callialasporites trilobatus*, *C. segmentatus* and *Ginkgoretectina vastus* (Dogra et al. 1994). The palynoflora of the Jabalpur Formation, as described by Kumar (1994), is predominately dominated by *Araucariacites–Callialasporites* along with *Cooksonites*, *Cyathidites*, *Triporoletes*, *Coptospora*, *Aquitriletes*, *Crybelosporites* and *Podosporites*.
(c) **Reports from different basins of India**: Palynoflora from Jabalpur Formation have been mostly reported from the Satpura Basin (Kumar 1994, 2011; Dogra et al. 1988, 1994; Prakash 2008; Tewari et al. 2009).

5 Palynomorphology and Their Palaeoclimatic Implications from the Indian Gondwana Sequences

The palaeoclimate of a particular geological period can be inferred using various indicators/proxies such as isotopic ratios of different elements, fossil remains (flora and fauna), lake sediments, ice cores, etc. Fossils play an essential role among palaeoclimatic indicators. Among them, animal fossils are more significant in marine sediments, while plant fossils are essential in terrestrial sediments. Among plant fossils (megaflora and palynoflora), palynomorphs helps in high precision palaeoclimate interpretations. There are two important prerequisites for interpreting palaeoclimate based on palynology: (i) evaluation of morphographic characters of dispersed spores and pollens, and (ii) changing composition of morphographic characters over time.

5.1 Evaluation of Morphographic Characters

Fossil spores and pollen exhibit different morphographic characters with specific configurations and structures, each with a precise function. Here, we will discuss different morphographic characters and their climatic sensitivity and behavioural patterns. The genera *Callumispora* and *Densipollenites* have been treated as characters for the description (Tiwari and Tripathi 1987). The relative abundance and distribution of different morphographic characters in Indian Gondwana sequences are presented in Fig. 4. The climatic conditions related to each morphographic feature are shown in Table 2.

(a) **Central body**: It is the central part of the pollen grain containing the cytoplasm and nucleus, which give rise to the microgametophyte after division. The central body is of two types based on the density of its wall: dense central body (Fig. 4(2)) and thin central body (Fig. 4(3)). The thickness of the central body increases in response to extreme climates to protect the delicate cell contents and vice versa. It is observed that pollen with a dense central body is dominant in Talchir and Karharbari assemblages (Fig. 4(2)) (Tiwari and Tripathi 1987). Though they are present until the Panchet Formation, they are on a decreasing trend.

(b) **Saccus**: The mode and extent of sexine with relation to nexine results in the formation of two types of saccate pollen: monosaccate and disaccate. Radial monosaccate aids in the flotation of pollen during dispersal and shields nexine by covering the majority of the subequatorial zone. Radial monosaccate pollen distribution (Fig. 4(1)) reveals that it is more prevalent in the Talchir and Karharbari formations and less prevalent in the Barakar, Barren Measures, and Raniganj formations. However, the Raniganj-Panchet boundary and Upper Raniganj Formation both exhibit a recurrence of radial monosaccate pollen.

In disaccate, there are two forms: haploxylonoid (Fig. 4(4)) and diploxylonoid (Fig. 4(5)). The shapes of these sacci are climate sensitive. While species

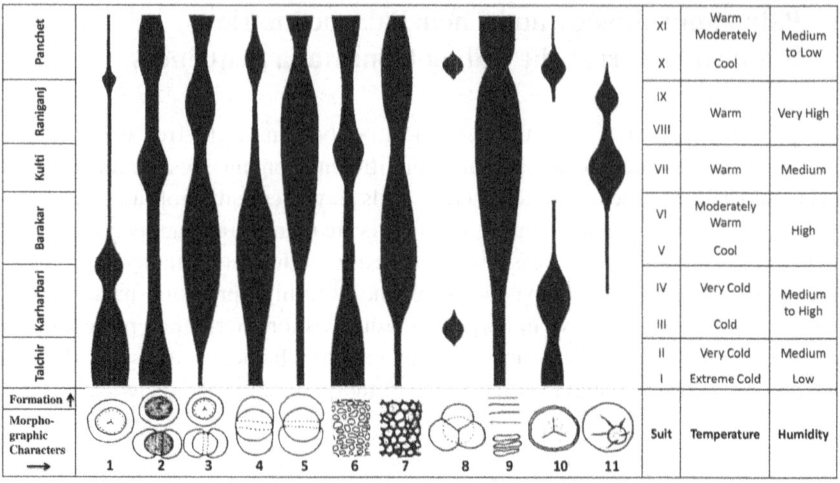

Fig. 4 Relative abundance and distribution of different morphographic characters in Indian Gondwana sequences (Tiwari and Tripathi 1987). 1. Radial monosaccate, 2. Dense central body, 3. Thin central body, 4. Haploxylonoid construction, 5. Diploxylonoid construction, 6. Leathery saccus, 7. Non-leathery saccus, 8. Tetrad of spores, 9. Striations and taeniae, 10. *Callumispora* and 11. *Densipollenites*

Table 2 Different morphographic characters and their respective climatic conditions (after Tiwari and Tripathi 1987)

Sl. No.	Morphographic characters	Climatic condition
1	Dense central body	Low humidity
2	Thin central body	High humidity
3	Radial monosaccate	Cold phase
4	Haploxylonoid construction	Cool phase
5	Diploxylonoid construction	Warm phase
6	Leathery saccus structure	Less humidity
7	Non-leathery saccus structure	Medium to high humidity
8	Striations and taeniae	Seasonal fluctuations
9	Tetrad of spores	Sudden change in temperature
10	*Callumispora*	Cool phase, low to medium humidity
11	*Densipollenites*	Warm phase, low to medium humidity

with diploxylonoid sacci grow in temperate zones, those with haploxylonoid sacci grow in cool temperatures (subpolar and subalpine zones) (Tiwari and Tripathi 1987). In the Lower Permian and Lower Triassic, the haploxylonoid state predominates over the diploxylonoid. Diploxylonoid conditions improve in the Upper Permian and remerge as a significant group in the later Lower Panchet.

(c) **Saccus infrastructure**: The saccus of Permian gymnospermous pollen is filled with alveolae (size varies from less than one micron to ten microns). This type of filled nature is known as pseudosaccate. The arrangement of endosexinal elements produces the alveolate construction, also known as an infrareticulate structure, which is made up of anastomosing ridges, or muri (thickness 0.5–3 μm). Gondwana pollen has also been shown to have an infra-punctate structure with a leathery, thick appearance similar to sexine. The leathery sacci can be differentiated from the non-leathery sacci, which have thin muri and a fin to medium-sized reticulate structure. Thus, the significance of leathery and non-leathery types is evident in the transition from extreme to pleasant climatic conditions (Tiwari and Tripathi 1987). The distribution of leathery pollen shows maximum abundance in Talchir palynoflora (Fig. 4(6)). Later, it declines gradually and is replaced with non-leathery types at the end of the Late Permian (Fig. 4(7)). The reappearance of leathery saccus again occurs in Triassic palynoflora.

(d) **Tetrad**: A tetrad releases four individual spores or pollen grains when the cementing substance dissolves. Their release depends on certain enzymes. There is a climatic condition causing the non-dispersal of four entities. Abnormal temperature behaviour can cause enzyme function to fail, leading to an explosion of tetrad synthesis. Thus, a higher number of tetrads in the palynoassemblage reflect the changing conditions in temperature. The tetrads are more frequent in Talchir and Lower Panchet (Fig. 4(8)) and scarce in other formations.

(e) **Striations and taeniae**: The striations (linear grooves) are present on the central body, and the taeniae (irregular ribbon-shaped stripes) are located on the proximal face of the central body, leaving thin, irregularly wide areas between them. The striations exhibit a clear evolution trend (complexity) during the Permian. Taeniae were the last to diversify in the Early Triassic (Fig. 4(9)). This quantitative complexity and abundance slowly increase from Talchir to Karharbari. It is at its peak in Barakar, Barren Measures, and Raniganj, but a decline is recorded at the Permian–Triassic boundary. The maximum diversification in striation types is evidently associated with the most luxuriant vegetation, which produced coal and thus signifies a warm humid climate.

(f) ***Callumispora***: It's a trilete bearing psilate spore with a structured, thick exine that shows layering in the optical section. This thick layering of the spore has been developed to protect the cell content from adverse conditions. These taxa are present in fair amounts in Lower Talchir and increase in Upper Talchir. It reaches its maximum abundance in Lower Karharbari. Later, it declines from Upper Karharbari to Lower Raniganj and reappears prominently in the Upper Raniganj-Lower Panchet palynoflora. The relative abundance of *Callumispora* is presented in Fig. 4(10).

(g) ***Densipollenites***: The monosaccus of this pollen completely encloses the central body from all sides. Due to their leathery saccus, *Densipollenites* are best suited for challenging climatic conditions. It shows the highest abundance in the Barren Measures Formation and subsequently declines in Raniganj. Later, it increases in Upper Raniganj and disappears in Lower Panchet (Fig. 4(11)).

5.2 Changing Composition of Morphographic Characters

There have been attempts to extrapolate climatic information from the unique combination of characteristics present in a certain palynoflora. Such combinations are known as 'suites'. Tiwari and Tripathi (1987) suggested eleven suites for the Indian Gondwana. These suites and their respective climatic inferences have been presented in Table 3.

Table 3 Palynocompositions of Indian Gondwana and their respective climatic inferences (after Tiwari and Tripathi 1987)

Formations	Morphographic characters		Palynozones	Temperature	Humidity
Panchet	Suite 11	Thin central body, diploxylonoid, leathery saccus, taeniae	D	Warm	Medium to low
			C		
			Composition VI		
	Suite 10	Haploxylonoid, dense central body, taeniae, non-leathery saccus, radial monosaccate, *Callumispora*	B	Moderately cool	
			A		
Raniganj	Suite 9	Thin central body, haploxylonoid, non-leathery saccus, striation, radial monosaccate, *Densipollenites*	D	Warm	Very high
			C		
			Composition V		
	Suite 8	Thin central body, diploxylonoid, non-leathery saccus, striation	B		
			A		
Barren Measures (Kulti)	Suite 7	Dense central body, diploxylonoid, leathery saccus, striation, *Densipollenites*	Composition IV A	Warm	Medium
Barakar	Suite 6	Thin central body, diploxylonoid, non-leathery saccus, striation	B	Moderately warm	High
			Composition III		
	Suite 5	Thin central body, haploxylonoid, non-leathery saccus	A	Cool	

(continued)

Table 3 (continued)

Formations	Morphographic characters		Palynozones	Temperature	Humidity
	Suite 4	Radial monosaccate, thin central body, haploxylonoid, non-leathery saccus	B	Very cold	
Karharbari			Composition II		Medium to high
	Suite 3	Thin central body, monosaccate, non-leathery saccus, haploxylonoid, *Callumispora*	A	Cold	
	Suite 2	Radial monosaccate, thin central body, leathery saccus, haploxylonoid, tetrad	B	Very cold	Medium
Talchir			Composition I		
	Suite 1	Radial monosaccate, dense central body, haploxylonoid, *Callumispora*	A	Extreme cold	Low

6 Palaeoclimate and Palynofloral Structuring of Gondwanaland

Tiwari and Vijaya (1995) illustrated palynofloral influence lines in south and north of equator during Early Permian (Fig. 5) and in Early Triassic (Fig. 6). They suggest the areas of influence of particular palynoflora, their intensity, and their gradient of paucity. 1S–4S represent the southern influence lines and 1N–4N shows the northern influence lines.

The Early Permian palynofloral structuring and distribution patterns of major morphos show that typical Gondwanan elements predominated between the 3-S and 4-S lines (Fig. 5). This region was influenced by glaciation and had a cool temperate climate. The palynomorphs mainly consist of girdling monosaccate pollen *Plicatipollenites*, *Parasaccites*, *Virkkipollertites*, *Potonieisporites*, rare striate and non-striate bisaccate pollen, and pteridophytic spores. This assemblage is primarily associated with the glacigene suites of the Talchir Formation in India.

The Late Permian period witnessed the nearly complete disappearance of glaciers from Gondwanaland, while a cool temperate climate with a warming trend prevailed. In India, striate (without monolete) and non-striate bisaccate pollen with pteridophytic spores predominate, with significant declining of monosaccates. The pattern of spatial distribution remains the same, as in Early Permian.

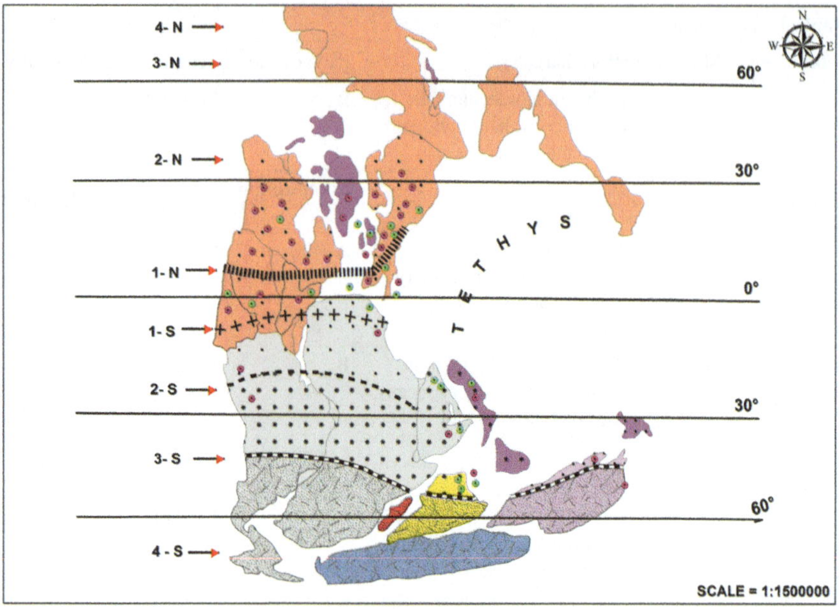

Fig. 5 Palynofloral influence lines in Early Permian. 1-S to 4-S in the southern hemisphere and 1-N to 4-N in the northern hemisphere, from equator towards poles, suggesting the areas of general impact of various palynofloras controlled by broad climatic zones during Early Permian time on Pangea. Fine and densely stippled: Glossopterid regime; coarsely stippled: decreasing effect of southern palynoflora towards equator; dot in green circle: tropical and highly wet equatorial palynoflora; dot in pink circle: Euromerian palynoflora. 1-N line marked by short oblique lines: beginning of drier and warmer zone northward; 1-S line, plus mark: southern limit of tropical zone. 2-S broken-line: northern limit of warm temperate with cooling effect from south; 3-S lines with alternate white and red bands: line of cool temperate to cold climate of the southern region. Rest of the arrows are levels of further changes in climate (after Tiwari and Vijaya 1995)

The Early Triassic period was marked by global warming and drier conditions. In India it was not a drier period as was thought earlier because diverse hygrophytic spores have been reported from the Early Triassic. The position of influence lines are shown in Fig. 6. The Gondwana palynoflora between 2-S and 3-S lines primarily consists of *Playfordiaspora*, *Krempipollenites*, *Satsangisaccates* and *Arcuatipollenites*, along with cavate and apiculate spores. The palaeoflora of the Middle Triassic did not differ significantly from that of the Early Triassic, but in the Late Triassic, more northerly elements entered the region of high latitude in the southern hemisphere. A comparison of the Permian and Triassic influence lines shows a major widening of the tropical wet season belt (1-S to 1-N) and that of the middle latitude and warm temperate belts (1-N to 2-N and 1-S to 2-S) in the Triassic.

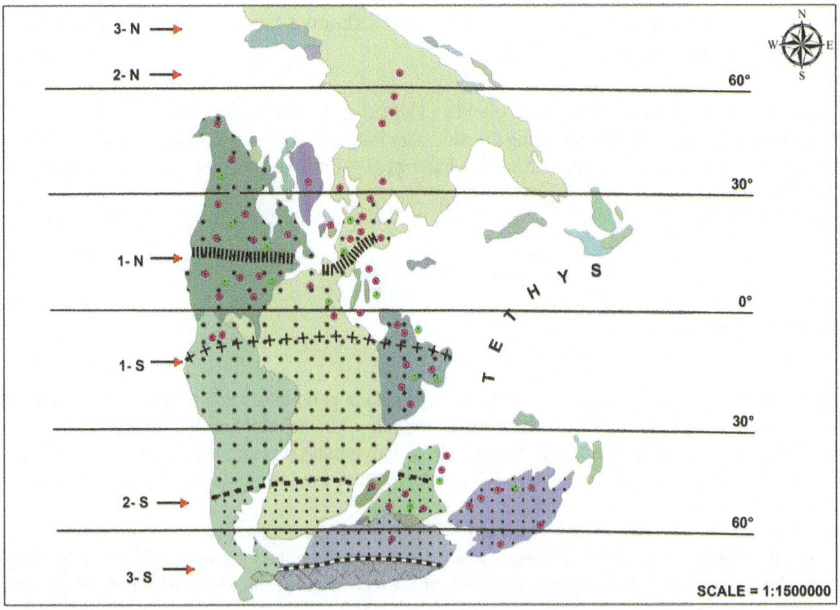

Fig. 6 Palynofloral influence lines in Early Triassic (symbols as in Fig. 5). Narrowing of southern cool temperate zone and widening of tropical and warm temperate areas, suggesting the beginning of uniformity in the climate. The Tethyan region has more of mixed palynoflora in Triassic than in the Permian (after Tiwari and Vijaya 1995)

7 Conclusion

The palynostratigraphic age of Gondwana is a challenging target, especially its significance in various palynozones. The present manuscript is a step towards the zonal framework of Indian Gondwana sequences. The biostratigraphic ages of different Gondwana sequences have been discussed with their respective palynoassemblage zones, along with FADs, LADs, and DODs. The climatic implications of these palynozones have been discussed in relation to evolution and the changing composition of the morphographic characters of fossil spores and pollen. Although a zonal framework for Gondwana has been erected, several gaps in the record need to be filled. The marine sequence of the Himalaya and Peninsular Indian Gondwana is not discussed. A palynological correlation with marine sections in the Himalayan region and peninsular India of Gondwanaland is necessary. The current paper is based on data and palynostratigraphy reports from the Indian Palaeozoic and Mesozoic sequences. There is an urgent need to prepare database software for different palynomorphs from the Indian Palaeozoic and Mesozoic to determine the age ranges of palynomorphs through the Gondwana sequence. The database information system is significant as it gives holistic information on palynomorphs, making the whole world of palynology a small village. This will aid new researchers in identifying and studying palynomorphs.

Acknowledgements The authors are greatly indebted to the Director, Birbal Sahni Institute of Palaeosciences, Lucknow for providing the necessary facilities. The author (SKS), INSPIRE Fellow is indebted to Department of Science and Technology (DST), Government of India for the research grant (IF200461).

References

Aggarwal N, Murthy S, Pillai SSK et al (2019) Artinskian palynoflora and palaeoclimate of Nand-Besur Block, Bandar Coalfield, Wardha Basin, India. J Palaeontol Soc India 64(2):241–255

Agnihotri D, Pillai SSK, Aggarwal N et al (2018) Palynomorphs from the Barakar Formation of Dhanpuri Open Cast Mine, Sohagpur Coalfield, Madhya Pradesh, India. Palaeobotanist 67:171–184

Backhouse J, Balme BE, Helby R et al (2002) Palynological zonation and correlation of the latest Triassic, Northern Carnarvon Basin. In: The sedimentary basins of Western Australia 3. Proceedings of the Petroleum Exploration Society of Australia Symposium, Perth, WA, pp 179–194

Bharadwaj DC (1962) The miospore genera in the coals of Raniganj Stage (Upper Permian), India. Palaeobotanist 9(1–2):68–106

Bharadwaj DC (1974) Palaeobotany of Talchir and Karharbari formations and Lower Gondwana glaciations. In: Surange KR, Lakhanpal RN, Bharadwaj DC (eds) Aspects and appraisal of Indian palaeobotany. Birbal Sahni Institute of Palaeobotany, Lucknow, pp 369–385

Bharadwaj DC, Saluja SK (1964) A sporological study of seam VII (Jote Dhemo colliery) in the Raniganj coalfield Bihar, India. Part I. Description of sporae dispersae. Palaeobotanist 12:181–215

Bharadwaj DC, Tiwari RS, Anand-Prakash (1978a) A Talchir mioflora from northern Satpura Basin, India. Palaeobotanist 25:62–69

Bharadwaj DC, Tiwari RS, Anand-Prakash (1978b) Palynology of Bijori Formation (Upper Permian) in Satpura Gondwana Basin, India. Palaeobotanist 25:70–78

Bhattacharya A, Nandi A, Dutta A (2001) Triassic mega and microplant fossils from the Kamthi Formation of Talcher Coalfield, Orissa with chronological significance. In: Proceedings of national seminar on recent advances in geology of coal and lignite basins of India, Calcutta, 1997, vol 54. Geological Survey of India Special Publication, pp 123–126

Bhattacharya S, Ankit Y, Murthy S et al (2021) Biotic response to environmental shift during the Permian-Triassic transition: assessment from organic geochemical proxies and palynomorphs in terrestrial sediments from Raniganj Sub-basin, India. Palaeogeogr Palaeoclimatol Palaeoecol 576:110483

Cashyap SM, Tewari RC (1987) Depositional model and tectonic evolution of Gondwana basins. Palaeobotanist 36:59–66

Casshyap SM, Tewari RC (1991) Depositional model and tectonic evolution of Gondwana basins. In: Venkatachala BS, Maheswari HK (eds) Indian Gondwana. Memoirs of the Geological Survey of India, vol 21, pp 95–206

Chakraborty C, Mandal N, Ghosh SK (2003) Kinematics of the Gondwana basins of peninsular India. Tectonophysics 377:299–324

Chandra A, Lele KM (1979) Talchir miofloras from South Rewa Gondwana Basin, India and their biostratigraphical significance. In: International palynological congress, Birbal Sahni Institute of Palaeobotany, Lucknow, vol 2, pp 117–151

Das DK (1958) On the microfossil content of Barakar Coal of Talcher Coalfield, Orissa. Q J Geol Min Metall Soc India 30(4):233–234

Dogra NN, Singh RY, Kulshrestha SK (1988) Palynological evidence on the age of Jabalpur and Lameta formations in the type area. Curr Sci 57(17):954–956

Dogra NN, Singh RY, Kulshreshtha SK (1994) Palynostratigraphy of infra-trappean Jabalpur and Lameta formations (Lower and Upper Cretaceous) in Madhya Pradesh, India. Cretac Res 15(2):205–215. https://doi.org/10.1006/cres.1994.1011

Dutta P (2002) Gondwana lithostratigraphy of peninsular India. Gondwana Res 5(2):540–553. https://doi.org/10.1016/S1342-937X(05)70742-5

Gautam S, Ram-Awatar TR et al (2016) Permian-Triassic palynoloral transition in Sohagpur Coalield, South Rewa Gondwana Basin, Madhya Pradesh, India. Palaeobotanist 65:109–129

Gautam S, Mendhe VA, Murthy S et al (2022) Palynoassemblages and depositional environment of the subsurface Permian sediments in Raniganj Coalfield, Damodar Basin, West Bengal, India. J Earth Syst Sci 131(4):1–21. https://doi.org/10.1007/s12040-022-01937-7

Ghosh P, Sengupta DP (2020) Geodynamics of Gondwanaland. In: Geodynamics of the Indian plate. Springer, Cham, pp 213–232

Holt KA, Bennett KD (2014) Principles and methods for automated palynology. New Phytol 203(3):735–742

Hyde HA, Williams DA (1944) Studies in atmospheric pollen. I. A daily census of pollens at Cardiff, 1942. New Phytol 43(1):49–61

Jha N (2006) Permian palynology from India and Africa-a phytogeographical paradigm. J Palaeontol Soc India 51(1):43–55

Jha N (2014) Upper Cisuralian palynology and palaeoclimate of Manuguru area Godavari basin, India and their global correlation. J Earth Syst Sci 123(7):1681–1692. https://doi.org/10.1007/s12040-014-0492-2

Jha N, Sarate OS (2001) Palynological analysis of Lower Gondwana coals from Ramagundam area, Godavari Valley Coalfield, Andhra Pradesh, India. Minetech 22:25–31

Jha N, Aggarwal N (2012) Permian-Triassic palynostratigraphy in Mailaram area, Godavari Graben, Andhra Pradesh, India. J Earth Syst Sci 121(5):1257–1285. https://doi.org/10.1007/s12040-012-0224-4

Jha N, Chary MB, Aggarwal N (2012) Permian Triassic palynofloral transition in Chintalapudi area, Godavari Graben, Andhra Pradesh, India. J Earth Syst Sci 121(5):1287–1303. https://doi.org/10.1007/s12040-012-0228-0

Jha N, Sabina KP, Aggarwal N et al (2014) Late Permian palynology and depositional environment of Chintalapudi sub basin, Pranhita-Godavari basin, Andhra Pradesh, India. J Asian Earth Sci 79:382–399. https://doi.org/10.1016/j.jseaes.2013.10.010

Kumar P (1994) Palynology of carbonaceous clays of Nimar Formation, Jhabua District, Madhya Pradesh, India. Geol Soc India 44(6):671–674

Kumar M (2011) Palynofloral and sedimentary organic matter analyses of Early Cretaceous sediments exposed along Sher River in Satpura Basin, Madhya Pradesh, India. Geophytology 41:9–23

Lele KM (1975) Studies in the Talchir flora of India–10. Early and Late Talchir miofloras from the West Bokaro Coalfield, Bihar. Palaeobotanist 22(3):219–235

McLoughlin S (2001) The breakup history of Gondwana and its impact on pre-Cenozoic floristic provincialism. Aust J Bot 49(3):271–300

Meena KL (2003) Sporae dispersae and correlation of Gondwana sediments in Talcher Coalfield, Son-Mahanadi Valley, Orissa. Geophytology 31:105–109

Meena KL, Murthy S (2011) Palynology of the Talchir Formation from the Ib-River Coalfield, Himgir basin, Odisha. J Indian Geol Congr 3:43–49

Meena KL, Pillai SSK, Murthy SK, Vethanayagam SM (2013) Early Talchir palynozone recorded from sediments of Baikunthpur Area, Chirimiri Coalfield, Chhattishgarh, India. Gondwana Geol Mag 28:45–52

Mishra DP, Singh VP, Saxena A et al (2022) Palaeoecology and depositional setting of an Early Permian (Artinskian) mire based on a multi-proxy study at the Jagannath coal mine (Talcher Coalfield), Mahanadi Basin, India. Palaeogeogr Palaeoclimatol Palaeoecol 601:111–124. https://doi.org/10.1016/j.palaeo.2022.111124

Mukhopadhyay G, Mukhopadhyay SK, Roychowdhury et al (2010) Stratigraphic correlation between different Gondwana basins of India. J Geol Soc India 76(3):251–266. https://doi.org/10.1007/s12594-010-0097-6

Murthy S (2017) Late Permian palynomorphs from the West Bokaro Coalfield, Damodar Basin, Jharkhand, India. Palaeobotanist 66(1–2):201–209

Murthy S, Vijaya Vethanayagam SM (2013) Palynostratigraphy of Permian succession in the Pench Valley Coalfield, Satpura basin, Madhya Pradesh, India. J Palaeontol Soc India 58(2):241–250

Murthy S, Sarate OS (2015) Early Triassic palynomorphs from Nand-Besur Coalfield, Nagpur District, Maharashtra, India. Geophytology 45(1):1–8

Murthy S, Tripathi A, Chakraborti B et al (2014) Palynostratigraphy of Permian succession from Binja Block, South Karanpura Coalfield, Jharkhand, India. J Earth Syst Sci 123(8):1895–1906. https://doi.org/10.1007/s12040-014-0505-1

Murthy S, Sarate OS, Pillai SSK et al (2017) Early Permian micro and megaspores from the Nand-Besur Block, Bandar Coalfield, Wardha Basin, Maharashtra, India. Palaeobotanist 66(2):177–189

Murthy S, Aggarwal N, Saxena A (2020a) Early Permian floral diversity and palaenvironment of the West Bokaro Coalfield, Damodar Basin, India. J Palaeontol Soc India 65:1–14

Murthy S, Saxena A, Chakraborti B (2020b) Palynostratigraphy of Permian and Mesozoic subsurface sediments of Brahmani Coalfield, Rajmahal Basin, India. J Palaeontol Soc India 65(2):149–161

Murthy S, Mendhe VA, Uhl D et al (2021) Palaeobotanical and biomarker evidence for wildfire in the Early Permian (Artinskian) of the Rajmahal Basin, India. J Palaeogeogr 10(1):1–21. https://doi.org/10.1186/s42501-021-00084-2

Patel R, Goswami S, Sahoo M et al (2021) Biodiversity of a Permian Temperate Forest: a case study from Ustali area, Ib River Basin, Odisha, India. Geol J 56(2):903–933. https://doi.org/10.1002/gj.3936

Pillai S, Meena KL, Tewari R et al (2016) Early Triassic palynomorphs from Kuraloi Block, Belphar area, Ib-River coalfield, Mahanadi basin, Odisha. J Geol Soc India 88(6):693–704. https://doi.org/10.1007/s12594-016-0536-0

Pillai SSK, Agnihotri D, Gautam S et al (2018) Glossopteris flora from the Pali Formation, Johilla coalfield, South Rewa Gondwana basin, Madhya Pradesh, India: palynological evidence for a Late Permian age. J Palaeontol Soc India 63(1):53–72

Pillai SSK, Mathews RP, Murthy S et al (2020) Palaeofloral investigation based on morphotaxonomy, palynomorphs, biomarkers and stable isotope from Lalmatia coal mine of Rajmahal Lower Gondwana Basin, Goddar District, Jharkhand, India: an inclusive empirical study. J Geol Soc India 96(1):43–57. https://doi.org/10.1007/s12594-020-1503-3

Prakash N (2008) Biodiversity and palaeoclimatic interpretation of Early Cretaceous flora of Jabalpur Formation, Satpura Basin, India. Palaeoworld 17(3–4):253–263

Prasad B, Pundir BS (2017) Gondwana biostratigraphy of the Purnea Basin (eastern Bihar, India), and its correlation with Rajmahal and Bengal Gondwana basins. J Geol Soc India 90(4):405–427. https://doi.org/10.1007/s12594-017-0735-3

Quattrocchio ME, Martínez MA, Pavisich AC et al (2006) Early Cretaceous palynostratigraphy, palynofacies and palaeoenvironments of well sections in northeastern Tierra del Fuego, Argentina. Cretac Res 27(4):584–602. https://doi.org/10.1016/j.cretres.2005.11.012

Rajanikanth A, Chinnappa CH (2016) Early Cretaceous flora of India—a review. Palaeobotanist 65(1–2):209–245

Ram-Awatar (1987) Palynological dating of supra Barakar formations in Son Valley Graben. Palaeobotanist 36:133–137

Rao AR, Dhar CL, Obergfell FA (1977) Badhaura Formation of Rajasthan: its stratigraphy and age. In: Laskar B, Rao CS (eds) Proceeding IV international Gondwana symposium, Calcutta, vol 2, p 281
Sahoo M, Aggarwal N, Goswami S (2020a) Palynological investigation of the Lower Gondwana outcrop near Gopalprasad, Odisha, India: an inference on the age, palaeovegetation and palaeoclimate. J Palaeontol Soc India 65(1):27–35
Sahoo M, Goswami S, Aggarwal N, Pillai SSK (2020b) Palaeofloristics of lower Gondwana exposure near Kumunda village, Angul District, Talcher Basin, Odisha, India: a comprehensive study on megafloral and palynofloral assemblages. J Geol Soc India 95(3):241–254. https://doi.org/10.1007/s12594-020-1422-3
Sarate OS, Murthy S (2017) Talchir palynoflora from Wardha Valley Coalfield, Central India: climatic implications. J Geosci Res 2(2):139–148
Saxena A, Singh KJ, Murthy S et al (2019) Early Permian macro and miofloral diversity from Singrauli Coalfield, Son-Mahanadi Basin, India. J Palaeontol Soc India 64(2):169–183
Shanker R, Singh G, Kumar G (1996) The sequence stratigraphy of Himalaya and its reflections in Gondwana sedimentation. Gondwana Nine, India 2:1159–1173
Shi G, Zhou Z, Xie Z (2010) A new Cephalotaxus and associated epiphyllous fungi from the Oligocene of Guangxi, South China. Rev Palaeobot Palynol 161(3–4):179–195. https://doi.org/10.1016/j.revpalbo.2010.04.002
Souza PA, Marques-Toigo M (2005) Progress on the palynostratigraphy of the Permian strata in Rio Grande do Sul State, Paraná basin, Brazil. An Acad Bras Ciênc 77:353–365
Souza PA, Perinotto JA, Félix CM et al (2015) Biostratigraphy and paleoecology of an unusual palynological record from the Aquidauana Formation, Late Pennsylvanian of Paraná Basin. An Acad Bras Ciênc 87:611–622
Srivastava SC (1973) Palynostratigraphy of the Giridih Coalfield. Geophytology 3:184–194
Srivastava SC, Jha N (1988) Palynostratigraphy of lower gondwana sediments in Godawari graben, Andhra Pradesh, India. Palaeobotanist 37(2):199–209
Srivastava SC, Sarate OS (1988) Palynology of the Talchir Formation from Betul Coalfield, Satpura Basin, India. J Palaeosci 37(1–3):81–84
Srivastava SC, Sarate OS (1989) Palynostratigraphy of the Lower Gondwana sediments from Shobhapur Block, Pathakhera Coalfield, Madhya Pradesh. Palaeobotanist 37:125–133
Tewari R, Kumar M, Prakash N (2009) Early Cretaceous megaspores from Sher River Section, Sehora, Satpura Gondwana Basin, Madhya Pradesh, India. Phytomorphology 59(1):7
Tiwari RS, Tripathi A (1987) Palynological zones and their climatic inference in the coal- bearing Gondwana of peninsular India. Palaeobotanist 36:87–101
Tiwari RS, Tripathi A (1992) Marker assemblage zones of spores and pollen species through Gondwana Palaeozoic and Mesozoic sequence in India. Palaeobotanist 40:194–236
Tiwari RS, Vijaya (1995) Differential morphographic identity of Gondwanic palynomorphs. Palaeobotanist 44:62–115
Tiwari RS, Kumar R (2002) Indian Gondwana palynochronology: relationships and chronocalibration. Palaeobotanist 51:13–30
Tiwari RS, Tripathi A, Jana BN (1991) Palynological evidences for upper Permian Raniganj coals in western part of Talchir Coalfield, Orissa, India. Curr Sci 61(6):407–410
Tiwari R, Tripathi A, Vijaya (1994) Organic-walled microfossils of doubtful origin in Permian and Triassic sequences on peninsular India. Palaeobotanist 43(1–3):1–38
Tripathi A (1993) Palynosequence in subsurface Permian sediments in Talcher Coalfield, Orissa, India. Geophytology 23(1):99–106
Tripathi A (1996) Early, Late Triassic palynoassemblages from subsurface Supra-Barakar sequence in Talcher Coalfield, Orissa, India. Geophytology 26(1):109–118
Tripathi A (2009) Palynology of the Brahmani River Section, Talcher Coalfield, Orissa, India. J Palaeontol Soc India 54(2):179–187
Tripathi A, Bhattacharya D (2001) Palynological resolution of Upper Permian sequence in Talcher Coalfield, Orissa, India. In: Proceedings of national seminar on recent advance in geology of

coal and lignite basins of India, Calcutta, 1997, vol 54. Geological Survey of India-Special Publication, pp 59–68

Tripathi A, Murthy S, Chakarborty B, Das DK (2012) Stratigraphic status of coal horizon in Tatapani-Ramkola Coalfield, Chhattisgarh, India. J Earth Syst Sci 121(2):537–557

Veevers JJ (1993) Gondwana facies of the Pangean supersequence: a review. Am Soc Mech Eng 513–520

Veevers JJ, Tewari RC (eds) (1995) Gondwana master basin of peninsular India between Tethys and the interior of the Gondwanaland province of Pangea. Geological Society of America

Venkatachala B, Tiwari R (1987) Lower Gondwana marine incursions: periods and pathways. Palaeobotanist 36:24–29 https://doi.org/10.54991/jop.1987.1556

Vijaya TA, Roy A et al (2012a) Palynostratigraphy and age correlation of subsurface strata within the sub-basins in Singrauli Gondwana Basin, India. J Earth Syst Sci 121(4):1071–1092. https://doi.org/10.1007/s12040-012-0213-7

Vijaya SM, Chakraborty B, Roy JS (2012b) Palynological dating of a subsurface coal bearing horizon in East Bokaro Coalfield, Damodar Basin, Jharkhand, India. Palaeontogr Abt B 288(1–4):41–63. https://doi.org/10.1127/palb/288/2012/41

Reappraisal of Permian and Early Triassic Palynoflora and Palynostratigraphy of Son-Mahanadi Basin and Their Climatic Implications

Srikanta Murthy, Anju Saxena, Sankar Suresh Kumar Pillai, and Suyash Gupta

Abstract Coal bearing Gondwana sequences of the peninsular India have been extensively studied for plant fossils since late nineteenth century. Over the years, ample amount of data has been generated enriching our understanding of past vegetation, its evolution, proliferation, decline and subsequent extinction during the Permian and Triassic periods. Palaeofloristic studies are imperative to understand the vegetation dynamics and biostratigraphy of Lower Gondwana and Triassic sediments as these are mostly devoid of widespread characteristics faunal remains except some rare occurrences. The present paper elucidates a reassessment of the palaeobotanical studies pertaining to Permian and Triassic sequences primarily focusing on palynological work with a glimpse of macrofloral studies and their implications in deducing the biostratigraphy, palaeoclimate and the palaeoenvironment of the Son-Mahanadi Basin. Like any other Gondawan basin, Permian Gondwana sequences of this basin are also characterized with the elements of Glossopteris flora that marked their first appearance during the earliest Permian (Talchir Formation) after the late Carboniferous deglaciation and gradually evolved and proliferated across the Permian encompassing Karharbari, Barakar, Barren Measures and Raniganj (= lower Kamthi formation) formations and ultimately got extinct during the early Triassic and succeeded by the appearance and diversification of *Dicroidium* flora during the Triassic. Attempts have been made to provide a detailed synthesis of the past vegetation and associated palaeoclimate prevalent during the deposition of each of the above mentioned formations and their biostratigraphic implications.

1 Introduction

Gondwana sediments in India, occur mainly along the five master Gondwana Basins in peninsular India. Almost 99% of coal resource for the country is from these Gondwana Basins. These basins are found along major river valleys either as separate

S. Murthy · A. Saxena (✉) · S. S. K. Pillai · S. Gupta
Birbal Sahni Institute of Palaeosciences, 53-University Road, Lucknow 226007, India
e-mail: anju_saxena@bsip.res.in; anju_saxena2002@yahoo.co.in

B. Samant and D. Thakre (eds.), *Applications of Palynology in Stratigraphy and Climate Studies*, Society of Earth Scientists Series, https://doi.org/10.1007/978-3-031-51877-5_3

body or combined with post-Permian sequences. These basins are named either after the major rivers namely, Damodar, Son, Mahanadi and Godavari or after the linear hill ranges like Satpura and Rajmahal. These basins are known as Damodar-Koel Basin, Son-Mahanadi Basin, Satpura Basin, Rajmahal Basin and Wardha-Godavari Basin. It is estimated that about 5 km thick strata has been deposited over 200 million years from Upper (earliest) Carboniferous to Early Cretaceous period (302–100 my) and are preserved in these basins are clubbed into Gondwana Super Group (Mukhopadhyay et al. 2010). The Gondwana Supergroup is subdivided into Permo-Carboniferous Lower Gondwana group and Mesozoic Upper Gondwana group.

Among the above mentioned basin, Son-Mahanadi Basin occupies the central and eastern part of the country and together constitutes the largest coal producing basin (Fig. 1). This basin comprises of eighteen coalfields and their adjoining areas which together constitute major Gondwana sedimentary deposits of the basin. These coalfields are Athgarh, Bisrampur, Chirimiri, Hasdo-Arand, Ib River, Jhilimili, Johilla, Korar, Korba, Lakhanpur, Mand-Raigarh, Sendurgarh, Sohagpur, Sonhat, Singrauli, Talcher, Tatapani-Ramkola, and Umaria. In general, Lower Gondwana (Permian) sequences are categorized in five formations namely, Talchir, Karharbari, Barakar, Barren Measures and Raniganj in stratigraphical order. Though, in the Son Basin, only Talchir, Barakar, Barren Measures and Raniganj formations are designated whereas in Mahanadi basin, Talchir, Barakar, Barren Measures and Kamthi (equivalent to Raniganj and early Triassic sequences) are recognized. Karharbari is demarcated in Damodar Basin only (Table 1). In the Son-Mahanadi Basin, from the palaeofloristic point of view, a plethora of work has been carried out in the last two centuries focusing on macrofloral assemblage. However, in comparison to macrofloral studies, palynological studies have received little late attention and got impetus during the latter half of the last century. Some of the pioneer work includes the work of Trivedi (1950), Bharadwaj and Sinha (1969a, b), Sinha (1969) and Tiwari (1969, 1971, 1973).

Focusing on macrofloral aspects extensively studied coalfields are Ib River Coalfield (Goswami et al. 2006a, b, 2018a, b; Singh et al. 2006; 2002; Patel et al. 2021a, b, c and references therein), Korba coalfield (Singh et al. 2011a, 2012), Singrauli Coalfield (Singh and Saxena 2015; Singh et al. 2016; Saxena et al. 2016a, b; Saxena et al. 2022 and references therein), Talcher Coalfield (Goswami and Singh 2013; Saxena et al. 2014; Mishra et al. 2022 and references therein) Tatapani-Ramkola (Saxena et al. 2019a; Singh et al. 2011b, Srivastava et al. 1997). More recently, to envisage a more holistic scenario of palaeovegetation both macro and miofloral studies are being carried out together.

The details about palynology, its basics and role in understanding the climate during the Gondwana sedimentation in India has been given in another chapter in this book itself by Sahoo et al. (2024) therefore, it is not dealt here in detail. The term ‘Palynology’ was coined by Hyde and Williams (1944) which is derived from the Greek word meaning ‘I Sprinkle’. Palynology is the study of microscopic bodies generally known as palynomorphs; pollen and spores and certain other microscopic structures, either of plant or of uncertain origin. The other common microscopic structures include acritarchs, dinoflagellates, algal and fungal spores etc. The size of palynomorphs ranges between 5 and 500 μm. The term ‘palynology’ was introduced

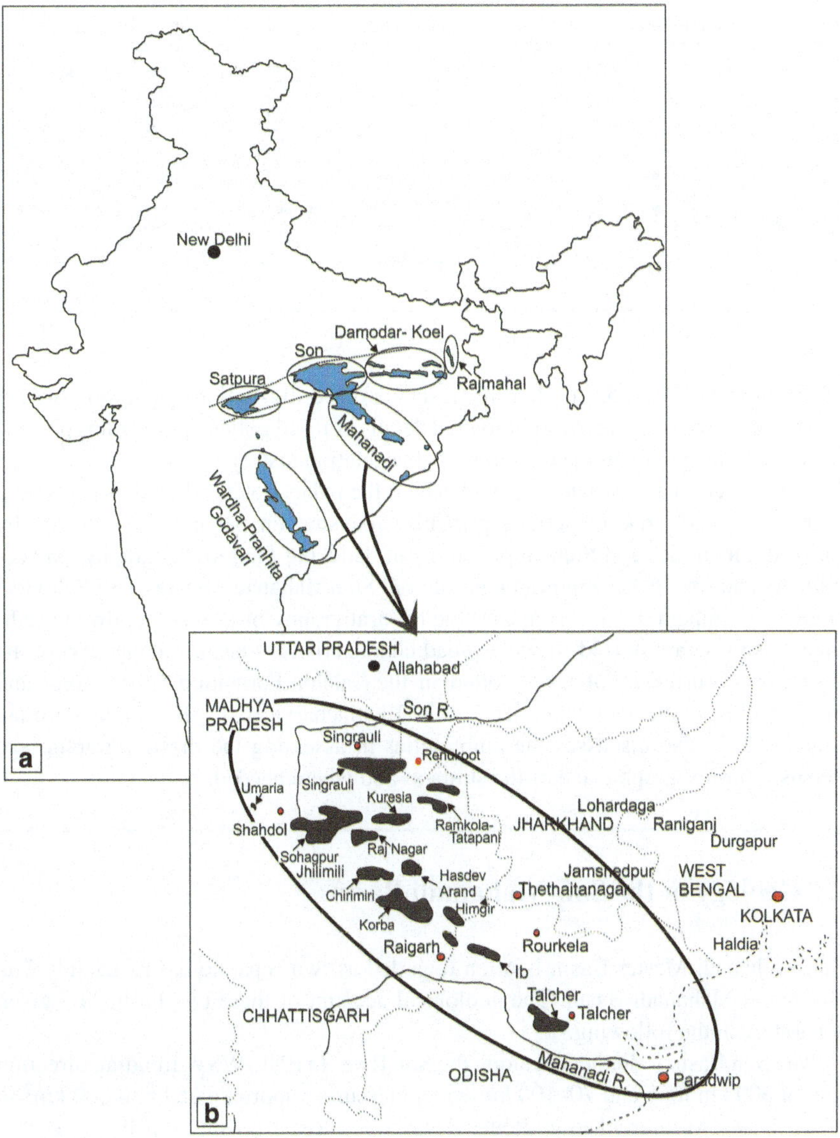

Fig. 1 **a** Coal-bearing Gondwana sedimentary basins of peninsular India. **b** Depiction of distribution of important coalfields in Son-Mahanadi Basin. The coalfields positions are drawn on the base map of India and are not to scale (after Saxena et al. 2021)

Table 1 Correlation of Late Palaeozoic Gondwana basins (after Mukhopadhyay et al. 2010)

Age	Gondwana Sedimentary Basins →				Damodar-Koel	Rajmahal	Southern Mahanadi	Son	Satpura	Godavari
251 Ma										
260 Ma	Time Slot III	Permian	Lopingian		Raniganj		Raniganj	Raniganj	Bijuri	Lr. Kamthi
			Guadalupian		Barren Measures		Barren Measures	Barren Measures	Motur	Barren Measures
271 Ma										
278 Ma	Time Slot II		Cisuralian	Kungurian	Up. Barakar	Up. Barakar	Up. Barakar	Up. Barakar	Up. Barakar	Up. Barakar
				Artinskian	Lr. Barakar/ Karharbari	Lr. Barakar	Lr. Barakar	Lr. Barakar	Lr. Barakar	Lr. Barakar
290 Ma	Time Slot I			Sakmarian Asselian	Talchir	Talchir	Talchir	Talchir	Talchir	Talchir
299 Ma										
302 Ma		Carboniferous		Gzhelian						

Non-deposition/Erosion

by Erdtman (1951, 1952) to describe a standardized illustration of the morphology of a pollen grain or spore. Morphological details of fossil pollen/spore grains provide significant insight about the evolution and vegetation history.

The present paper provides a synthesis of the palaeobotanical studies pertaining to Permian and Triassic periods primarily focusing on palynology followed by macrofloral studies and their implications in deducing the biostratigraphy, palaeoclimate and the palaeoenvironment of the Son-Mahanadi Basin. Palynological studies are imperative to understand the biostratigraphy of Lower Gondwana sediments as these are devoid of widespread characteristics faunal remains except the restricted occurrences of those found in the Talchir Formation (*Eurydesma* and other molluscs) and abundant trace fossils (Bhattacharya et al. 2021) in the Permian sequences. It also discusses the ambiguities in assessing the biostratigraphic age versus lithostratigraphic age of the strata as and where needed.

2 Geology of the Son-Mahanadi Basin

Son-Mahanadi Master Basin is often regarded as two separate basins namely Son Basin and Mahanadi Basin. The geological account of these two basins are given separately in the following.

The Son Basin is delineated along the Son River in ENE-WSW trending direction (about 300 km long and 70–100 km wide, measuring approximately 30,000 km^2 in area). It is also referred to as Rewa Basin/South Rewa Basin (as it lies in earlier Rewa state). It forms the continuity between Koel-Damodar Basin and the Satpura Gondwana Basin and occurs at the junction of Mahanadi Basin. Son Basin is broadly situated within the Proterozoic mobile belt known as the Central Indian Tectonic Zone (CITZ). This tectonic zone consists of two parallel E-W trending structural domains: the Son-Narmada mega-lineament to the north and the Sausar mobile belt to the south. All the major tectonic discontinuities are confined to the CITZ, where northern limits are bounded by Son-Narmada Fault and southern limits by Central Indian Shear Zone. The Gondwana sequences of Son Basin ranges from Permian to Early Cretaceous. This sequence is further divided into different formations namely Talchir and Barakar

(Early to middle Permian), Barren Measures and Raniganj (Late Permian), Pali (Late Permian-? Early Triassic), Parsora (Late Triassic/Early Jurassic), and Chandia Bed (Early Cretaceous). The major coal bearing Permian sequences are primarily exposed along the periphery of the basin, while a thick sequence of Mesozoic strata covers a large portion of the basin encompassing in the western and central regions.

The Son Basin exhibits asymmetrical profile in both its longitudinal and transverse profiles. It is a composite basin consisting of distinct sub-basins viz. Tatapani-Ramkola Basin in the east, Umaria Basin in the west, Singrauli Basin (comprising Moher and Main sub-basins) in the north and Sohagpur Basin (comprising Jhilimili, Sonhat, Sohagpur, and Johilla sub-basins) in the south (Mukhopadhyay and Bhattacharyya 2019). The generalized stratigraphic sequences of some of the coalfields of Son Basin are given in Tables 2 and 3. The Singrauli Coalfield is located at the northernmost boundary of the Son-Mahanadi Basin that extends from the centre of Peninsular India to the east coast. Towards the northern part of this coalfield, no sediments of Gondwana period occur. It is bounded between 23°47′ and 24°12′ N and 81°48′ and 82°52′ E and lies in the drainage area of Son and Rihand rivers. Structurally, this coalfield can be divided into two tectono-sedimentary sub-basins: (i) Moher sub-basin on the north-eastern side and (ii) the Singrauli main sub-basin to the west. The Permian sequences along with coal bearing strataare exposed in the Moher sub-basin whereas, the Triassic sediments of Panchet (Pali) and Mahadeva (Parsora) formations are mostly limited to the Singrauli main sub-basin. The Tatapani–Ramkola (TR) Coalfield is a north–south belt of coal-bearing rocks. On the northern part is Tatapani and on southern part is Ramkola and is approximately 4–5 km in width and 65 km in length. The coalfield lies between 23°30′–23°55′ N and 83°00′–83°40′ E (Raja Rao 1983). The Mahanadi Basin extends along the Mahanadi River with a vast extent of Lower Gondwana formations and covers Surguja, Korba, Bilaspur, and Raigarh districts of Chhattisgarh state and Angul, Jharsuguda, Sundargarh, and Sambalpur districts of Odisha state. Important coalfields of the region are: Mand-Raigarh, Korba, Hasdo-Arand, Talcher and Ib River. Talcher Coalfield or sub-basin is the southernmost member of this basin and is connected with other coalfields through a narrow belt of the sediments of Talchir Formation. Mand-Raigarh and Ib River coalfields lies within the catchment area of Mahanadi River, whereas, Hasdo-Arand and Korba coalfields extend over the drainage area of the Mahanadi and Son rivers, but owing to geological similarities and continuity, both these coalfields are classified under Mahanadi Basin. Talcher Basin lies within the catchment area of the Brahmani River, and is also grouped within the Mahanadi Basin.

Like other Gondwana rift valleys, Mahanadi Basin also exhibits half-graben configuration, where sub basins are asymmetrical in shape, showing the deposits of the thickest sedimentary piles closer to the faulted margin. The sedimentation in these basins is considered to be commenced in the sag basins and later in the rift basin setting. The Talcher basin, located west of the Hasdo-Arand and Korba basins, has better control over E-W trending faults and lineaments. The rest of the basin is dominated by NW-SE trending structures. Some intra-basinal longitudinal faults, like the E-W trending fault along the central part of the Talcher basin have throws of

100 s of meters, leading to the repletion of Barakar and Barren Measures sequences (Roychowdhury 2019).

Sedimentation in the Mahanadi Basin, began after the latest Carboniferous deglaciation as the case in other major Gondwana basins. The lithology of the Mahanadi Basin changes from N-S to NNE-SSW across the Mand-Raigarh basin. In the eastern side, Mand-Raigarh basin has dominance of shale facies, and in the western part is shows dominance of arenaceous facies. Lithological architecture of Barakar and Barren Measure formations, shows continuity with the Talcher and Ib River basins. In contrast, the lithological framework of the Mand-Raigarh Basin is more akin to that of the Hasdo-Arand and Korba coalfields. The generalized stratigraphic nomenclature of the Mahanadi Basin is given in Table 4. Geographical spreads of significant coalfields/basins are given in Table 5.

This coalfield is surrounded by Precambrian metamorphic rocks forming the basement, on the three sides, over which the Gondwana sediments were deposited. Chirimiri coalfield, is also a part of Son basin and lies between 23°8′–23°15′ N and

Table 2 The stratigraphy of the Permian sequences of Mahanadi Basin, Odisha, India (after Goswami et al. 2006b)

Age	Formation	Lithology	Area
Upper Permian	Lower Kamthi	Fine to medium grained well sorted sandstone, siltstone, clay bed, coal, shale and sandstone	Ib River and Talcher basins
Middle Permian	Barren Measures	Clay and ironstone nodules, shale, sandstone and coal	Ib River and Talcher basins
Late Lower Permian	Barakar	Shale, sandstone and coal	Ib River, Talcher, Gaisilat, Katringia and Athmalik basins
Middle Lower Permian	Karharbari	Shale, sandstone and coal	Ib River, Talcher, Gaisilat, Katringia and Athmalik basins
Early Lower Permian	Talchir	Boulder bed, needle shale, greenish sandstone, marl and rhythmites	Ib River, Talcher, Gaisilat, Katringia and Athmalik basins

Table 3 Geographical spreads of important coalfields/basin of Mahanadi Basin, India (after Roychowdhury 2019)

Coalfield/basin	Area (sq. km)	Latitude	Longitude	State
Hasdo-Arand	1200	22°39′–22°57′	82°210–83°39′	Chhattisgarh
Korba	520	22°15′–22°30′	82°15′–82°55′	Chhattisgarh
Mand-Raigarh	900	21°45′–22°42′	83°01′–83°44′	Chhattisgarh
Ib River	1375	21°30′–22°06′	83°37′–84°10′	Odisha
Talcher	1800	20°53′–21°12′	84°20′–85°23′	Odisha

82°17′–82°25′ E. Talchir Formation is exposed along rivulets flowing near the Chirimiri town and is characterized by olive green, thinly laminated shales and medium grained sandstone succeeded by alternate beds of shale and sandstones (Raja Rao 1983).

Table 4 General stratigraphic succession of Singrauli Coalfield, Son Basin (after GSI unpublished report, in Vijaya et al. 2012)

Age	Formation/group	Thickness	General lithology
Recent			Alluvium
Cretaceous	Basic intrusive		Dolerite dykes and sills
Late Triassic	Parsora (Mahadeva)	500 m +	Medium-to-coarse grained ferruginous quartzose sandstone
Early Triassic	Pali (Panchet)	700 m +	Greenish yellow to reddish yellow, medium-to-coarse grained sandstone with variegated siltstone and clay
Late Permian	Raniganj	215–400 m	Fine to medium grained dirty to buff coloured subarkosic to feldspathic wacke with alternation of thin lamination of grey and carbonaceous shale along with impersistent coal seams
Middle Permian	Barren Measures	110– 300 m	Dark brown to brownish yellow to greenish grey, medium-to-coarse grained flaggy sandstone with thin grey clay bands in between
Early Permian	Barakar	325–550 m	Dirty white fine-to-coarse grained sub-arkosic to arkosic sandstone along with siltstone, shale, carbonaceous shale and coal seams
Early Permian	Talchir	75–230 m	Dark greenish grey to grey shale, fine grained sandstone diamictite, siltstone pebbly sandstone and boulder bed
Unconformity			
Precambrian	Mahakoshal		Granite, gneiss, quartzite, phyllite, schist and pegmatite

Table 5 General stratigraphic succession of Chirimiri Coalfield, Son Basin (Raja Rao 1983)

Age	Formation	Lithology
Upper Cretaceous to lower Eocene (?)	Deccan Trap	Basic flows, dykes and sills
Lower Permian	Barakar	Essentially sandstone with subordinate shales and coal seams (230–435 m)
Upper Carboniferous to Early Permian	Talchir	Predominantly olive green shales and medium to fine grained sandstone
Unconformity		
Precambrian basement		Granite, gniesses and quartzite

3 Palynofloristics and Palaeoenvironment During the Deposition of Different Formations in Son-Mahanadi Basin

3.1 Palynofloristics and Vegetation of Talchir Formation (late Carboniferous to Early Permian)

Gondwana sedimentation in peninsular India commences with melting of Late Carboniferous ice that paved way for the deposition of thick sedimentary sequences resting unconformably over the Precambrian Basement. The lowermost formation is designated as the Talchir Formation, after the Talcher Coalfield in Mahanadi Basin. The type locality of Talchir Formation is exposed in the eastern part of Talcher Coalfield in and around Sarang Village. Basal most unit of Talchir Formation is of glacio-fluvial nature and commonly referred as Boulder Bed or Tillite. The overlying units comprises of conglomerate, sandstone and shales. The shales are peculiar needle shales typically of buff coloured. The associated sedimentary structures such as ripple marks, cross-bedded sandstone, hummocky cross-stratification, and pebbly conglomerate suggests that deposition took place in fluviatile dominated setting.

The basal most units of this formation are hitherto devoid of any macrofloral or palynological remains and are considered as of glaciogenic origin, whereas, the middle and upper units of Talchir Formation is mostly considered as the continental facies deposit as is evident by the occurrence of plant megafossils in the overlying needle shales (Chandra and Singh 1996; Singh et al. 2005). A good number of macrofloral assemblages has been recovered form the overlying units of boulder beds; primarily comprising of *Gangamopteris* and *Neoggerathiopsis* leaves remains along with occasional records of *Euryphyllum*, conifers, and bryophytic remains (Chandra and Srivastava 1982; Chandra 1995; Chandra and Singh 1996; Singh et al. 2005; Saxena et al. 2022 and references therein). Plant fossils together with sedimentary features indicate that deposition took place in a water body with vast expansion. The stable isotopic analyses recently carried out on the carbonate nodules present in the siltstone bed of the Talchir Formation exposed near Bedasar Village near Angul Town indicate that this phase of Talchir sedimentation occurred in a fresh-water environment, most probably in lakes formed by the melted glacial waters derived from the adjoining hills (Bhattacharya et al. 2002).

Talchir palynoflora: Like other Gondwana basins of India, the Talchir palynoflora is characterized by the dominant occurrences of radial monosaccate palynotaxa namely *Parasaccites* and *Plicatipollenites* and *Virikkipollenites* along with spores *Callumispora* and *Jayantisporites*. Some common taxa are shown in Fig. 2. Some initial palynological studies were carried out by Surange and Lele (1954) where winged spores were reported from the 'Talchir Stage' of Giridih coalfield; Lele (1964) from Talchir needle shale of South Rewa Gondwana basin and Chandra and Lele (1979) from South Rewa Gondwana Basin. Bharadwaj and Srivastava (1973) reported Talchir miofloras from the borecore of Korba Coalfield. A detailed and more recent study carried out by Vijaya et al. (2012) based on the borecore sediments of Singrauli Coalfield, reveals the Talchir palynoassemblage having abundance of *Parasaccites*- and *Plicatipollenites* associated with *Callumispora* and *Jayantisporites*. Spores-pollen diversity is shown by the presence of *Potonieisporites magnus*, *Tuberisaccites tuberculatus*, *Crucisaccites* sp., *Microbaculispora tentula*, *Lacinitriletes minutus*, *Brevitriletes unicus*, *Circumstriatites* spp., *Faunipollenites* sp., *Striatopodocarpites* sp., *Crescentipollenites fuscus*, *Tiwariasporis* sp., *Platysaccus* sp., and the alete forms *Tetraporina*, and *Maculatasporites*. This assemblage has been compared with the *Parasaccites korbaensis* Palynozone in Tiwari and Tripathi (1992), and dated as earliest Permian in age. However in a more recent study carried out by Murthy et al. (2020), the Talchir sediments from the bore core procured from Rajmahal Basin were reassessed for the age and lithostratigraphic versus biostratigraphic palynodating limitations in peninsular Gondwana sequences. They have identified two palynoassemblages. Palynoassemblage-I which is marked by the dominance of monosaccate pollen grains—*Potonieisporites*, *Plicatipollenites* and *Parasaccites* (*Cannanoropollis*), and is corroborated to the *Potonieisporites neglectus* Assemblage Zone of Tiwari and Tripathi (1992) of the lower Talchir Formation and a late Pennsylvanian age has been assigned. Palynoassemblage-II shows prominence of monosaccate pollen grains with subdued occurrences of bisaccate pollen grains and trilete spores and has been equated with *Plicatipollenites gondwanensis* Assemblage Zone (Tiwari and Tripathi 1992) of the upper Talchir Formation and for that an early Cisuralian age is assessed.

Similarly, in another study carried out by Tripathi et al. (2012), on bore core samples of Tatapani-Ramkola Coalfield, lithologically designated Talchir sediments have yielded the palynoflora correlatable to Lower Barakar palynoflora (*Scheuringipollenites barakarensis* Assemblage Zone of Tiwari and Tripathi 1992). Meena and Murthy (2011) reported a Talchir palynoassemblage from the outcrop samples exposed near the Chilkutia rivulet section in Ib River Basin. This palynoflora is rich in radial monosaccate pollen taxa viz. *Parasaccites* (*P. korbaensis*, *P. obscures*, *P. densicorpus*), *Plicatipollenites* (*P. gondwanensis*, *P. densus*, *P. indicus*) along with *Tuberisaccites tuberculosus*, *Tuberisaccites lobatus*, *Potonieisporites*, *Sahnites*, *Striatites*, *Faunipollenites*, *Lophotriletes* and assigned an earliest Permian (Asselian-Sakmarian) age to these sediments. Later, Meena et al. (2013) reported palynoassemblage from the sediments of Baikunthpur area, Chirimiri Coalfield, Chhattishgarh. The assemblage includes dominance of *Plicatipollenites*, *Parasaccites* and the other palynomorphs are *Potonieisporites*, *Tuberisaccites*, *Lophotriletes*, *Imparitriletes*, *Vesicaspora*, *Pachysaccus*, *Virkkipollenites*), *Callumispora*, *Valisaccites* and *Caheniasaccites*. It is placed in the *Plicatipollenites gondwanensis* assemblage zone of Tiwari and Tripathi (1992) and early Early Permian age is assigned.

These palynoflora of Son-Mahanadi basins are comparable with other known palynoassemblages of different Talchir exposures of India as reported by various workers such as Lele and Karim (1971) from Jayanti Coalfield, Bharadwaj et al. (1979) from Manendragarh area, Srivastava et al. (1988) from Betula Coalfield, Satpura Basin, Lele (1975) from West Bokaro Coalfield, Murthy et al. (2020) from

Fig. 2 Frequently encountered Palynotaxa in the Permian and Triassic sequences of Son-Mahanadi Basin. 1. *Microfoveolataspora bokaroensis* Tiwari (1965), 2. *Didecitriletes ericianus* Venkatachala and Kar (1965), 3. *Callumispora barakarensis* Bharadwaj and Srivastava (1969), 4. *Jayantisporites pseudozonatus* Lele and Makada (1972), 5. *Parasaccites korbaensis* Bharadwaj and Tiwari (1964), 6. *Parasaccites talchirensis* Lele and Makhada (1972), 7. *Plicatipollenites indicus* Lele (1964), 8. *Plicatipollenites gondwanaensis* (Balme and Hennel) Lele (1964), 9. *Potonieisporites neglectus* Potonié and Lele (1961), 10. *Crucisaccites latisulcatus* Lele and Maithy (1964), 11. *Barakarites indicus* Bharadwaj and Tiwari (1964), 12. *Scheuringipollenites tentulus* Tiwari emend. Tiwari (1973), 13. *Scheuringipollenites maximus* (Hart) Tiwari (1973), 14. *Scheuringipollenites barakarensis* Tiwari emend. Tiwari (1973), 15. *Lueckisporites virkkiae* Potonie and Klaus, 16. *Faunipollenites varius* Bharadwaj (1962), 17. *Rhizomaspora indica* Tiwari (1965), 18. *Striatopodocarpites decorus* Bharadwaj and Salujha (1964), 19. *Lunatisporites ovatus* (Goubin) Maheshwari and Banerji (1975), 20. *Tiwariaspora simplex* (Tiwari) Maheshwari and Kar (1967)

Rajmahal Basin, Sarate and Murthy (2017), Sabina et al. (2021) from Wardha Valley Coalfield.

3.2 *Palynofloristics and Vegetation of Karharbari Formation (Early Permian, Sakamarian-Artinskian)*

Karharbari as a Formation, has been debatable as it had not been deposited uniformly in all the basins and lack sharp delineation contact with underlying strata although having distinct lithocharacters and bioassemblage (mainly macroflora). Karaharbari Formation has gradational contact with overlying Barakar Formation. Sensu stricto, Karharbari Formation is recognized in Damodar Basin. It was first delineated in the Giridih Coalfield. Although it is reported to occur in patches in some of the basins namely, Son-Mahanadi and Satpura (primarily on the basis of floral assemblages) whereas in other basins it is considered as the basal most unit of overlying Barakar Formation by various workers (ref). It is the lowermost coal bearing horizon of Permian sequences in India. Lithologically, this formation comprises of grey to brown mottled carbonaceous sandstones, grit and conglomerates, polymictic pebble bed at the base and occasional coal seams (Sastry et al. 1977). Petrologically, it has dominance of reworked Talchir material in the matrix and distinct heavy mineral assemblage (Ghosh and Basu 1969). Coal seams are non-laminated and coal is dull in nature (Pareek 1969). Age bracketing of Karharbari is considered to be ranging from late Sakmarian to Artinskian.

The status of Karharbari as a definite chronostratigraphic unit in between Talchir and Barakar formations is contentious. However, on the basis of having typical macrofloral assemblage it is considered as a 'Biozone' by many palaeobotanists (see Banerjee, 1987). The characteristics macrofloral assemblage include *Botrychiopsis, Buriadia, Euryphyllum, Gangamopteris, Noeggerathiopsis, Ottokaria* and *Rubidgia* (Feistmantel 1886; Maithy 1966; Banerjee 1987; Singh et al. 2006 and references therein). The typical miofloral elements include *Callumispora, Parasaccites, Plicatipollenites*, brackish water acritarchs and fungal spores.

First palynological study on the Karharbari sediments was carried out by Sen (1953), in the type area i.e. Giridih Coalfield where lower Karharbari assemblage was identified on the basis of dominance of trilete spores (?*Callumispora*) and dominance of monosaccate pollens over bisaccates; later dominant in the overlying younger seam. Later on, number of palynological studies were carried out by Guhasarkar (1956), Maithy (1965, 1969), Bharadwaj (1966, 1971, 1974), Srivastava (1973), Tiwari (1974), Maithy and Misra (1984) from the type area. Karaharbari palynological assemblage is characterized by *Callumispora-Parasaccites-Plicatipollenites* assemblage. Two distinct palynoassemblages were identified, lower assemblage having the dominance of *Callumispora* (a trilete spore) and subdominance of *Parasaccites, Plicatipollenites, Brevitriletes* whereas upper assemblage has dominance of *Parasaccites* and *Plicatipollenites* and subdominance of *Callumispora* and *Brevitriletes.*

In the other coalfields of Son-Mahanadi Basin, the presence of Karharbari is demarcated by Tiwari (1967) who had reported Karharbari palynoassemblage from Ib River Coalfield having dominance of monosaccate pollen and trilete spores. Carbonaceous shale samples from the Ib seam have yielded a varied Karharbari palynoassemblage having the dominance of monosaccate pollens and trilete spores, the composition is designated as Ib-2 palynoassemblage (Navale and Tiwari 1968). The assemblage has prominent occurrences of *Parasaccites* and *Brevitriletes*, and subdominance of *Potonieisporites*, *Striatopodocarpites*, *Plicatipollenites*, *Punctatisporites*, (*Callumispora*) and *Virkkipollenites* and spores namely, *Horriditriletes* and *Indotriradites*, along with other genera. The assemblage is dated as Artinskian in age, and is s correlatable to the early Permian miofloral assemblages of the Korba (Tiwari 1965), Sohagpur (Navale and Tiwari 1967; Meena 2000; Ram-Awatar et al. 2004) and Talcher coalfields (Navale and Tiwari 1966). Presence of Karharbari Formation is also established in the bore core sequence of Talcher Coalfield by Tripathi and Bhattacharya (2001) on the basis of palynological studies. The assemblage is characterized by the abundance of *Parasaccites* followed by *Microbaculispora*, *Brevitriletes*, *Callumispora*, *Striatites*, *Faunipollenites* and *Rhizomaspora.* An upper Karharbari palynoassemblage was reported by Kar and Srivastava (2003), on the basis of sediment samples of a bore core raised from upstream of Deona rivulet and also samples recovered from an outcrop section exposed in Sukhnaiya rivulet; their Palynozone-1 shows the dominance of radial monosaccate pollen *Parasaccites* and sub-dominance of *Callumispora* and is corroborated with coeval palynoassemblage *Parasaccites-Callumispora* Assemblage Zone of Tiwari and Tripathi (1992). A Karharabri palynoassemblage has been identified in the bore core sediments of Mand-Raigarh Coalfield by Jana et al. (2002), although lithologically Karharbari Formation was not delineated in the borecore. Recently, Aggarwal et al. (2022) demarcated an upper Karharbari Palynoassemblage in the sediments of Nandira colliery, Talcher Coalfield based on the dominance of *Parasaccites*, and *Plicatipollenites* and subdued ocuurence of *Potonieisporites.* Based on floral entities, lithological and petrological observations Banerjee (1987) envisaged that the deposition of Karaharbari sediments took place in the same environment as that of Talchir, but deposition of lower Karharbari was under distinct microenvironment.

3.3 Palynofloristics and Vegetation of Barakar Formation (late Early Permian = Artinskian-Kungurian)

Barakar Formation is the most well developed formation existing in all the Gondwana basins and is the store house of more than 90% coal of India. Its lithological architecture contains thick sequences of repetitive cycles of shale, coal and sandstone deposition, comprising of fine to coarse grained sandstones, siltstones and dark grey to grey carbonaceous shale, sandy shale, fireclay and interbedded thick coal seams. Coal deposition within the Barakar Formation had possibly occurred in two phases the first phase took place above the Karharbari/Lower Barakar sediments where some of the thickest seams were deposited followed by the second phase that occurred during the deposition of Upper Barakar sediments, and this second phase is confined in the eastern part of peninsula (Mukhopadhyay et al. 2010). Palaeofloristically, it is the most extensively studied formation, having highly varied and rich macro and miofloral records. Plenty of macrofloral elements have been reported from this formation from the various coalfields of Son-Mahanadi Basin specially from Korba, Ib River, Singrauli, Talcher, and Tatapani-Ramkola Coalfield (Singh et al. 2011a, b, 2012; Goswami and Singh 2013; Goswami et al. 2018a, b; Saxena et al. 2016b; 2019a, b and reference therein). On the basis of macrofloral assemblages, two floral zones have been identified in Ib River basin (Singh et al. 2007; Goswami et al. 2018a), the older zone (lower Barakar Formation—Artinskian) shows dominance of genera *Gangamopteris–Noeggerathiopsis–Euryphyllum–Glossopteris* complex along with the occurrences of *Buriadia, Cordaites, Euryphyllum, Noeggerathiopsis, Kawizophyllum, Gangamopteris, Ottokaria, Vertebraria,* and some of the characteristic key Karharbari *Glossopteris* species namely, *G. angusta, G. communis, G. decipiens, G. karharbariensis, G. giridihensis, G. recurva, G. taenioides, G. zeilleri* (Goswami et al. 2006a). The assemblage seems to be the extension of Karharbari vegetation. The younger (upper Barakar Formation—Kungurian) zone is characterized by genera *Cyclodendron, Paracalamites, Trizygia, Schizoneura, Raniganjia, Benlightfootia, Bengalia, Neomariopteris, Dichotomopteris, Ginkgoites, Glossopteris, Surangephyllum, Palaeovittaria, Scutum, Senotheca* and *Vertebraria* (Goswami et al. 2018a).

Palynologically, also two palynofloral assemblages are recognized, the Lower Barakar palynoassemblage i.e. *Scheuringipollenites barakarensis* Assemblage Zone and Upper Barakar *Faunipollenites varius* Assemblage Zone (Tiwari and Tripathi 1992). In the Lower Barakar, non-striate pollens are dominant, and in the Upper Barakar, striate bisaccate pollens are more abundant, replacing the non-striate pollens. Some common Barakar palynomorphs are shown in Fig. 2.

The Lower Barakar palynoassemblages are recorded from the Jagannath coal mine of Talcher Basin (Mahanadi Basin) by Mishra et al. (2021, 2022) wherein the palynofloral composition reveals dominance of *Scheuringipollenites* followed by *Faunipollenites, Parasaccites, Callumispora, Microbaculispora* and *Rhizomaspora* along with low occurrences of other palynotaxa. The assemblage is compared to *Scheuringipollenites barakarensis* Assemblage Zone (Tiwari and Tripathi 1992). However, from the same area, lithologically and macrofloristically dated Lower Barakar sediments have yielded an upper Permian palynoflora (Patel et al. 2021a). The other palynological studies carried out recently on Barakar sediments in Mahanadi Basin are by Patel et al. (2021b, c), Sahoo et al. (2020a, b) although yielded the palynoflora having Late Permian affinity. Earlier works include palynological analyses of the sediments of Himgir-Rampur coal mine done by Navale and Tiwari

(1968) and Tiwari (1967) that had yielded a bisaccate-monolete rich miofloral assemblage. The palynoassemblage has dominance of *Sulcatisporites* (= *Scheuringipollenites*), *Brijrajisporites*, *Rhizomaspora*, *Primuspollenites*, followed by *Lahirites*, *Ibisporites*, *Latosporites*, *Cuneatisporites* and *Platysaccus* etc. and is designated as the upper Barakar flora (late Early Permian age). Similarly, Meena (1999, 2004), Tiwari et al. (1995), Meena and Pillai (2011) reported palynoassemblage from Chaturdhara rivulet section, Basundhara rivulet section, Chaturdhara rivulet section and bore-core IBKA-2 of Kuraloi Block in the Ib river Basin respectively. These palynoassemblage indicate late Early Permian age by having the characterized palynotaxa *Faunipollenites*–*Scheuringipollenites* along with *Rhizomaspora*, *Scheuringipollenites*, *Distriatites*, *Densipollenites*, *Striatopodocarpites* etc. From the borecore sediments of Mand-Raigarh, five distinct palynoassemblages have been demarcated which include the palynoassemblages having Karharabari and Raniganj affinity, though lithologically these two formations were not delineated in the borecore sediments (Jana et al. 2002). They argued that Barakar palynoflora indicate a time-regressive as well as time-transgressive nature of sedimentation during Barakar period in Mand-Raigarh area. From the Son Basin, Kar and Srivastava (2003), also reported Lower Barakar palynoflora from the Sukhnaiya rivulet outcrop section and their Palynozone II has abundance of *Scheuringipollenites* and subdominance of *Primuspollentites*, *Ibisporites* and *Rhizomaspora* and is corroborated to the *Scheuringipollenites maximus* Assemblage Zone of Tiwari and Tripathi (1992) and an Early Permian (Artinskian) age has been interpreted. In another study carried out in an open cast coal seam sequence of Bina colliery (Turra Seam) in Singrauli Coalfield, a Lower Barakar palynoassemblage has been identified (Singh et al. 2017). The assemblage is characterized with the dominance of *Scheuringipollenites* and *Parasaccites* over *Faunipollenites* along with the other age marker species namely, *Indotriradites*, *Microfoveolatispora*, *Rhizomaspora* and *Densipollenites* and is equitable to *Scheuringipollenites barakarensis* Assemblage-Zone of Tiwari and Tripathi (1992). This palynocomposition put these beds at the transition of Karharbari and Lower Barakar formations and are dated as late Early Permian (Artinskian) in age. This assemblage has also yielded peculiar occurrences of naked, fossil spore tetrads having affinities with the dispersed miospore genera *Indotriradites*, *Microbaculispora* and *Microfoveolatispora* which are the pioneer record of these tetrads from any Artinskian strata in the world; these were studied in detail for their significance in assessing the climate (Saxena et al. 2016a).

An Early Permian Lower Barakar palynoflora is reported from the lithologically designated Raniganj strata of Jhingurdah bottom seam of Singrauli coalfield, revealing an ambiguity in lithological and biostratigraphic age assessment (Singh et al. 2017). Referred as Palynoassemblage-III, the palynocomposition revealed dominance of genera *Scheuringipollenites* followed by *Faunipollenites*, *Verticipollenites*, *Barakarites*, *Crescentipollenites*, *Corriasaccites*, *Laevigatosporites* and is comparable with the *Scheuringipollenites barakarensis* Assemblage Zone-V of Tiwari and Tripathi (1992) and an Early Permian (Artinskian) age (Lower Barakar) was assigned to studied strata. In an another study from Son Basin, a well preserved Lower Barakar palynoflora was recorded from Dhanpuri Open Cast Mine, Sohagpur

Coalfield, Madhya Pradesh (Agnihotri et al. 2018a). Palynomorphs in this study include megaspores and spores/pollen grains. Megaspores are represented by 6 genera and 13 species, genera comprising *Bankisporites, Barakarella, Biharisporites, Bokarosporites, Jhariatriletes* and *Talchirella.* The microspore assemblage shows dominance of monosaccate pollen grains; *Parasaccites obscures, P. korbaensis* and *Plicatipollenites indicus*, have dominance over non-striate bisaccate taxa *Scheuringipollenites barakarensis* and *S. maximus.* Other palynomorphs include *Callumispora barakarensis, Faunipollenites* sp. and *Striatopodocarpites multistriatus, Ibisporites diplosaccus and Platysaccus ovatus*, along with *Tiwariasporis gondwanensis* and *Vitreisporites* sp. A Lower Barakar palynoflora has also been reported from Umaria Coalfield by Srivastava and Anand Prakash (1984).

The Upper Barakar palynoflora is reported in the borecore samples of Tatpani Ramkola Coalfield (Son Basin) by Tripathi et al. (2012), and designated as palynoassemblage II characterized by the dominance of *Striatopodocarpites, Faunipollenites* and *Scheuringipollenites* followed by marker taxa *Densipollenites densus, Cyclobaculisporites gondwanensis, Horriditriletes curvibaculosus, Verticipollenites crassus* and *Striatopodocarpites ovatus.* This assemblage is correlated with *Faunipollenitis varius* Assemblage Zone of Tiwari and Tripathi (1992), and is assessed as late Early Permian in age. Similar palynocmposition is recorded by Singh et al. (2017) from the Purewa Bottom Seam of Singrauli Coalfield having *Faunipollenites* as the dominant genus followed by *Scheuringipollenites, Densipollenites, Barakarites, Parasaccites, Striatopodocarpites* and *Rhizomaspora* corroborated with the *Faunipollenites varius* Palynoassemblage Zone-VI of Tiwari and Tripathi (1992) and a late Early Permian (Kungurian) age was assigned. Palynoassemblage III of Kar and Srivastava (2003) (Sukhnaiya rivulet section) reveals the dominance of *Faunipollenites* and *Striatopodocarpites* followed by *Striatites, Distriatites*, and *Lahirites.* It is also assessed as Upper Barakar palynoflora. In an another study carried out at igneous intrusion affected sediments by Sarana and Kar (2011), from Dhamani block of Tatapani-Ramkola Coalfield, has yielded palynocmposition having the dominance of striate bisaccates i.e. *Faunipollenites* and *Striatopodocarpites* and subdued occurrence of non-striate pollen *Scheuringipollenites* this palynoassemblage is correlatable to the *Faunipollenites-Scheuringipollenites* Assemblage Zone and is Early Permian in age.

The lower and upper Barakar palynofloras are comparable with the contemporaneous sequences of Indian Gondwana sediments of Damodar Basin (Tiwari and Tripathi 1992; Murthy 2010; Murthy et al. 2010), Rajmahal Basin (Murthy et al. 2021) and Satpura Basin (Murthy et al. 2013).

3.4 Palynofloristics and Vegetation of Barren Measures (middle Permian = Guadalupian)

As the name itself suggests, this formation is non-coaliferous and represents break in coal deposition. It is sandwiched between the underlying coaliferous Barakar Formation and overlying coaliferous Raniganj/Kamthi Formation. Lithologically, it comprises of grey shales, carbonaceous shales and ferruginous and buff colored sandstones. Owing to its typical lithology, the formation is also known as 'Ironstone Shale' or Kulti Formation in Damodar Basin and as 'Motur Formation' in Satpura Basin. Previously this formation was recognized only in the Damodar and Koel basins, whereas in other basins it was referred as post-Barakar sequences or merged with Barakar Formation (Raja Rao 1983). The type area for this formation is in Jharia Coalfield, Damodar Basin. Except for a few reports from Damodar and Pranhita–Godavari Basin, the formation is by and large devoid of megafossil remains. Srivastava and Tewari (2001) have recorded a rich *Glossopteris* assemblage from the Barren Measures Formation of Jharia Coalfield, Damodar Basin and Choparappu et al. (2019) have described *Glossopteris-Gangamopteris-Noeggerathiopsis* assemblage from Pranhita–Godavari Basin. From Son-Mahandi Basin, hitherto no published macrofloral records are available. Although, recently, elements of Glossopteris flora been recovered from the non-coaliferous 'red beds' above the Barakar Formation of Singrauli Coalfield (unpublished data).

Only one palynozone has been identified in this formation namely, *Densipollenites indicus* Assemblage Zone (Tiwari and Tripathi 1992) based on the sediments of Damodar Basin. However, in comparison to rare macrofloral records from Son-Mahanadi Basin, some palynological reports are available, most of them are from Tatapani-Ramkola (TR) Coalfield and are based on borecore samples. Although, in many instances, the retrieved palynological ages do not conform to the lithologically designated Barren Measures sediments. In the bore-core study from Dhamani Block of Tatapani-Ramkola Coalfield, Sarana and Kar (2011) recovered a palynoassemblage comprising of abundance of *Faunipollenites*, *Striatopodocarpites*, *Striatites* and *Strotersporites* followed by the subdominance of *Densipollenites* (mainly *D. invisus*) and was compared to *Striatopodocarpites-Densipollenites* Assemblage Zone of Tiwari and Tripathi (1992) and was equated with the Barren Measures Palynozone (Middle Permian age). Similarly, the presence of Barren Measures Formation was recognized on the basis of palynological study (Kar 2003), done on the bore core specimens of Meghuli Block, TR Coalfield. Four distinct palynoassemblages were identified in that study, of which Palynoassemblage III abundance of characteristic palynotaxa of Barren Measures Formation and is correlatable with *Faunipollenites–Striatopodocarpites-Densipollenites* Assemblage Zone. Earlier, in this coalfield, Barren Measures sediments were not demarcated by Raja Rao (1983). However, there presence as lithological units of Barren Measures Formation were identified later in 1996 by Geological Survey of India.

In another detailed study of Tripathi et al. (2012) from this coalfield, lithologically designated Barren Measures strata did not conform to palynological dates. The recorded palynoflora from bore core TRBD-3, from the part of the sequence demarcated as Barren Measures; is identified as Palynoassemblage-III and is similar to *Gondisporites raniganjensis* Assemblage zone, correlated with palynoflora of the Raniganj Formation and dated as latest Permian in age. The assemblage has dominance of *Striatopodocarpites* followed by *Crescentipollenites, Densipollenites, Verticipollenites, Distriatites, Faunipollenites, Weylandites* and marker species of *Distriatites bilateris, Gondisporites reticulates*, and *Verticipollenites oblongus*. Similarly, lithologically designated as Barren Measures sediment samples procured from the upper part (19.00–54.80 m) of this core has yielded palynoassemblage of latest Permian age and has the typical elements of Raniganj Formation comprising of *Densipollenites*, followed by *Brevitriletes, Laevigatosporites, Striatopodocarpites, Striatites, Verticipollenites, Distriatites* and marker species—*Lundbladispora brevicula, Arcuatipollenites diffuses, Krempipollenites indicus*. This assemblage is correlatable to *Densipollenites magnicorpus* Assemblage zone of Tiwari and Tripathi (1992).

Similarly, in another bore core (TROD-1), lower part of the Barren Measures Formation sequence has yielded a palynoflora having characteristic presence of *Striatopodocarpites* followed by *Brevitriletes, Cyclogranisporites, Microbaculispora, Densipollenites, Faunipollenites, Barakarites*, and marker taxa- *Arcuatipollenites diffuses, Densipollenites magnicorpus, Gondisporites reticulates*, and *Krempipollenites indicus*. This assemblage is comparable to *Densipollenites magnicorpus* Assemblage Zone of Tiwari and Tripathi (1992) and dated as latest Permian in age. However, the upper part of Barren Measures Formation sediments of this core has abundance of *Krempipollenites indicus* and is determined as Early Triassic in age. The palynological findings and lithological designations are not in accordance at places in these adjacent areas and needs to be redefined in time span (Saxena et al. 2019b). The palynoflora of Barren Measures Formation of Son-Mahanadi Basin (Kar 2003; Sarana and Kar 2011) is comparable with the Barren Measures palynoflora of Damodar Basin (Bharadwaj et al. 1965; Kar 1966, 1968). In a latest study carried out by Murthy et al (2023) based on the borecore samples of Pench area, Satpura Basin, lithologically designated Motur Formation sediments have yielded palynocomposition dominated by *Straitopodocarpites* followed by *Faunipollenites* and other taxa and is equatable to *Gondisporites raniganjensis* palynozone of Tiwari and Tripathi (1992) and are dated as Guadalupian Tiwari and Tripathi (1992), more precisely this assemblage zone is dated as Wordian-Capitanian in age (Prasad and Pundir 2017).

3.5 Palynofloristics and Vegetation of Raniganj/Lower Kamthi Formations (Late Permian, Lopingian)

Raniganj Formation is named after its type locality in Damodar Basin, and is known as the Bijori Formation in Satpura Gondwana Basin and is equivalent of lower Kamthi

Formation in Mahanadi Basin. After the hiatus in coal formation during the deposition of Barren Measures Formation, the amelioration in climate during Raniganj times supported the luxuriant growth of Glossopteris flora which resulted in accumulation of thick coal seams. Lithologically, it comprises of fine to coarse grained sandstones which are often pebbly, cross-bedded and ferruginous in nature, massive sandstones, siltstones, carbonaceous and grey shales and claystones along with intermittent coal seams. Raniganj coal seams are thinner in comparison to Barakar coal seams. Macrofloristically, Raniganj/Kamthi formation is well studied. In Son Basin, a good assemblage of Raniganj macroflora is reported from the Jhingurdah coal seam, Singrauli Coalfield (Singh and Saxena 2015) and Tatapani-Ramkola Coalfield (Singh et al. 2012; Saxena et al. 2019a). A significant finding of female fructification genus *Umkomasia* that belongs to the Triassic element *Dicroidium* was made from Late Permian sequence exposed along Behra Rivulet near Karaonda village in Tatapani-Ramkola Coalfield, Chhattisgarh (Chandra et al. 2008) which is the first record of this genus from the Late Permian sequence of Indian Gondwana. The genus was represented by two species, viz., *Umkomasia polycarpa* and *U. uniramia.* Similar reports of the occurrence of *Dricoidium* foliage (4 species) are known from Upper Permian sequences of Jordan (Kerp et al. 2006; Abu Hamad et al. 2008). These finding proposes that the corystospermales genus *Dicroidium* might have appeared in the Late Permian but reached its acme in the Middle–Upper Triassic.

Late Permian flora of Mahanadi Basin is more rich and diversified than Son Basin flora. In Ib River Coalfield Kamthi Formation flora is well studied by Goswami et al. (2006a, c, 2018a, b, 2022b and references therein), Patel et al. (2021a), Singh et al. (2007) and in Talcher Basin by Goswami and Singh (2013), Saxena et al. (2014), Goswami et al. (2018b), Sahoo et al. (2020a, b), Patel et al. (2021b, c), Aggarwal et al. (2022) and Mishra et al. (2022). Raniganj and Kamthi macroflora is characterized with the abundance of Glossopteris, Pteridophytes and gymnosperms and it includes maximum number of leaf forms, fructifications and woods. Pteridophytes and gymnosperms reached their maximum development both in quantity as well as diversity (Chandra 1992).

In Raniganj/lower Kamthi Formation, palynoflora is characterized by the dominance of striate bisaccate pollens mainly *Striatopodocarpites* along with increased frequency of trilete spores viz. *Microfoveolatispora*, *Microbaculispora*, *Cyclogranisporites* and *Horriditriletes* is observed. Four distinct palynoassemblages are identified based on marker taxa. The older one is marked by the dominance of *Striatopodocarpites*, the next level marked by the spore genus *Gondisporites*, followed by monosaccate genus *Densipollenites* and the bisaccate genus *Crescentipollenites* signifies the youngest level. Increase in radial monosaccate pollens is also observed at the last phase of the Raniganj palynocomposition. The first appearance of *Playfordia cancellosa*, *Lunatisporites pelluciudus* and spore taxa of *Lunbladispora* spp. in the Raniganj flora suggest the end of Permian period. Some of these palynotaxa are shown in Fig. 3.

Palynologically, Raniganj sediments are well studied in the Son Basin namely from Tatapani-Ramkola Coalfield (Srivastava et al. 1997; Kar and Srivastava 2003); Mand-Raigarh (Murthy et al. 2014) Singrauli Coalfield (Bharadwaj and Sinha 1969b;

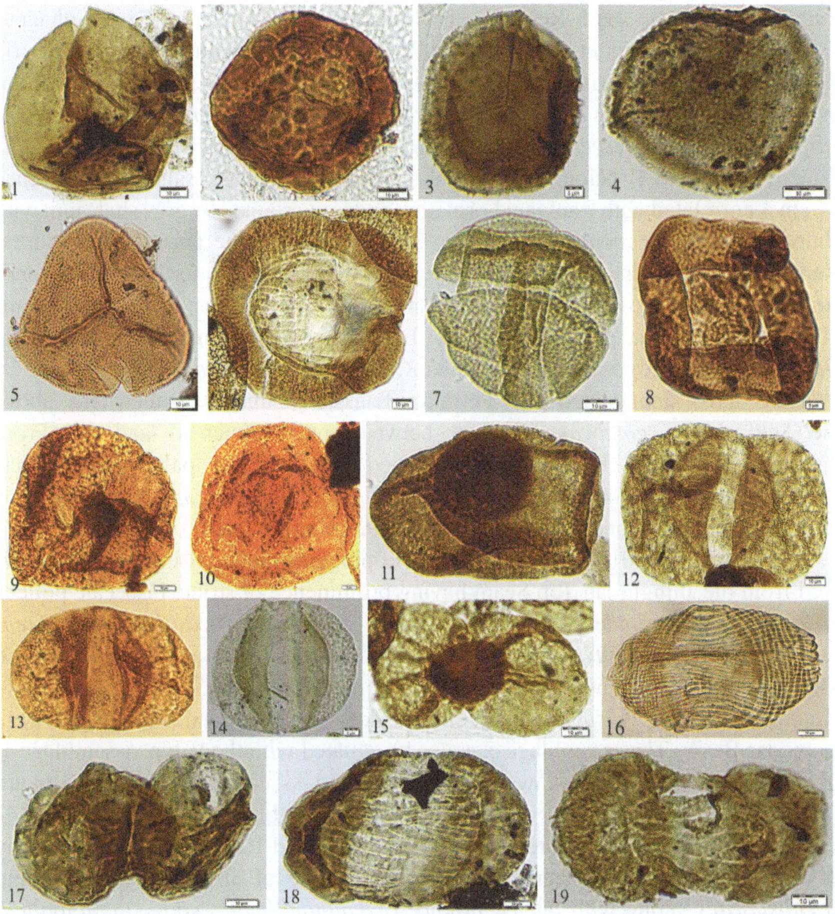

Fig. 3 Frequently encountered palynotaxa in the Permian and Triassic sequences of Son-Mahanadi Basin. 1. *Microfoveolataspora raniganjensis* Bharadwaj emend. Tiwari and Singh (1981), 2. *Verrucosisporites densus* Bharadwaj and Tiwari (1977), 3. *Lundbladispora raniganjensis* Tiwari and Rana (1981), 4. *Gondisporites raniganjensis* Bharadwaj (1962), 5. Microbaculispora *gondwanensis* Bharadwaj (1962), 6. *Striomonosaccites ovatus* Bharadwaj (1962), 7. *Guttulapollenites hannonicus* Goubin (1965), 8. *Crucisaccites monoletus* Maithy embed. Tiwari et al. (1989), 9. *Densipollenites invisus* Bharadwaj and Salujha (1964), 10. *Densipollenites magnicorpus* Tiwari and Rana (1980), 11. *Densipollenites densus* Bharadwaj and Srivastava (1969). 12. *Crescentipollenites fuscus* (Bharadwaj) Bharadwaj et al. (1974), 13. *Alisporites indicus* Bharadwaj and Srivastava (1969), 14. *Alisporites ovalis* Kumar (1973), 15. *Rhizomaspora indica* Tiwari (1965), 16. *Weylandites indicus* Bharadwaj and Srivastava (1969), 17. *Verticipollenites gibbosus* Bharadwaj (1962), 18. *Striatopodocarpites magnicorpus* Bharadwaj and Tiwari (1964), 19. *Lunatisporites pellucidus* (Goubin.) Tiwari and Vijaya (1995)

Singh et al. 2017), South Rewa Basin (Tiwari and Ram-Awatar 1986), Sohagpur Coalfield (Ram-Awatar et al. 2004). In a Permian–Triassic sequence exposed in Iria Rivulet, TR coalfield, Srivastava et al. (1997) identified three assemblage zones based on palynocmposition. In stratigraphical order these are—(I) *Densipollenites magnicarpus* Zone, (II) *Crescentipollenites fuscus* Zone, and (III) *Falcisporites stabilis* Zone. Zone I and II belong to the coal, carbonaceous shale, sandstone sequence which is correlated with Raniganj Formation and is overlained by grey shale unit with abundance of *Schizoneura gondwanensis* (a prominent Late Permian macrofloral element). These beds are overlain by greyish/olive-green splintery shale and sandstone of Triassic age and designated as Panchet Formation. The third palynoassemblage zone is characterized with Early Triassic palynotaxa namely *Falcisporites stabilis*, *Playfordiaspora cancellosa*, and *Klausipollenites schaubergeri*.

They have demarcated the Permian–Triassic boundary somewhere above the *Schizoneura gondwanensis* bed between palynozones II and III. From the same coalfield, Kar (2003), identified four distinct palynoassemblages. Among these, Palynoassemblage IV—*Faunipollenites–Striatopodocarpites–Striatites* Assemblage Zone is assigned as Late Permian in age. From Singrauli coalfield (Singh et al. 2017) a Late Permian (Lopingian) palynoassemblage is recovered from the Jhingurdah Top Seam having prominence of striate bisaccate pollens *Striatopodocarpites* and *Faunipollenites* along with marker taxa *Alisporites*, *Arcuatipollenites*, *Striasulcites*, *Playfordiaspora*, *Klausipollenites* and *Densipollenites*. This assemblage (Palynoassemblage IV) is correlatable with *Striatopodocarpites–Faunipollenites–Gondisporites* Assemblage Zone of Tiwari and Tripathi (1992) in Damodar Basin. The assemblage can also be equated with the Palynoassemblage-II of Sohagpur coalfield (Ram-Awatar et al. 2004) and Assemblage-III of East Bokaro coalfield (Vijaya et al. 2012). A Raniganj palynoassemblage (Palynoassemblage 1) is also reported from the borecore sediments of Mand-Raigarh coalfield (Murthy et al. 2014) and is characterized by the dominance of *Striatopodocarpites* and *Gondisporites*, along with fair occurrence of *Faunipollenites* spp., *Guttulapollenites hannonicus*, *Arcuatipollenites* spp., *Lundbladispora microconata* and is equated with *Gondisporites raniganjensis* Palynoassemblage of Damodar Basin.

In Mahanadi Basin, the lower Kamthi Formation (= Raniganj) is exposed in the Belpahar area of Ib River Basin and some areas of the adjacent Sundargarh District of this basin. A Late Permian palynoassemblage was first reported by Maiti (1994) from the subsurface samples of Sundargarh district. Kamthi sediments procured from the borecore of Belpahar area were studied by Meena (1998) yielded rich assemblage of palynomorphs having abundance of striate bisaccate pollen *Striatopodocarpites* and *Crescentipollenites* along with gradual decline of *Densipollenites* and appearance of marker palynotaxa namely *Lunatisporites*, *Goubinispora*, *Falcisporites*, *Chordasporites* and *Satsangisaccites* in the assemblage correlatable to Early Triassic (Induan) age to this assemblage. A late Late Permian (Lopingian-Induan) age was established for the studied sediments of Kamthi Formation. In another study based on borecore samples (Meena 2000) of Ib River Coalfield, the palynoassemblage-2 revealed the prominence of striate bisaccate pollen *Striatopodocarpites* followed by *Faunipollenites*, *Crescentipollenites*, *Distriatites* and presence of marker palynotaxa viz. *Densipollenites magnicorpus*, *Klausipollenites and Falcisporites*, assemblage is correlated with *Densipollenites magnicorpus* assemblage zone suggesting Late Permian age (Lopingian). Based on the palynocompositions derived from three different bore-cores raised in Tangadih area (Jharsuguda District) of the Ib-River basin, two palynoassemblages were identified; Assemblage-1 is characterized by the abundance of striate bisaccate pollen grains (*Striatopodocarpites*, *Faunipollenites* and *Crescentipollenites*), and subdominance of non-striate bisacccate pollen grains, along with the appearance of *Arcuatipollenites* and significant percentage of trilete spores suggesting a Late Permian age (Meena and Goswami 2004).

In addition to miospores, a good assemblage of megaspores has also been reported from the Late Permian sediments of the Kuraloi area in the Ib River Basin (Tewari et al. 2009). The assemblage includes *Banksisporites indicus*, *B. utkalensis*, *Biharisporites spinosus*, *Biharisporites* sp. A. *Biharisporites* sp. B, *Bokarosporites rotundus*, *Jhariatriletes baculosus*, *J. srivastavae*, *J. damudicus*, *Manumisporites distinctus*, *Ramispinatispora indica*, *R. nautiyalii*, *R. mahanadiensis* and *Ramispinatispora* sp., *Singhisporites baculatus*, *S. radialis* and *S. surangei*. However, later palynological investigations of the sediments of the Kuraloi Block, revealed an Early Triassic age (lower part of the Upper Kamthi Formation). The above palynoassemblage has yielded cingulated, zonate, taeniate, nonstriate palynotaxa which having an early Triassic affinity and shows the dominance of *Klausipollenites*, *Gondwanipollenites* and *Weylandites* (Pillai et al. 2016).

In a detailed palynological study carried out on the borecore sediments (TP-8) of Talcher Coalfield, Tripathi (1997) identified four assemblages (I–IV). Assemblage-I recovered from the lithologically classified Barakar Formation had yielded Late Permian palynotaxa showing dominance of *Striatopodocarpites* spp, common occurrence of *Faunipollenites*, *Gondisporites*, *Callumispora* and *Densipollenites* spp., along with *Falcisporites*, *Lunatisporites* and *Satsangisaccites*. Later, from the same bore Permian–Triassic transition was delineated by Tripathi (2001) on the basis of two distinct palynoassemblages—(i) *Densipollenites magnicorpus* of Late Permian and (ii) *Krempipollenites indicus* of Early Triassic age. More recently, in a multiproxy study carried out on the sediments of Jagannath coalmine, Talcher coalfield revealed a Late Permian (Wuchiapingian-Changshingian) palynoflora recovered from the lithologically and macrofloristically assigned lower Barakar sediments. The palynocomposition has domination of *Striatopodocarpites* spp. followed by sub-dominance of *Faunipollenites* spp. along with a fair representation of *Densipollenites* spp. (Patel et al. 2021b). In another study from Hingula area, Talcher coalfield, Late Permian (Wordian-Capatanian) palynoflora is recorded although on the basis of single productive sample, although macrofloristically these sediments are dated as Early Permian lower Barakar in age (Patel et al. 2021c). Raniganj palynoflora of Son Basin and equivalent palynoflora of lower Kamthi Formation has strong affinities with the palynoflora of Damodar Basin which is hitherto considered as the reference basin for inter-basinal correlation among peninsular Gondwana basins. In addition to the miospore assemblages, acritarchs and ichofossils of both fluvial and marine facies have also been recognized in the Lower Kamthi Formation and older horizons of the Talcher Basin (Goswami and Singh 2013 and references therein) suggesting intermittent marine incursions at various time interval during the Permian.

3.6 Palynofloristics of Panchet Formation and Equivalent Sequences of Triassic Period

Permian–Triassic sediments have been analyzed worldwide to assess the cause, time and impact of great mass extinction event on biota recovery and radiation (Cascales-Miñana and Cleal 2014; Cascales-Miñana et al. 2016; Benton 2018). Primary cause for the severe loss of biodiversity are considered as high level of CO_2 and raised global temperature because of large basaltic lava flows of Siberian traps (Visscher et al. 1996; Shen et al. 2019). In Southern hemisphere (Gondwanan land) glossopterids were abundant during the Permian, but they were reduced at Permian–Triassic boundary and were replaced by peculiar *Dicrodium* flora (Retallack 1995; Dobruskina 1994).

In India these Triassic sediments are known as 'Panchet Formation' in Damodar Basin, 'Kamthi Formation' in Mahanadi Basin, 'Upper Kamthi Formation' in Godavari Basin, 'Panchmarhi Formation' in Satpura Basin and 'Pali Formation' in Son Basin (Mukhopadhyay et al. 2010). The Mesozoic successions in Son Basin are identified as Pali, Tiki and Parasora formations ranging in age from Late Permian to Late Triassic/Early Jurassic. Lithologically, the Pali Formation is further grouped into three members—Lower, Middle and Upper. The Lower and Middle members are considered equivalent to Barren Measures (middle Permian) and Raniganj Formation (Late Permian) of the Damodar Basin respectively, while Upper member exhibits Permian–Triassic transition.

One of the most common macrofloral element of Triassic sediments of Indian Gondwana is the foliage of corystosperm genus *Dicrodium*. The occurrence of *Dicroidium*-in the Son/South Rewa Basin was reported by Lele (1955, 1961). One of the prominent locality is 'Nidpur Locality' in the western part of Singrauli coalfield, Sidhi District, (M.P.) and is known for prolific yield of middle Triassic flora including *Dicroidium* and the beds are referred as 'Nidpur Beds'. Dicroidium flora is well studied by Bose and Srivastava (1971), Pant and Basu (1973, 1979a), Bhowmik and Das 2012) and Srivastava (1974) who reported *Pteruchus indicus*, the male fertile organ of *Dicroidium*. Subsequently, Pal (1984) described five species of *Dicroidium* from the Tiki Formation exposed near Harai and Giar villages, Shadol District, M.P. The occurrence of this flora is also known from the Triassic beds of Talcher Coalfield, Odisha (Pal and Ghosh 1997; Bhattacharya et al. 2001). A detailed analysis of genus *Dicroidium* and its distribution in Indian Gondwana has been carried out by Pal et al. (2014), Singh et al. (2012) and Saxena et al. (2019b) have reported existence of few elements of Glossoptreis Flora from the Triassic beds (Panchet Formation) of Iria Rivulet section, Tatapani—Ramkola Coalfield. These elements include *Paracalamites*, *Schizoneura*, *Dizeugotheca*, *Glossopteris* and *Vertebraria* although in subdued occurrences along with *Dicroidium* foliage and other Triassic elements namely *Linguifolium, Heidiphyllum* and *Rissikia* (Saxena et al. 2019b).

The age of the Pali Formation in South Rewa/Son Basin has always been contentious. Different workers have assigned differed ages ranging from middle late Permian to Late Triassic (see Pillai et al. 2018). A well preserved macroplant fossil assemblage was procured from an outcrop sequence of Johilla coalfield, belonging to Pali Formation, assemblage comprises of *Neomariopteris hughesii*, along with fifteen species of *Glossopteris, Vertebraria indica, Utkaliolepis indica* (scale leaf) and dispersed seeds namely *Cordaicarpus zeilleri, Rotundocarpus striatus, Samaropsis goraiensis* and *S. feistmantelii* (Pillai et al. 2018). Though, on the basis of palynology and macrofloral remains, this assemblage is dated as Late Permian in age and is comparable with the coeval non–coaliferous beds of Kamthi Formation in Mahanadi Basin and Bijori Formations of Satpura Basin, and has limited comparison with Pachhwara Formation of the Rajmahal Basin. Palynological assemblage of this study is characterized by dominance of striate bisaccate pollen grains mainly *Faunipollenites* and *Striatopodocarpites* along with with *Crescentipollenites, Verticipollenites, Densipollenites, Striatites, Lahirites, Distriatites, Scheuringipollenites, Arcuatipollenites, Decussatisporites* and age marker taxa such as *Guttulapollenites, Alisporites* and *Weylandites* suggests a late Permian age.

In Son Basin, in palynological context meager works have been done on Triassic sediments. Pant and Basu (1979b) described some Triassic megaspores from the Nidpur beds of Singrauli Coalfield. Tiwari and Ram-Awatar (1987) recovered an assemblage of miospore containing *Striatopodocarpiles, Faunipollenites, Densipollenites, Lundbladispora* and *Lunatisporites* along with age marker taxa like *Playfordiaspora, Lueckisporites, Falcisporites* and *Nidipollenites* from the outcrop sediments exposed near Johilla River between Dargaon and Salaia villages. It has been concluded that the studied bed showed Permain/Triassic transition phase, however, this assemblage can be broadly correlated with the Lower Triassic assemblage of Damodar Basin. Subsequently, Tiwari and Ram-Awatar (1990) also recovered a miospore assemblage of Early Triassic age. Lithologically designated as the Middle Pali Member, Ram-Awatar (1993) has recovered palynological assemblage from the bore core sediments (SPB-17) of Sohagpur Coalfield comprising a total of 41 spores and pollen genera. The common forms are *Densipollenites, Scheuringipollenites Faunipollenites, Striatopodocarpites, Crescentipollenites, Striatites* and *Lahirites*. Although, the recovered palynoassemblage is correlatable with Raniganj Formation of Darmodar Basin. In another bore-core sequence from the Middle Member of the Pali Formation, three palynoassemblages (I, II, III) were identified by Ram-Awatar (1996a). The older two assemblages (II and III) reveals abundance of striate-disaccate pollen, i.e., *Faunipollenites* and *Srtriatopodocarpites* and subdominance of *Crescentipollenites* along with *Gondisporites, Crucisaccites, Lunatisporites* and *Chordasporites* indicating a Late Permian age. Presence of Leiospherids at the top of the sequence (Assemblage-I) suggests a probable marine influence in the area.

Ram-Awatar (1996b) studied a palynological assemblage recovered from the Supra-Barakar sediments from the Umaria Coalfield comprising mainly striated bisaccate pollens along with other significant taxa namely *Lundbladispora*, *Playfordiaspora*, *Densoisporites*, *Falcisporites*, *Brachysaccus*, *Lycopodiacidites*, *Klausipollenites*, *Goubinispra* and *Uvaesporis*. This assemblage indicates the Permian–Triassic transition within the Supra-Barakar sediments.

Upper Pali sediments (Chundi River Section) were studied by Ram-Awatar (1997) and two palynoassemblages were identified. Assemblages are characterized by the presence of non-striate bisaccate pollen *Goubinispora*, *Brachysaccus*, *Densoisporites*, *Foveosporires*, *Lundbladispora*, *Nidipollenites*, *Playfordiaspora*, *Satsangisaccites*, *Staurosaccites*, *Todisporites*, *Converrucosisporires* and *Kamthisaccites*. These assemblages suggest that the sediments of Chundi River Section were deposited during Late Permian to early-middle Triassic times. In another study of sediments belonging to Upper Member of Pali Formation, Son-Chundi River Section in Shahdol Disuict, M.P., were analyzed for palynofloral composition (Ram-Awatar 2003). The recovered assemblage comprises 33 genera with 45 species. Significant palynotaxa are *Densoisporites playfordii*, *Brachysaccus ovalis*, *Falcisporites stabilis*, *Krempipolleniles vistitus*, *Titkisporites bahnaei*. *Lunddbladispora brevicula*, *Chordasporites sp.*, *Staurosaccites marginalis*, *Satsangiaccites nidpurensis*, *Samaropollenites speciosus* and *Goubinispora morondavensis* suggesting Early to Middle Triassic age.

Recently, Gautam et al. (2016) investigated and identified three palynoassemblage zones of late Permian, Early and Late Triassic age respectively for the borehole SNB-1 from the Sohagpur Coalfield, western part of the Rewa Basin. The palynoassemblage-I reveals abundance of striate bisaccate pollen mainly *Striatopodocarpites ovatus*, *Faunipollenites varius* and *Crescentipollenites fuscus* along with *Alisporites ovalis*, *Lunatisporites pellucidus*, *Densipollenites indicus* and is of late Permian age. Palynoassemblage II has prominence of cavate/cingulate spores, e.g. *Lundbladispora microconata*, *Densoisporites complicates*, followed by subdued occurrences of *Kraeuselisporites rallus*, *Lapposisporites lapposus*, *Playfordiaspora crenulata*, *Goubinispora indica*, *Klausipollenites schaubergeri*, *Limatulasporites fossulatus* and *Reduviasporonites chalastus* suggesting an early Triassic age whereas Palynoassemblage III shows abundance of *Alisporites astigmosus* along with *Falcisporites nuthallensis*, *Dubrajisporites unicus*, *Enzonalasporites densus*, *Leschikisporites aduncus*, *Lunatisporites rhaeticus*, *Lueckisporites virkkiae*, *Tikisporites balmei*, *Playfordiaspora vellata*, *Polycingulatisporites crenulatus*, *Cingulizonates indicus*, *Neoraistrickia taylori* and *Aratrisporites fischeri*. This palynoassemblage is dated as late Triassic in age. Miospore assemblage from the Tiki Formation sediments were studied by Maheshwari and Kumaran (1979), near Tharipathar and Ghiar villages in Son River section and dated as Upper Triassic.

In comparison to the macrofloral studies, palynological studies of Triassic sediments in Mahanadi Basin are meagre, and are mostly confined to Talcher Coalfield and Ib River Coalfield. Based on the studies of bore core sequence of TP-8

(307–367 m depth) from Talcher Coalfield, Tripathi (1996) and Tripathi and Bhattacharya (2001) recorded an Early Triassic palynofloral assemblage comprising *Striatopodocarpites, Satsangisaccites, Falcisporites, Weylandites muraticavea, Playfordiaspora, Lundbladispora, Arcuatipollenites* and *Alisporites*. From the same borehole TP-8 (151–239 m depth), they also reported Late Triassic assemblage which is characterized by the presence of *Brachysaccus, Rimaesporites, Samaropollenites* and *Callialasporites*. Furtehr, borehole samples in the Chhendipada–Patrapara region of Talcher Coalfield were analysed by Meena (2003, borehole TP-9, 259–280 m depth) and Bhattacharya et al. (2001), and recorded Early Triassic assemblage. Goswami and Singh (2013) have assessed and reviewed in detail the floral biodiversity of Talcher Coalfield during Permian and Triassic times.

4 Palaeoclimate

4.1 Palaeoclimate During the Deposition of Talchir Formation

Lithologically, Talchir Formation is made up of Talchir glacigene sediments—tillite, striated pavements, varves followed by shale units, which reveals that the cold climate was prevailing during the deposition of Talchir sediments (Earliest Permian). According to Tiwari and Tripathi (1988), on the basis of morphological characters of palynomorphs; Talchir Formation is categorized in two suits namely, suite-1 (lower part of Talchir Formation) and suite-2 upper part. The morphological characters of palynomorphs in suite-1 indicates an extreme cold climate and has the dominance of *Plicatipollenites-Parasaccites* complex with less diversification in other palynomorphs and in suite-2, the upper part of the Talchir Formation, the association of palynomorphs *Parasaccites-Plicatipollenites* complex with higher degree of variation in diversity of other palynomorphs is observed, it revealed amelioration of climate from frigid to very cool condition with medium humidity. This climate is also supported by macrofloral studies of Talchir Formation envisaging sparse vegetation and with small and stunted trees (Chandra 1992; Saxena et al. 2022) and mineralogical studies as less cool (Singh 1976) and as cold semi arid (Suttner and Dutta 1986).

4.2 Palaeoclimate During the Deposition of Karharbari Formation

Tiwari and Tripathi (1988) have categorized Karharbari Formation into two suites (suite-3 and suite 4). Based on morphological character of palynomorphs of suite

-3, it is inferred that climate amelioration took place which was conducive for luxuriant vegetation growth and is also evidenced by the presence of coal seams in this formation. Similarly, the morphological characters and dominance of palynomorphs *Parasaccites* and *Callumispora* in the suite-4 (composition IIB) reveals recurrence of very cold phase with medium to high humidity. The macroplant remains of the Karaharbari Formation are quite distinct than those of underlying Talchir and overlying Barakar formations. The floral assemblage comprises of dominance of *Gangamopteris*, *Noeggerathiopsis* and *Glossopteris* plants along with the occurrences of *Buriadia* and *Botrychiopsis* (*Gondwanidium*). These floral findings led to envisage the climate amelioration during Karharbari times (Chandra 1992).

4.3 Palaeoclimate During the Deposition of Barakar Formation

The Karharbari-Barakar transition was gradual as peculiar plant fossils of Karharbari Formation are also recorded from the lower Barakar sequences and this transition is also marked by the deposition of some of the thickest coal seams. It is envisaged that cooler climatic conditions during Karharbari times were continued to exist during the lower Barakar times also (Artinskian). However, discernible changes in the floral composition of Upper Barakar Flora, indicates prevalence of moderately warm and humid climate during that time.

The morphological characters representing suite-5 are associated with dominance of *Scheuringipollenites* (composition IIIA) pointing towards a favorable and cooler climate in the beginning which gradually becomes moderately warm as revealed by suite-6 with dominance of *Faunipollenites* (composition IIIB), it indicated high humidity throughout the Barakar Formation, also high humidity indication was confirmed by presence of thick coal strata and abundance of pteridophytic spores in the Barakar (Tiwari and Tripathi 1988). However, Singh (1976) has interpreted prevalence of subtropical climate and warm temperature in the lower part and tropical humid climate in the upper part of the Barakar Formation based on the sedimentological study. Similarly, Dutta and Suttner (1986) have confirmed for temperate humid climate during the Barakar based on oxygen isotope analysis (Fig. 4).

The formation of coal swamps during Barakar times in almost all the sedimentary Gondwana basins of India, evidently suggests existence of dense vegetation and sufficient rainfall in this region during Barakar (Artinskian-Kungurian) (Chandra and Chandra 1988; Chandra 1992). It is also supported by the fact that most of the recorded *Glossopteris* leaves of this formation are usually lanceolate in shape, medium to large size, have acute apices and acute-cuneate or tapering bases. Leaves have entire margins and narrow elongated, polygonal meshes. These morphological features have been interpreted to indicate the prevalence of warm and humid climatic conditions. Acute apices and narrow tapering or acute cuneate bases of the leaves suggest that enough sunlight was available to the plants. Therefore, leaves did not have to expand

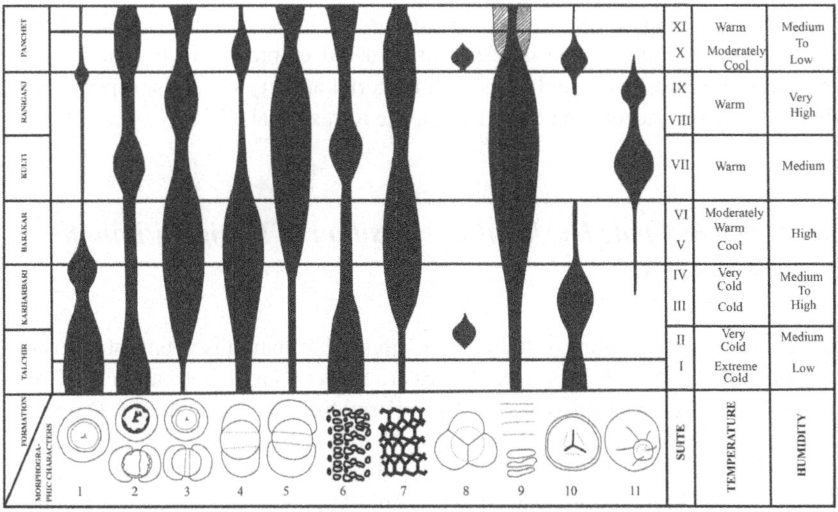

Fig. 4 Trends in the distribution of morphographic characters considered to be important from Talchir to Lower Panchet, in terms of relative abundance. 1. Radial monosaccate; 2. Dense central body; 3. Thin central body; 4. Haploxylonoid construction; 5. Diploxylonoid construction; 6. Leathery saccus; 7. Non-leathery saccus; 8. Tetrad of spores; 9. Solid black striations and oblique lines Taeniae; 10. *Callumispora*; 11. *Densipollenites* (after Tiwari and Tripathi 1988)

their apices for absorption of more sunlight. Based on the floral composition of upper and lower Barakar sequences, a cool temperate climate with high humidity and sufficient light intensity is deduced for the lower Barakar sequences whereas moderately warm climate with high humidity is inferred for upper Barakar Formation (Goswami and Singh 2013; Goswami et al. 2018b, 2022a).

4.4 *Palaeoclimate During the Deposition of Barren Measures*

The richness of *Densipollenites* (composition IV) in Barren Measures (Suite-7) reveals conditions are different than the Barakar Formation. Barren Measures times had experienced warm climate with medium to low humidity (Tiwari and Tripathi 1988). Bharadwaj (1973), Kar (1976), Lele (1976), Chandra and Chandra (1988) have concluded that the diversity in palynomorphs does not support views for dry arid climate during the Barren Measures depositional period. Singh (1976) mineralogical and lithological studies have shown that the dominance of red facies and absence of coal seams in Barren Measures Formation is the result of depositional conditions of environment rather than climate. Kar et al. (1964) argued that warm and humid climate was prevailing during the deposition of Barren Measures Formation owing to the similarity of heavy mineral compositions in Barakar and Barren Measures formations. Though, macrofloral records from this formation are meagre,

Chandra (1992) has drawn similar climatic inferences. Presence of siderite with higher concentration of Iron is responsible of red coloration in Barren Measures and siderite presence also indicates high bacterial activity which is responsible for absence of carbonaceous matter in Barren Measures (Shah 1976).

4.5 Palaeoclimate During the Deposition of Raniganj/Kamthi Formation

In the Permian successions of India, the Raniganj Formation is youngest lithostratigraphic unit, its equivalents in other basins are Lower Kamthi Formation in Wardha-Godavari and Mahanadi basins, the Bijori Formation in Satpura Basin. During the deposition of Raniganj/lower Kamthi (Late Permian) sequences, climate was relatively warm with high humidity that supported the luxuriant plant growth, thus dense forests as evident by occurrences of ample pteridophytic and Glossopterid remains. According to Tiwari and Tripathi (1988) in suite-8 and suite-9 the morphological characters in palynological composition-V indicate warm and highly humid climate throughout the span of Raniganj Formation (= Kamthi Formation) and this climate is also supported by the presence of thick coal deposits in Raniganj Formation. The palynological studies suggests a change in lowering of temperature and decrease in the humidity.

However, there is drastic change in lithologies of this time frame when Damodar and Son-Mahanadi basins are compared. The preponderance of the red bed facies of the ferruginous sandstones in the Mahanadi Basin, points out existence of seasonal inconsistency of dry spells and semi arid conditions. These red bed facies could be attributed to local substrate oxidization. This could be the reason for the absence of coal seams in the lower Kamthi Formation of Ib-River Basin in Mahanadi Basin, despite the occurrence of of large quantities of vegetable matter as evident by ample macrofloral records (Goswami et al. 2018a). Therefore, it can be argued that though, climatic conditions were substantially favorable for the growth and development of vegetation cover, substrate conditions prevented formation of coal. At the same time, thick coal seams with good quality coal were formed in the adjacent Damodar Basin which is not the case in the Son Basin. Moreover, Raniganj Formation of Damodar basin has yielded highly rich and diversified macro and miofloral remains of Glossopteridales and Pteridophytes. The occurrence of pteridophytic remains in the Kamthi beds of the Ib-River Basin is quite low as compared to the adjacent Talcher Basin. It led to envisage that climatic conditions were a little drier in the Kamthi region during that time (Goswami et al. 2018b). Thus, floristic diversity derived from the Kamthi exposures of Ib-River Basin indicate that these Late Permian forests were growing under warm and highly humid conditions as in Raniganj Formation of Son or Damodar basins (Tiwari and Tripathi 1988). The faunal records revealed prevalence of temperate climate (Singh 1976). Dutta and Suttner (1986) and Suttner and Dutta (1986) have generalized the climate experiencing gradual change from warm and semi-arid based on isotope and mineralogical study from Raniganj to Panchet formations.

4.6 Palaeoclimate During the Deposition of Panchet Formation

The transition from the Permian to Triassic is characterized by major mass extinction event known as 'Great Dying' and Siberian volcanic eruptions are considered as a one of the major causative factor. In the Gondwanan floral realm it is marked by the decline and extinction of 'Glossopteris Flora' and advent of altogether new 'Dicroidium dominant flora' followed by 'Ptilophyllum Flora'. However, the floral changeover across the Permian–Triassic was not as severely affected as was the case with faunal entities. Several studies suggest that the flora was not altogether extinct rather suffered a gradual decline at this time interval (Cascales-Miñana and Cleal 2014; Novak et al. 2019).

As for as the changeover in plant domain, from Raniganj Formation (Lopingian) to Panchet Formation (Triassic) in Son-Mahnadi Basin is concerned, different opinions exist. Based on change in lithology, i.e. from grey shales of Raniganj to olive and red shales of Panchet Formation, shift from warm and humid climate at the end of Permian to prevalence of arid/semi-arid climate at the beginning of Triassic is envisaged. Earlier macrofloral evidences had advocated widespread aridity with reduced rainfall during the onset of Triassic (Fox 1940; Lele 1976). However, based on good macrofloral assemblage from the early Triassic sediments, subsequent workers have suggested prevalence of warm medium to low humid climate with intermittent semi arid spells (Goswami et al. 2018a; Saxena et al. 2019b).

Contrasting views based on palynological data were expressed on the basis of palynological studies. Bharadwaj (1975) envisaged a glacial phase during P–T interval. Tiwari and Tripathi (1988), argued that occurrence of abundant and diverse palynomorphs from lower Triassic sediments and their morphological characters indicate cool and medium to low humid climate at the beginning of Triassic which gradually became warmer towards the end of Triassic as indicated by Pollen Suite-11 and the morphological characters of pollens and spores of this suite. Based on evolution of striations in pollens, Vijaya (1989) argued that during the Early Triassic time, definite climate change was there, but ruled out the existence of arid conditions. This inference was based on the occurrence of rich diversification in spores and pollens and their abundance suggesting that luxuriant vegetation was thriving. It was further substantiated by diminishing trend of striation-complexity and diversification in taeniae characters in the Early Triassic saccate pollen group. She further opined that thickening of several strips in the form of taeniae is linked with the warming climate. Mineralogical and oxygen isotope studies also revealed a gradual shift in climate that changed to warmer and moderately semi-arid in Upper Panchet with seasonal fluctuations rather than extreme aridiry (Suttner and Dutta 1986). Singh

(1976) discussed that red colour of the Panchet sediments is considered to be due to the presence of siderite and is facies controlled, and these sediments were deposited under the influence of subtropical climate with prolonged seasonal perturbations. On the basis of palaeofloral evidences gathered from the Son Basin, Saxena et al. (2019b) also discussed and opined that at Permian Triassic boundary event, plant life was significantly transformed but did not become completely extinct. The debate whether there was global mass extinction in the plant realm was there at P/T or not is still going on (Fig. 5).

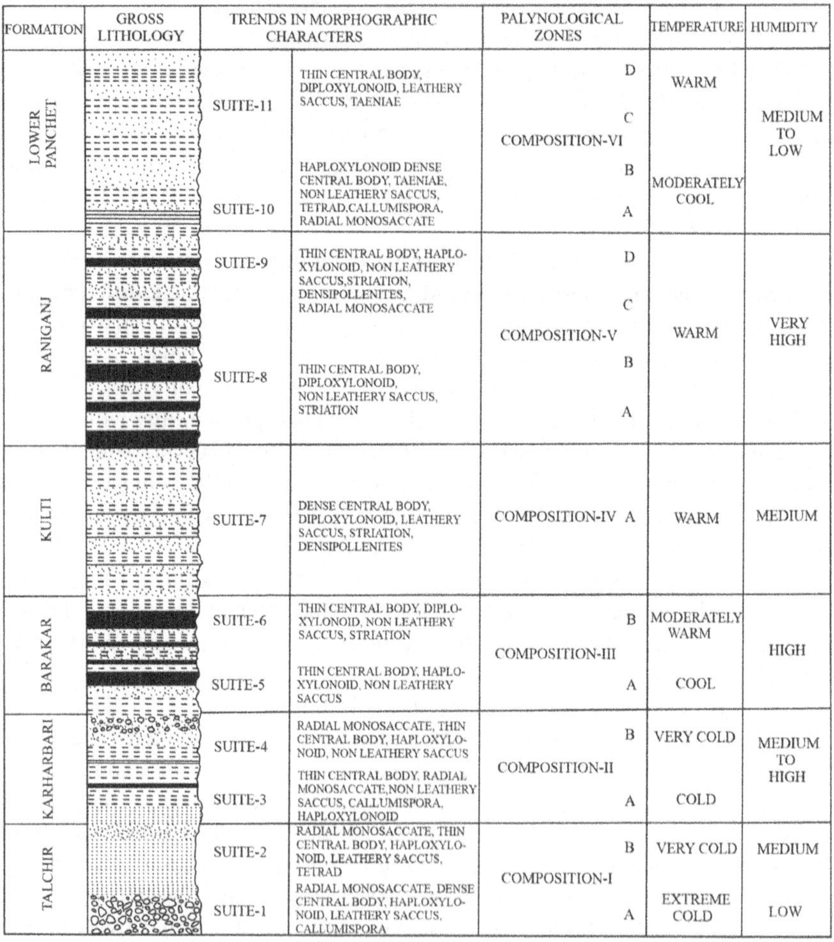

Fig. 5 Synthesis of lithological and palynological compositions to evaluate palaeoclimate (after Tiwari and Tripathi 1988)

5 Botanical Affinities

Based on the recovered miospores entities from the coalfield and outcrop sediments of the Son-Mahanadi Basin, the affinities of spores and pollens with their parent plant groups are depicted in the following (Table 6).

6 Concluding Remarks

Morphotaxonomic characters of Permian and Triassic palynomorphs have been studied in detail, for their evolutionary tendency and as indicators to climatic variabilities by many palynologists. However the most exhaustive work was done by Tiwari and Tripathi (1992, 1988) for lower Gondwana formations which is hitherto followed by Indian palynologists. They established that the morphological characters change with time adapting to climatic perturbations and exhibit the new characters replacing older ones in the characteristic assemblages. Thus these assemblages indicate the horizon as well as demarcating the formational boundaries. In Indian Gondwana basins, in most of the cases biostratigraphic delineations coincides with those of the lithostratigraphic demarcations/boundaries which causes a constraint in precise age determination. Further the ambiguities in the usage of nomenclature of some of the key palynotaxa in Indian and global perspective poses a serious concern in establishing correlations among the coeval assemblages. Efforts are being made to provide new insights and correlation among the Indian Gondwana basins based on robust palynological data (Murthy et al. 2020; Prasad and Pundir 2017). Therefore, it is matter of investigation to re-evaluate the palynoflora for their biozones versus corresponding lithological boundaries. Moreover, several palynotaxa which has been shown as the first occurrences in Early Permian of continental Gondwana sequences are known to have their first occurrences in much older sequences of Himalayas and other parts of the world (e.g. Agnihotri et al. 2018b; Gupta et al. 2022). Based on the hitherto available records, the characteristic palynotaxa and their corresponding formations and age are given in brief in the followings.

The Talchir palynoflora is uniformly characterized by the dominance of radial monosaccate pollens. Within this formation, dominance of generic constituent *Plicatipollenites* is identified as older assemblage; the younger assemblage has *Parasaccites* rich palynocomposition. The older flora is less diversified whereas younger flora exhibit higher diversity among the radial monosaccate and bisaccate palynotaxa.

Palynoflora of the upper Talchir Formation continues to the lower Karharbari horizon and is characterized by higher frequency of radial monosaccates pollens. The older assemblage of Karharbari has the dominance of *Callumispora* and first appearance of *Crucisaccites* and *Stellapollenites* pollens.

The high diversity of striate and non-striate bisaccate pollen are marked in the Barakar palynoflora. The Barakar palynocomposition is divisible in two zones. The

Table 6 Botanical affinities of the significant palynotaxa recovered from the sediments of Son-Mahandi Basin (based on Balme 1995; Lindström and McLoughlin 2007; diPasquo and Grader 2012; Mishra et al. 2017)

Palynotaxa	Botanical affinities
Equisetopsids	**Filicales**
Calamospora aplata	*Brevitriletes unicus*
Calamospora exila	*Brevitriletes communis*
Verrucosisporites densus	*Callumispora fungosa*
Verrucosisporites surangei	*Callumispora gretensis*
Verrucosisporites triassicus	*Cyclotriletes granulates*
Cordaitales	*Cyclogranisporites triletes*
Barakarites indicus	*Cyclogranisporites leopoldi*
Barakarites crassus	*Cyclobaculisporites minutus*
Barakarites densicorpus	*Cyclobaculisporites baculatus*
Densipollenites invisus	*Dentatispora crasa*
Densipollenites indicus	*Dentatispora gondwanensis*
Densipollenites magnicorpus	*Dentatispora mammoida*
Densipollenites densus	*Horriditriletes horridus*
Divarisaccus lelei	*Horriditriletes curvibaculus*
Divarisaccus scorteus	*Microbaculispora barakarensis*
Divarisaccus strengeri	*Microbaculispora tentula*
Parasaccites korbaensis	*Microbaculispora villosa*
Parasaccites obscurus	*Microbaculispora gondwanensis*
Parasaccites bilateralis	*Microfoveolatispora barakarensis*
Parasaccites ovatus	*Microfoveolatispora raniganjensis*
Parasaccites densicorpus	*Microfoveolatispora fovelata*
Plicatipollenites gondwanensis	**Glossopteridales**
Plicatipollenites indicus	*Circumstriatites obscurus*
Plicatipollenites trigonalis	*Circumstriatites ovatus*
Striomonosaccites ovatus	*Circumstriatites talcherirensis*
Striomonosaccites circularis	*Limitisporites rectus*
Tuberisaccatus lobatus	*Limitisporites hexagonalis*
Lycopsidales	*Faunipollenites varius*
Gondisporites raniganjensis	*Faunipollenites bharadwajii*
Gondisporites reticulates	*Faunipollenites perexigus*
Indospora clara	*Faunipollenites singrauliensis*
Indotriradites korbaensis	*Striatopodocarpites magnicorpus*
Indotriradites sparsus	*Striatopodocarpites rarus*
Indotriradites surangei	*Striatopodocarpites decorus*

(continued)

Table 6 (continued)

Palynotaxa	Botanical affinities
Jayantisporites pseudozonatus	*Striatopodocarpites brevis*
Jayantisporites conatus	*Striatopodocarpites diffusus*
Lundbladispora raniganjensis	*Striatopodocarpites magnificus*
Lundbladispora microconata	*Striatopodocarpitesrotundus*
Lundbladispora baculata	*Striatopodocarpites subcircularis*
Scheuringipollenitesbarakarensis	*Potonieisporites barrelis*
Scheuringipollenites maximus	*Rhizomaspora indica*
Scheuringipollenitestentulus	*Rhizomaspora costa*
Striasulcites ovatus	*Rhizomaspora singula*
Striasulcites tectus	*Rhizomaspora monosulcata*
Verticipollenites gibbosus	*Primuspollenites lelei*
Verticipollenites oblongus	*Primuspollenites densus*
Verticipollenitessecretus	*Distratites bilateris*
Verticipollenites subcircularis	*Distriatites bilateralis*
Weylandites bilateralis	*Distriatites indicus*
Weylandites circularis	*Corisaccites alutas*
Weylandites irregularis	*Corisaccites indicus*
Weylandites minutes	*Dicappipollenites crussus*
Coniferales	*Distriamonocolpites ovalis*
Chordasporites sp.	*Hamiapollenites saccatus*
Lunatisporites (Arcuatipollenites) pellucidus	*Hamiapollenites insolitus*
Lunatisporites ovatus	*Sahnites panchetensis*
Lunatisporites damudicus	*Sahnites barrelis*
Guttulapollenites hannonicus	*Caheniasaccites ovatus*
Gutulapollenites balmei	**Coniferales/Peltaspermales**
Crescentipollenites fuscus	*Alisporites indicus*
Crescentipollenites ampulus	*Falcisporites zapfei*
Crescentipollenites globosus	*Klausipollenites schaubergeri*
Crescentipollenites gondwanensis	**Unknown affinity**
Crescentipollenites densus	*Dicappipollenites crassus*
Potonieisporites lelei	*Tiwariasporis indicus*
Potonieisporites neglectus	*Tiwariasporis novas*
Potonieisporites magnus	*Tiwariasporis flavatus*

Lower Barakar palynoassemblage i.e. *Scheuringipollenites barakarensis* Assemblage Zone and Upper Barakar *Faunipollenites varius* Assemblage Zone (Tiwari and Tripathi 1992). In the Lower Barakar, non-striate pollens are dominant, and in the Upper Barakar, striate bisaccate pollens are more abundant, replacing the non-striate pollens. The older assemblage is characterized by abundance of *Scheuringipollenites* while younger assemblage shows striate pollen genus *Faunipollenites*. The distribution of trilete spores shows richness in the Lower Barakar and decline in the Upper Barakar. First occurrences of cingulated zonate spores viz. *Indotriradites* and *Dentatispora* along with varied apiculate taxa and monosaccate pollen exhibits very low frequency and significant increase in diversity of bisaccate pollens is observed first time during Barakar times.

The dominance of striate bisaccate pollen mainly the genus *Striatopodocarpites* continued in the Barren Measures palynoflora along with abundance of monosaccate pollen genus *Densipollenites*. Less frequency of trilete spores is marked in the Barren Measures. Only one palynozone has been identified in this formation namely, *Densipollenites indicus* Assemblage Zone (Tiwari and Tripathi 1992) and a middle Permian age is inferred for this formation.

In Raniganj/lower Kamthi, palynoflora is characterized by the dominance of striate bisaccate pollens mainly *Striatopodocarpites* along with increased frequency of trilete spores viz. *Microfoveolatispora*, *Microbaculispora*, *Cyclogranisporites* and *Horriditriletes* is observed. Four distinct palynoassemblages are identified based on marker taxa. The older one is marked by the dominance of *Striatopodocarpites*, the next level marked by the spore genus *Gondisporites*, followed by monosaccate genus *Densipollenites* and the bisaccate genus *Crescentipollenites* signifies the youngest level. Increase in radial monosaccate pollens is also observed at the last phase of the Raniganj palynocomposition. The first appearance of *Playfordia cancellosa*, *Lunatisporites pelluciudus* and spore taxa of *Lunbladispora* spp. in the Raniganj flora suggest the end of Permian period.

Floral changeover across the Permian–Triassic transition is considered to be not as dramatically affected by the extinction event as happened in the marine faunal populations. Several palaeofloristic studies of Son-Mahanadi basin suggest that plant life was significantly transformed but did not become completely extinct at Permian Triassic transition as evident by occurrence few *Glossopteris* species and other allied taxa along with several pollen taxa from the Panchet sediments as well, which were continued from the Raniganj sediments (Saxena et al. 2019b). Similar inferences are also made from the Mahanadi Basin, by Goswami and Singh (2013), that many groups of plants (including pollen and spores) have disappeared in a step-wise manner during the Permian–Triassic interval (Raniganj/Panchet) and, similarly, in steps, many new fore-runners appeared at the onset of Triassic.

Acknowledgements Authors are thankful to Director, BSIP for providing facilities and support and granting permission (RDCC/9/2023-24) to publish this work.

References

Abu Hamad A, Kerp H, Vörding B et al (2008) A Late Permian flora with *Dicroidium* from the Dead Sea region, Jordan. Rev Palaeobot Palynol 149:85–130

Aggarwal N, Mathews RP, Ansari AH et al (2022) Palaeoenvironmental reconstruction for the Permian (lower Gondwana) succession of the Godavari Valley Coalfield in southern India based on a combined palynofacies, carbon isotope, and biomarker study. J Palaeogeogr 11(1):123–144

Aggarwal N, Patel R, Goswami S (2022) Palaeoclimate, Palaeoecology and Palaeovegetation in and around Nandira Colliery, Talcher Basin, Odisha, India during Early Permian: Inferences from Typical Karharbari Palynofloral and Palynofacies Analysis. J Geol Soc India 98(9):1244–1252

Agnihotri D, Pillai SSK, Aggarwal N et al (2018a) Palynomorphs from the Barakar Formation of Dhampuri Open Cast Mines, Sohagpur Coalfield, Madhya Pradesh, India. Palaeobotanist 67:171–184

Agnihotri D, Pandita SK, Tewari R et al (2018b) Palynology and detrital zircon geochronology of the Carboniferous Fenestella Shale Formation of the Tethyan realm in Kashmir Himalaya: implications for global correlation and floristic evolution. J Asian Earth Sci 157:348–359

Balme BE (1995) Fossil in situ spores and pollen grains: an annotated catalogue. Rev Palaeobot Palynol 87:81–323

Banerjee M (1987) Karharbari: a formation or biozone. Palaeobotanist 36:37–50

Benton MJ (2018) Hyperthermal–driven mass extinctions: killing models during the Permian–Triassic mass extinction. Philos Trans R Soc A: Math, Phys Eng Sci, A 376:20170076. https://doi.org/10.1098/rsta.2017.0076

Bharadwaj DC (1962) Miospore genera in the coals of Raniganj Stage (Upper Permian) India. Palaeobotanist 9:68–106

Bharadwaj DC (1966) Distribution of spores and pollen grains dispersed in the Lower Gondwana Formations of India. In: Symposium. Floristics and stratigraphy of Gondwana land, pp 69–84

Bharadwaj DC (1971) Palynostratigraphy of lower Gondwana successions in India. In: International Gondwana symposium, Annals of Geology Department, Aligarh Muslim University, Aligarh, vols 5 and 6, pp 30–419

Bharadwaj DC (1973) Palynology in biostratigraphy and palaeoecology of Indian Lower Gondwana Formations. Palaeobotanist 22(1–3):150–157

Bharadwaj DC (1974) Palaeobotany of Talchir and Karharbari formations and Lower Gondwana glaciation. In: Aspects and appraisal of Indian palaeobotany, pp 369–385

Bharadwaj DC (1975) Palynology in biostratigraphy and palaeoecology of Indian Lower Gondwana formations. Palaeobotanist 22:150–157

Bharadwaj DC, Sinha V (1969a) Some new miospores from Lower Gondwana Coals. In: Santapau H et al (eds) Journal Sen memorial volume. Botanical Society of Bengal, Calcutta, pp 7–16

Bharadwaj DC, Sinha V (1969b) Sporological succession and age of Jhingurdah Seam, Singrauli Coalfield, M.P. India. Palaeobotanist 17:275–287

Bharadwaj DC, Srivastava SC (1973) Sub surface palynological succession in Korba Coalfield, M.P. India. Palaeobotanist 20:137–151

Bharadwaj DC, Salujha SK (1964) Sporological study of seam VIII in Raniganj Coalfield, Bihar (India). Part II–Distribution of Sporae dispersae and correlation. Palaeobotanist 13:57–73

Bharadwaj DC, Sah SCD, Tiwari RS (1965) Sporological analysis of some coal and carbonaceous shales from Barren Measure Stage (Lower Gondwana) of India. Palaeobotanist 13:222–226

Bharadwaj DC, Srivastava SC (1969) Some new miospore from Barakar Stage, Lower Gondwana, India. Palaeobotanist 17:220–228

Bharadwaj DC, Tiwari RS (1964) The correlation of Coal seams in Korba Coalfield, Lower Gondwanas, India. CR 5th International Congress of Carboniferous, Paris (1963):1131–1143

Bharadwaj DC, Tiwari RS (1977) Permian-Triassic miofloras from the Ranging Coalfield, India. Palaeobotanist 24(1):26–49

Bharadwaj DC, Tiwari RS, Anand-Prakash (1979) Permo-Triassic palynostratigraphy and lithological characteristic in Damodar Basin, India. Biol Membr 4:49–82

Bharadwaj DC, Navale GKB Anand Prakash (1974) Palynostratigraphy and Petrology of Lower Gondwana coals in Pench-Kanhan Coalfield, Satpura Gondwana Basin, M.P, India. Geophytology 4(1):7–24

Bhattacharya A, Nandi A, Dutta A (2001) Triassic mega and micro-plant fossils from the Kamthi Formation of Talcher Coalfield, Orissa with chronological significance. In: Proceedings of national seminar on recent advances in geology of coal and lignite basins of India, Calcutta, 1997, vol 54. Geological Survey of India, Special Publication, pp 123–126

Bhattacharya B, Bhattacharjee J, Banerjee S et al (2021) Palaeogeographic implications of ichnotaxa assemblages from early Permian fluvio–marine Barakar Formation, Raniganj Basin, India. J Earth Syst Sci 130:1–25

Bhattacharya SK, Ghosh P, Chakrabarti A (2002) Isotopic analysis of Permo-Carboniferous Talchir sediments from East-Central India: signature of glacial melt–water lakes. Chem Geol 188(3–4):261–274

Bhowmik N, Das N (2012) On three new species of Pteruchus Thomas from the Triassic of Nidpur, M.P. India. Palaeontographica 28:5–39

Bose MN, Srivastava SC (1971) The genus Dicroidium from the Triassic of Nidpur, M.P., India. Palaeobotanist 19(1):41–51

Cascales-Miñana B, Cleal CJ (2014) The plant fossil record reflects just two great extinction events. Terra Nova 26:195–200

Cascales-Miñana B, Diez JB, Gerrienne P et al (2016) A palaeobotanical perspective on the great end-Permian biotic crisis. Hist Biol 28:1066–1074

Chandra A, Lele KM (1979) Talchir miofloras from South Rewa Gondwana Basin, India and their biostratigraphical significance. In: International palynological conference, pp 1976–1977

Chandra A, Srivastava AK (1982) Plant fossils from the Talchir and coal–bearing formation of South Rewa Gondwana Basin, India and their biostratigraphical significance. Palaeobotanist 30:143–167

Chandra S (1992) Changing patterns of the Permian Gondwana vegetation. Palaeobotanist 40:73–100

Chandra S (1995) Bryophytic remains from the Early Permian sediments of India. Palaeobotanist 43:16–48

Chandra S, Chandra A (1988) Vegetational changes and their climatic implications in coal–bearing Gondwana. Palaeobotanist 36:74–86

Chandra S, Singh KJ (1996) Plant fossils from the type locality of Talchir Formation and evidence of earliest plant/animal activity in Gondwana of India. In: Mitra ND (ed) Gondwana ninth international Gondwana symposium, Hyderabad, vol I, pp 397–414

Chandra S, Singh KJ, Jha N (2008) First report of the fertile plant genus *Umkomasia* from Late Permian beds in India and its biostratigraphic significance. Palaeontology 51:817–826

Chopparapu C, Annamraju R, Pillai S et al (2019) Glossopteris flora from Barren Measures, Pranhita-Godavari Basin, India. J Geol Soc India 94(4):405–410

Di Pasquo MM, Grader GW (2012) The palynology of the lower Permian (Asselian–? Artinskian) Copacabana Formation of Apillapampa, Cochabamba, Bolivia. Palynology 36:264–276

Dobruskina IA (1994) Triassic floras of Eurasia. Österr Akad Wiss 10:1–422

Dutta PK, Suttner LJ (1986) Alluvial sandstone composition and palaeoclimate II. Authigenic mineralogy. J Sedimentol Petrol 56:346–358

Erdtman G (1951) On the '*Tricolporites protrudens* problem'. Sven Bot Tidskr 45:355

Erdtman G (1952) On pollen and spore terminology. Palaeobotanist 1:169–176

Feistmantel O (1886) The fossil flora of the Gondwana system. The fossil flora of some coalfields in Western Bengal. Palaeontol Indica Ser 12 4(2):1–71

Fox CS (1940) The climates of Gondwanaland during the Gondwana Era in the Indian region. In: 17th International Geological Congress, Moscow. Symposium on palaeozoic and precambrian climates, vol 6, pp 187–211

Gautam S, Ram-Awatar TR et al (2016) Permian-Triassic palynofloral transition in Sohagpur Coalfield, South Rewa Gondwana Basin, Madhya Pradesh, India. Palaeobotanist 65:109–129

Ghosh PK, Basu A (1969) A note on the classification of the Lower Gondwanas of India. Rec Geol Surv India 97:168–171

Goswami S, Singh KJ, Chandra S (2006a) Pteridophytes from Lower Gondwana Formations of Ib–River Coalfield, Orissa, and their diversity and distribution in the Permian of India. J Asian Earth Sci 28:234–250

Goswami S, Das M, Guru BC (2006b) Permian biodiversity of Mahanadi Master Basin, Orissa, India, and their environmental countenance. Acta Palaeobot–Krakow 46(2):101–118

Goswami S, Singh KJ, Chandra S (2006c) Palaeobotany of Gondwana Basins of Orissa, India: a bird's eye view. J Asian Earth Sci 28:218–233

Goswami S, Singh KJ (2013) Floral biodiversity and geology of the Talcher Basin, Orissa, India during the Permian-Triassic interval. Geol J 48:39–56

Goswami S, Saxena A, Singh KJ et al (2018a) An appraisal of the Permian palaeobiodiversity and geology of the Ib–River Basin, eastern coastal area, India. J Asian Earth Sci 157:283–301

Goswami S, Das K, Sahoo M et al (2018b) Biostratigraphy and floristic evolution of coal swamp floras of a part of Talcher Basin, India: a window on a Permian temperate ecosystem. Arab J Geosci 11:1–14

Goswami S, Singh KJ, Saxena A et al (2022a) Witnessing floral evolution: a case study from Barakar formation in Lajkura colliery, Ib river coalfield, Mahanadi Basin, India. Hist Biol 34:30–41

Goswami S, Sahoo M, Patel R et al (2022b) Record of scale leaves from Lower Kamthi Formation of Talcher Basin, Odisha, India. Geophytology 50:127–135

Goubin N (1965) Description et repartition des principaux pollenites permicns, triassiques et jurassiques des sondages du basin de Morondava (Madagascar), Revue Institut Francais du Petrole 20:1415–1461

Guhasarkar T (1956) Note on the microflora contents of the Karharbari, Barakar Coals of the Giridih Coalfield. Q Geol Soc, India 28(4):141–142

Gupta S, Saxena A, Shabbar H et al (2022) First record of late Carboniferous palynoassemblage from Ganmachidam Formation, Spiti Valley: implications for age assessment and extent of Glossopterid elements in the Tethyan realm. Geol J 57:2160–2178

Hyde HA, Williams DA (1944) Studies in atmospheric pollen. I. A daily census of pollens at Cardiff, 1942. New Phytol 43:49–61

Jana BN, Bhattacharyya AP, Chakraborti B et al (2002) Permian palynological succession from Mand Raigarh Coalfield, Chhattisgarh. J Geol Soc India 59:537–546

Kar P, Banerjee AK, Banerjee SP et al (1964) Hevey mineral assemblage and their significance in Lower Gondwana sediments of west Bokaro Coalfield. In: Proceedings of 22nd International Geological Congress, New Delhi, vol 9, pp 290–302

Kar R (2003) Palynological recognition of Barren Measures sediments (Middle Permian) from Tatapani-Ramkola Coalfield, Chhattisgarh, India. Gondwana Geol Mag 6:239–244

Kar R, Srivastava SC (2003) Palynological delimitation of the coal–bearing Lower Gondwana sediments in the southern part of Tatapani-Ramkola Coalfield, Chhattisgarh, India. J Geol Soc India 61:557–564

Kar RK (1966) Palynology of the Barren Measures Sequence from Jharia Coalfield, Bihar, India. 1. Summary and discussion. In: Symposium on floristics and stratigraphy of Gondwana world, Lucknow, pp 121–127

Kar RK (1968) Palynology of the Barren Measures Sequence from Jharia Coalfield, Bihar, India. 2. General palynology. Palaeobotanist 16:115–140

Kar RK (1976) Miofloristic evidence for climatic vicissitudes in India during Gondwana. Geophytology 6:230–244

Kerp H, Abu Hamad A, Vörding B et al (2006) Typical Triassic Gondwanan floral elements in the Upper Permian of palaeotropics. Geology 34:265–268

Kumar P (1973) The sporae dispesae of Jabalpur Stage, Upper Gondwana, India. Palaeobotanist 20:90–126

Lele KM (1955) Plant fossils from Parsora in the south Rewa Gondwana basin, India. Palaeobotanist 4:23–34

Lele KM (1961) Studies in the Indian Middle Gondwana Flora: 1. On Dicroidium from the South Rewa Gondwana Basin. Palaeobotanist 10(1–2):48–68
Lele KM (1964) Studies in the Talchir flora of India. 2: Resolution of the spore genus *Nuskoisporites* Pot. and Kl. Palaeobotanist 12(2):147–168
Lele KM (1975) Studies in the Talchir flora of India–10. Early and Late Talchir miofloras from the West Bokaro coalfield, Bihar. Palaeobotanist 22:219–235
Lele KM (1976) Palaeoclimatic implications of Gondwana floras. Geophytology 6:207–229
Lele KM, Karim R (1971) Studies in the Talchir flora of India. 6. Palynology of the Talchir boulder beds in the Jayanti Coalfield, Bihar. Palaeobotanist 19:52–69
Lele KM, Maithy PK (1964) An unusual monosaccate spore from the Karharbari stage, Giridih Coalfield, India. Geophytology, 12:307–312.
Lele KM, Makada R (1972) Studies in the Talchir flora of India–7. Palynology of the Talchir Formation in the Jayanti Coalfield, Bihar. Geophytology 2:41–73
Lindström S, McLoughlin S (2007) Synchronous palynofloristic extinction and recovery after the end-Permian event in the Prince Charles Mountains, Antarctica: implications for palynofloristic turnover across Gondwana. Rev Palaeobot Palynol 145:89–122
Maheshwari HK, Kar RK (1967) Tiwarisporis gen. nov., a new spore genus from the Permian of Cong and India. Curr Sci 36:369–370
Maheshwari HK, Banerji J (1975) Lower Triassic palynomorphs from the Maitur Formation, West Bengal, India. Palaeontographica 152 B:149–190
Murthy S, Saxena A, Khnagar R, Pillai SS, Uhl D, Singh VP, Gupta S, Borkar N (2023) Palynofloristics and wildfire evidence from Permian deposits of the Satpura Gondwana Basin, India: a multiproxy approach. Hist Biol 1–22. https://doi.org/10.1080/08912963.2023.2272690
Maheshwari HK, Kumaran KPN (1979) Upper Triassic sporae dispersae from the Tiki Formation–1: miospores from the Son River section between Tharipathar and Ghiar, South Rewa Gondwana Basin. Palaeontographica 171B:137–169
Maithy PK (1965) Studies in the Glossopteris flora of India–20. Noeggerathiopsis and allied remains from the Karharbari beds, Giridih Coalfield, Bihar. Palaeobotanist 13:89–103
Maithy PK (1966) Studies in the Glossopteris flora of India—26. Glossopteridales from the Karharbari beds, Giridih Coalfield, India. Palaeobotanist 13:248–263
Maithy PK (1969) Studies in the Glossopteris flora of India–39. On some new plant fossils from the Karharbari beds, Giridih coalfield, India. Palaeobotanist 18(1–3):167–172
Maithy PK, Misra PK (1984) A new Cyanophycean remain from the Karharbari Formation, Giridih Coalfield, Bihar. Palaeobotanist 32(1–3):130–133
Maiti A (1994) The palynoassemblage zones of Raniganj Formation (Upper Permian) as revealed from the sub–surface study in Ib River Coalfield, Sundargarh District, Orissa, India. In: 9th international Gondwana symposium, Hyderabad, Abstract, p 12
Meena KL (1998) Palynological dating of sub–surface Kamthi sediments in Ib River Coalfield, Orissa. Geophytology 27:107–110
Meena KL (2000) Palynodating of subsurface sediments of bore–hole IBH–6 in Ib River Coalfield, Orissa, India. Geophytology 2:111–113
Meena KL (2003) Sporae dispersae and correlation of Gondwana sediments in Talcher Coalfield, Son-Mahanadi valley, Orissa. Geophytology 31(1):105–109
Meena KL (1999) Late Barakar flora from the Chaturdhara Nala section, Ib–River Coalfield, Sundargarh, Orissa, India. Palaeobotanist 48(1–3):141–145
Meena KL (2004) Palynology of surface sediments exposed in Basundhara Nala, Ib–River Coalfield, Sundargarh District, Orissa, India. Minetech 25(3):12–17
Meena KL, Goswami S (2004) Palynostratigraphic studies of Late Permian sediments from Tangadih area, Ib River Basin, Orissa, India. Minetech 25:18–26
Meena KL, Pillai SSK (2011) Palynodating of sub-surface sediments from Kuraloi Block, Ib–River Coalfield, Jharsuguda, Odisha, Son-Mahanadi Graben, India. Palaeobotanist 60(1–2):335–343
Meena KL, Murthy S (2011) Palynology of the Talchir Formation from the Ib–River Coalfield, Himgir basin, Odisha. J Indian Geol Congr 3:43–49

Meena KL, Pillai SSK, Murthy S et al (2013) Early Talchir palynozone recorded from sediments of Baikunthpur Area, Chirimiri Coalfield, Chhattisgarh, India. Gondwana Geol Mag 28

Mishra DP, Murthy S, Pandey B et al (2021) Palaeobotanical evidence for Artinskian wildfire in the Talcher Coalfield, Mahanadi Basin, India. J Palaeontol Soc India 66(2):303–314

Mishra DP, Singh VP, Saxena A et al (2022) Palaeoecology and depositional setting of an Early Permian (Artinskian) mire based on a multi-proxy study at the Jagannath coal mine (Talcher Coalfield), Mahanadi Basin, India. Palaeogeogr Palaeoclimatol Palaeoecol 601:111124

Mishra S, Aggarwal N, Jha N (2017) Palaeoenvironmental change across the Permian-Triassic boundary inferred from palynomorph assemblages (Godavari Graben, South India). Palaeobiodivers Palaeoenviron 98(2):177–204

Mukhopadhyay G, Mukhopadhyay SK, Roychowdhury M et al (2010) Stratigraphic correlations between different Gondwana basins of India. J Geol Soc India 76:251–266

Mukhopadhyay S, Bhattacharyya S (2019) Son Valley Basin: sedimentation, structure, and tectonics. In: Developments in structural geology and tectonics, vol 4. Elsevier, pp 73–84

Murthy S (2010) Palynostratigraphy of the Permian succession in Borehole RJS-2, Raniganj Coalfield, Damodar Basin, West Bengal. J Indian Geol Congr 2(2):83–90

Murthy S, Chakraborty B, Roy MD (2010) Palynodating of subsurface sediments, Raniganj coalfield, Damodar Basin, West Bengal. J Earth Syst Sci 119(5):701–710

Murthy S, Vijaya, Vethanayagam SM (2013) Palynostratigraphy of Permian succession in the Pench Valley Coalfield, Satpura Basin, Madhya Pradesh, India. J Palaeontol Soc India 58(2):241–250

Murthy S, Ram-Awatar GS (2014) Palynostratigraphy of Permian succession in the Mand-Raigarh coalfield, Chhattisgarh, India. J Earth Syst Sci 123(8):1879–1893

Murthy S, Kavali PS, Di Pasquo M et al (2020) Late Pennsylvanian and Early Cisuralian palynofloras from the Rajmahal Basin, eastern India, and their chronological significance. Hist Biol 32(2):143–159

Murthy S, Mendhe VA, Uhl D et al (2021) Palaeobotanical and biomarker evidence for wildfire in the early Permian (Artinskian) of the Rajmahal Basin, India. J Palaeogeogr 10:1–26. https://doi.org/10.1186/s42501-021-00084-2

Navale GKB, Tiwari RS (1966) A preliminary sporological analysis of some coals from Talchir Coalfield, India. Palaeobotanist 15:47–51

Navale GKB, Tiwari RS (1967) Petro-palynological study of Churcha Seam, Sohagpur Coalfield (MP), India. Geol Soc India 8:68–74

Navale GKB, Tiwari RS (1968) Palynological correlation of coal seams, their nature and formation in Rampur Coalfield, Lower Gondwana (India). Rev Palaeobot Palynol 6:155–169

Novak H, Schneebeli-Hermann E, Kustatscher E (2019) No mass extinction for land plants at the Permian-Triassic transition. Nat Commun 10:384

Potonié R, Lele KM (1961) Studies in Talchir Flora of India–1. Sporae dispersae from the Talchir beds of south Rewa Gondwana Basin. Palaeobotanist 8:22–31

Prasad B, Pundir BS (2017) Gondwana biostratigraphy of the Purnea Basin (eastern Bihar, India), and its correlation with Rajmahal and Bengal Gondwana basins. J Geol Soc India 90(4):405–427

Sahoo M, Murthy S, Saxena A, Pillai SSK, Kumar S (2024) Significance of palynology in understanding age, palaeoclimate and correlation of Indian Gondwana sediments. In Samant, B. and Thakre, D. (Eds.) Application of Palynology in stratigraphy and climate studies. Springer Nature Publishing (Accepted)

Pal PK (1984) Triassic plant megafossils from the Tiki Formation, South Rewa Gondwana Basin, India. Palaeobotanist 32:253–309

Pal PK, Ghosh AK (1997) Megafloral zonation of Permian-Triassic sequence in the Kamthi Formation, Talcher Coalified, Orissa. Palaeobotanist 46:81–87

Pal PK, Ghosh AK, Kar R et al (2014) Reappraisal of the genus Dicroidium Gothan from the Triassic sediments of India. Palaeobotanist 63:137–155

Pant DD, Basu N (1973) *Pteruchus indicus* sp. nov. from the Triassic of Nidpur, India. Palaeontographica 144B:11–24

Pant DD, Basu N (1979a) Some further remains of fructifications from the Triassic of Nidpur, India. Palaeontographica 168B:129–146

Pant DD, Basu N (1979b) On some megaspores from the Triassic of Nidpuri, India. Rev Palaeobot Palynol 28:203–221

Pareek HS (1969) Petrological study of coals from Singrauli coalfield, M.P. Rec Geol Surv India 97:87–90

Patel R, Goswami S, Aggarwal N et al (2021a) Lower Gondwana megaflora, palynoflora, and biomarkers from Jagannath Colliery, Talcher Basin, Odisha, India, and its biostratigraphic significance. Geol J 57:986–1004

Patel R, Goswami S, Aggarwal N et al (2021b) Palaeofloristics of lower Gondwana exposure in Hingula area, Talcher Basin, Odisha, India: an inclusive study on biomarkers, megafloral and palynofloral assemblages. Hist Biol 1–17

Patel R, Goswami S, Sahoo M et al (2021c) Biodiversity of a Permian temperate forest: a case study from Ustali area, Ib River Basin, Odisha, India. Geol J 56(2):903–933

Pillai SSK, Meena KL, Tewari R et al (2016) Early Triassic Palynomorphs from Kuraloi Block, Belphar Area, Ib–River Coalfield, Mahanadi Basin, Odisha. J Geol Soc India 88(6):693–704

Pillai SSK, Agnihotri D, Gautam S at al (2018) Glossopteris flora from the Pali Formation, Johilla Coalfield, South Rewa Gondwana Basin, Madhya Pradesh, India: palynological evidence for a Late Permian age. J Palaeontol Soc India 63(1):53–72

Raja Rao CS (1983) Coalfields of India, vol. III; coal resources of Madhya Pradesh, Jammu and Kashmir. Bull Geol Surv India, Ser A 45:75–80

Ram-Awatar (1993) Palynological dating of subsurface sequence of Middle Pali Member in Sohagpur Coalfield (M.P.), India. Geophytology 23:107–114

Ram-Awatar (1996a) Palynozonation of Middle Pali Member in Sohagpur Coalfield, Madhya Pradesh. Palaeobotanist 43:96–101

Ram-Awatar (1996b) Palynostratigraphy of Supra–Barakar sediments (Pali, Tiki and Parsora formations) and their stratigraphic positions in South Rewa Basin, Madhya Pradesh. In: Guha PKS et al (eds)Proceedings of 9th international Gondwana symposium, Hyderabad, vol 1, pp 439–454

Ram-Awatar (1997) Palynological evidence for the Permian–Triassic boundary in Sohagpur Coalfield, India. Palaeobotanist 46:101–106

Ram-Awatar (2003) A Triassic palynoflora from Pali Formation, South Rewa Gondwana Basin, Madhya Pradesh, India. Palaeobotanist 52:49–52

Ram-Awatar, Mukhopadhyay A, Adhikari S (2004) Palynostratigraphy of sub surface Lower Gondwana, Pali sediments, Sohagpur Coalfield, Madhya Pradesh, India. Palaeobotanist 53:51–59

Retallack GJ (1995) Permian-Triassic life crisis on land. Science 267:77–80

Roychowdhury M (2019) Mahanadi Valley Basin structure and tectonics. In: Developments in structural geology and tectonics, vol 4. Elsevier, pp 95–102

Sabina PK, Roy A, Di Pasquo M et al (2021) Upper Pennsylvanian age of the Lower Talchir Formation in the Wardha Basin, Central India, based on guide palynomorphs present in radiometrically-dated palynozonations in South America, Africa, and Australia. Ameghiniana 58(4):318–344

Sahoo M, Aggarwal N, Goswami S (2020a) Palynological investigation of the Lower Gondwana outcrop near Gopal prasad, Odisha, India: an inference on the age, palaeovegetation and palaeoclimate. J Geol Soc India 65(1):27–35

Sahoo M, Goswami S, Aggarwal N et al (2020b) Palaeofloristics of lower Gondwana exposure near Kumunda village, Angul District, Talcher Basin, Odisha, India: a comprehensive study on megafloral and palynofloral assemblages. J Geol Soc India 95(3):241–254

Sarana S, Kar R (2011) Effect of igneous intrusive on coal microconstituents: study from an Indian Gondwana coalfield. Int J Coal Geol 85(1):161–167

Sarate OS, Murthy S (2017) Talchir Palynoflora from Wardha Valley Coalfield, Central India: climatic implications. J Geosci Res 2(2):139–148

Sastry MVA, Acharya SC, Shah SC et al (1977) Stratigraphy Lexicon of Gondwana formations of India. Geol Surv India Misc Publ 36:1–170

Saxena A, Singh KJ, Goswami S (2014) Advent and decline of the genus *Glossopteris* Brongniart in the Talcher Coalfield, Mahanadi Basin, Odisha, India. Palaeobotanist 63:157–168

Saxena A, Singh KJ, Murthy S et al (2016a) Spore tetrads, possible indicators of intense climatic regimes: case study from an early Permian stratum of Singrauli Coalfield, Son-Mahanadi Basin, India. Geol Mag 153:426–437

Saxena A, Singh KJ, Shabbar H et al (2016b) Macrofloral assemblage from the Early Permian Barakar Formation of Singrauli Coalfield, Son-Mahanadi Basin, India. Palaeobotanist 65:139–150

Saxena A, Singh KJ, Murthy S et al (2019a) Early Permian floral diversity deduced from the Barakar Formation of Singrauli Coalfield, Son-Mahanadi Basin, India. J Palaeontol Soc India 64(2):169–183

Saxena A, Singh KJ, Cleal CJ et al (2019b) Development of the *Glossopteris* flora and its end Permian demise in the Tatapani-Ramkola Coalfield, Son-Mahanadi Basin, India. Geol J 54(4):2472–2494

Saxena A, Gupta S, Singh KJ et al (2021) Diversity of the Genus *Gangamopteris McCoy* in the Early Permian Sequences of Singrauli Coalfield, Son-Mahanadi Basin, India. J Palaeontol Soc India 66(1):23–34

Saxena A, Khan MM, Raychowdhury N et al (2022) Early Permian macrofloral diversity in Indian Gondwana: evidence from Talchir Formation of Singrauli coalfield, Son-Mahanadi valley basin, central India. J Earth Syst Sci 131(2):1–19

Sen J (1953) Principles and problems of microfloral correlation of Indian coal seams with special reference to Karharbari. Bull Natl Inst Sci, India 2:129–140

Shah SC (1976) Climate during Gondwana Era in peninsular India: faunal evidences. Geophytology 6:186–206

Shen J, Chen J, Algeo TJ et al (2019) Evidence for a prolonged Permian-Triassic extinction interval from global marine mercury records. Nat Commun 10:1563

Singh IB (1976) Mineralogical evidence for climate vicissitudes in India during Gondwana times. Geophytology 6:174–185

Singh KJ, Saxena A (2015) End Permian (Lopingian) floral diversity in Singrauli Coalfield: evidences from Jhingurdah Colliery, Son-Mahanadi Basin, India. J Paleontol Soc India 60:97–112

Singh KJ, Chandra A, Chandra S (2005) Evaluation of earliest Permian flora of India and its equivalents in other Gondwana continents. Palaeobotanist 54:107–113

Singh KJ, Goswami S, Singh G (2011a) Palaeodiversity in the genus Glossopteris from the Lower Gondwana rocks of the Korba Coalfield, Chhattisgarh State, India. J Palaeontol Soc India 56:45–64

Singh KJ, Chandra S, Saxena A (2011b) *Tatapania* gen. nov., a possible cone of *Schizoneura gondwanensis* Feistmantel from the Late Permian in the Tatapani-Ramkola Coalfield, India. Palaeobotanist 60:251–263

Singh KJ, Goswami S, Chandra S (2006) The genus Glossopteris from Lower Gondwana formations of Ib–River Coalfield, Orissa, India. J Palaeontol Soc India 51:81–107

Singh KJ, Goswami S, Chandra S (2007) Occurrence of Cordaitales from lower Gondwana sediments of Ib-River Coalfield, Orissa, India: an Indian scenario. J Asian Earth Sci 29(5–6):666–684

Singh KJ, Murthy S, Saxena A et al (2017) Permian macro- and miofloral diversity, palynodating and palaeoclimate implications deduced from the coal–bearing sequences of Singrauli coalfield, Son-Mahanadi Basin, central India. J Earth Syst Sci 126(2):1–16

Singh KJ, Naugolnykh SV, Saxena A (2012) Permian and Triassic plant assemblages from the Tatapani–Ramkola Coalfield (India). In: Naugolnykh SV, Kodrul TM, Kokina OA (eds) Palaeontology and evolution of the biodiversity in the Earth history. Journal of the Russian Academy of Sciences, Moscow, pp 98–109

Singh KJ, Saxena A, Goswami S (2016) In-situ occurrence of *Vertebraria* roots in the Raniganj Formation of Singrauli Coalfield and its palaeoecological significance. Curr Sci 110:299–301

Sinha V (1969) Some "Acritarchs" and other microfossils from Barakar Stage of Lower Gondwana India. Palaeobotanist 17:326–331

Srivastava AK, Tewari R (2001) Lower Gondwana plant fossils from the Barren Measures Sequence from Jharia Coalfield, Bihar, India. Geol Surv India, Spec Publ 54:127–134
Srivastava SC (1973) Palynostratigraphy of the Giridih coalfield. Geophytology 3(2):184–194
Srivastava SC (1974) Pteridospermic remains from the Triassic of Nidpur, M.P. Sci Cult 41:211–212
Srivastava SC, Anand-Prakash KR (1997) Palynology of Permian-Triassic sequence in Iria Nala, Tatapani-Ramkola Coalfield, India. Palaeobotanist 46:75–80
Srivastava SC, Anand-Prakash SOS (1988) Palynology of the Talchir Formation from Betul Coalfield, Satpura Basin, India. Palaeobotanist 37:81–84
Srivastava SC, Anand-Prakash (1984) Palynological succession of the Lower Gondwana sediments in Umaria Coalfield, Madhya Pradesh. Palaeobotanist 32:26–34
Surange KR, Lele KM (1954) Studies in the Glossopteris flora of India–3. Plant fossils from the Talchir Needle shales from Giridih Coalfield. Palaeobotanist 4:153–157
Suttner LJ, Dutta PK (1986) Alluvial sandstone composition and palaeoclimate, I frame work mineralogy. J Sedimentol Petrol 56:329–345
Tiwari RS (1965) Miospore assemblage in some coals of Barakar Stage (Lower Gondwana) of India. Palaeobotanist 13:168–214
Tiwari RS (1967) Palynological investigations of some coal seams in the Ib–River Coalfield (Orissa), India. Palaeobotanist 16(1–3):222–242
Tiwari RS (1969) Sporological succession in Purew Seam, Singrauli Coalfield, M.P. In: Sen memorial volume. Botanical Society of Bengal, Calcutta, pp 93–100
Tiwari RS (1971) Sporological succession in Kota and Turra seams, Singrauli Coalfield, (M.P.), India. Palaeobotanist 18(3):264–269
Tiwari RS (1973) Palynological succession in the Barakar type area. Geophytology 3:166–183
Tiwari RS (1974) Inter-relationships of palynofloras in the Barakar stage (–Lower Gondwana), India. Geophytology 4(2):111–129
Tiwari RS, Ram-Awatar (1986) Late Permian palynofossils from the Pali Formation, South Rewa Basin, Madhya Pradesh. Bull Geol, Min Metall Soc India 54:250–255
Tiwari RS, Rana V (1980) A middle Triassic mioflora from India. Biol Memo 5(1):30–55
Tiwari RS, Rana V (1981) Sporae dispersae of some Early and Middle Trias¬sic sediments from Damodar Basin, India. Palaeobotanist 27:190–220
Tiwari RS, Singh V (1981) Morphographic study of some dispersed trilete miospores (sub–Infraturma–Varitrilete) from the Lower Gondwana of India. Palaeobotanist 27:253–296
Tiwari RS, Srivastava SC, Tripathi A, Vijaya (1989) Morphographic study of Permian palynomorphs: Callumispora, Parasaccites, Crucisaccites and Faunipollenites. Palaeobotanist 37:215–266
Tiwari RS, Tripathi A (1988) Palynological zones and their climate inference in the coal bearing Gondwana of Peninsular India. Palaeobotanist 36:87–101
Tiwari RS, Ram-Awatar (1990) Palyno–dating of Nidpur beds, Son Graben, Madhya Pradesh. In: Jain KP, Tiwari RS (eds) Proc. symposium vistas in Indian palaeobotany. Palaeobotanist 38:105–121
Tiwari R, Mehrotra NC, Meena KL et al (2009) Permian megaspores from Kuraloi area, Ib–River Coalfield, Mahanadi Basin, Orissa. J Geol Soc India 74:669–678
Tiwari RS, Tripathi A (1992) Marker assemblage zones of spore and pollen species through Gondwana Palaeozoic and Mesozoic sequence in India. Palaeobotanist 40:194–236
Tiwari RS, Tripathi A, Vijaya (1995) Organic-walled microfossils of doubtful origin in Permian and Triassic sequences on Peninsular India. Palaeobotanist 43(1–3):1–38
Tiwari RS, Vijaya (1995) Differential morphographic identity of Gondwana palynomorphs. Palaeobotanist 44:62–115
Tripathi A (1996) Early and Late Triassic palynoassemblage from subsurface Supra-Barakar sequences in Talchir Coalfield, Orissa, India. Geophytology 26(1):109–118
Tripathi A (1997) Palynostratigraphy and palynofacies analysis of subsurface Permian sediments in Talcher coalfield, Orissa. Palaeobotanist 46(3):79–88

Tripathi A (2001) Palynological expression about Permian-Triassic transition in the Talcher coalfield, Orissa, India. Palaeobotanist 50(1–3):247–253
Tripathi A, Bhattacharya D (2001) Palynological resolution of Upper Permian sequence in Talcher Coalfield, Orissa, India. In: Proceedings of national seminar on recent advance in geology of coal and lignite basins of India, Calcutta, 1997, vol 54. Geological Survey of India, Special Publication, pp 59–68
Tripathi A, Murthy S, Chakarborty B et al (2012) Stratigraphic status of coal horizon in Tatapani-Ramkola Coalfield, Chhattisgarh, India. J Earth Syst Sci 121(2):537–557
Trivedi BS (1950) Megaspores from lower Gond-Wana of Singrauli coalfield, District Mirzapur. Curr Sci 19(4):126
Venkatachala BS, Kar RK (1965) Two new triletes spore genera from the Permian of India. Palaeobotanist 13:337–340
Vijaya (1989) Evolutionary pattern of striations and taeniae in the Indian Gondwana saccate pollen. Palaeobotanist 38:83–91
Vijaya, Tripathi A, Roy A et al (2012) Palynostratigraphy and age correlation of subsurface strata within the sub-basins in Singrauli Gondwana Basin, India. J Earth Syst Sci 121:1071–1092
Visscher H, Brinkhuis H, Dilcher D et al (1996) The terminal Paleozoic fungal event: evidence of terrestrial ecosystem destabilization and collapse. Proc Natl Acad Sci 93:2155–2158

An Overview of Upper Gondwana Rajmahal Flora and Its Significance

Neelam Das and Anju Saxena

Abstract The flora of the Rajmahal Formation is typically known as 'Rajmahal Flora'. It is Early Cretaceous (Barremian to Aptian) in age and is one of the most interesting Mesozoic floral succession characterized by prolific and diversified floral entities. Distinctive amalgamation of volcanic and sedimentary sequences (intercalations of basaltic lava flows and intertrappean sediments) is a unique feature of this formation which had led to the excellent preservation of plant body remains along with anatomical details and it is one of the national geological monuments of India. A number of floral elements of Rajmahal Flora have been under continuous taxonomic reassessment, particularly enigmatic macroremains of angiosperms, though definite pollen records of angiospermid affinity are known from this formation. The present paper provides an updated synthesis and analysis of the palaeofloritics of the Rajmahal Formation and underlying Dubrajpur Formation and its relevance in tracing the angiosperm evolution.

Keywords Dubrajpur Formation · Rajmahal Flora · Ptilophyllum Flora · Upper Gondwana · Mesozoic

1 Introduction

Rajmahal Flora is the most enthralling and intriguing among all the Gondwana floras known from India. This flora is also known as 'Ptilophyllum Flora' and is characterized by prolific and diversified plant fossil assemblage of Mesozoic period. Peculiar volcano-sedimentary nature of sediments of Rajmahal Basin makes it distinct from the lithoarchitecture of other Gondwana basins. This unique amalgamation of sediments of non-marine origin followed by the deposition of periodic lava flows during the Cretaceous had resulted in the excellent store-house of Mesozoic plant assemblages. Stratigraphically, the sedimentary succession in the Rajmahal Basin begins with the deposition of Lower Permian strata belonging to Talchir and Barakar

N. Das · A. Saxena (✉)
Birbal Sahni Institute of Palaeosciences, 53-University Road, Lucknow 226007, India
e-mail: anju_saxena@bsip.res.in

B. Samant and D. Thakre (eds.), *Applications of Palynology in Stratigraphy and Climate Studies*, Society of Earth Scientists Series, https://doi.org/10.1007/978-3-031-51877-5_4

formations (Barakar, the coal bearing formation) preceding a big hiatus and are unconformably superimposed over the basement rocks of Archean granitic complex (Murthy et al. 2020a). Late Permian and Triassic sequences are missing as exposures (though palynologically recorded) in the region, and the Lower Permian strata are overlained by infratrappean Dubrajpur Formation (late Triassic-Jurassic) followed by superimposition of Rajmahal Formation (traps and intertrappean beds) during early Cretaceous. The Mesozoic exposures are spread across the basin whereas the coal bearing Permian successions are restricted to western part of the basin. The intertrappean beds are very thin as compared to the thickness of individual lava flow in which thickness of intertrappean beds vary from a few centimetres to about few meters whereas in subsurface these are up to 100 m thick (Raja Rao 1987). Traditionally, the Rajmahal traps (lava flows) are recognized as one of the manifestations of the disruption of Indian plates and the subsequent adjustment of adjacent plate boundaries (Sen Gupta 1988). It is considered that Rajmahal lava flows were poured out as fissure-type eruptions, although isolated volcanic vents have been recognised in some areas (Sarbadhikari 1968).

The Rajmahal Formation is one of India's best-known Upper Gondwana sequences, and its plant fossil-bearing intertrappean beds have been declared as 'National Geological Monuments of India' by the Geological Survey of India (GSI). Rajmahal Hills have attracted attention of geologists and have always been a cause of curiosity for palaeobotanists since 1850 when the area was first explored by McClelland. Initially, the 'Rajmahal Flora' was predominantly known by impressions until the discovery of petrified plant remains from Nipania locality in the Rajmahal Hills by Mr. Hobson of the Geological Survey of India in 1928 (Hobson 1928). The pioneer work on the geology of Rajmahal Basin with a brief introduction on plant fossils was carried out by McClelland (1850), followed by the detailed geological studies by Oldham (1854, 1860) and systematic study of plant fossils by Oldham and Morris (1862). Significant work on palaeobotanical aspect of Rajmhal Flora was carried out by Feistmantel (1876a, b, 1877a, 1881). Later on, almost at the gap of 40 years, the Rajmhal Basin got attention of Prof. Birbal Sahni (Father of Indian Palaeobotany), who had done considerable work in this region along with his co-workers during 1920 to 1948, and some of his monumental work from this region include reconstruction of synthetic group of plants i.e. Pentoxylales and reconstruction of whole plant of *Williamsonia sewardiana*. He opined that the Rajmahal Flora is of middle Jurassic age. It would not be wide of the mark to assume that it was probably the richness and exceptionality of flora of Rajmahal area that served as one of the impetuses and impelled Prof. Sahni to fulfil his dream of establishing an institute dedicated to palaeobotany in India wayback in 1946, which is currently known as Birbal Sahni Institute of Palaeosciences (formerly Birbal Sahni Institute of Palaeobotany). Reconstruction of Pentoxylon with *Nipaniophyllum* leaves and female cone *Carnoconites* made by Prof. Sahni was taken as the emblem of the institute. Further significant palaeobotanical work on the Rajmahal Flora was carried out by many workers (e.g. AR Rao, Vishnu Mittre, Sukh Dev, MN Bose, BD Sharma, Sen Gupta, J Banerji and others); details about some of the significant contributions made over the years have been provided and discussed in the subsequent section 'Flora of Rajmahal' of this chapter. Sen Gupta (1988) did a detailed synthesis and discussion of stratigraphical and palaeobotanical aspects of Rajmahal Basin. An overview of this flora reveals that of all the upper Gondwana sequences of India, Rajmahal Hills and surrounding areas have yielded the richest, well-preserved and highly diversified plant fossils in the form of impressions/compressions and petrified remains of the Mesozoic era; and have the dominance of gymnosperms (bennettitales, cycads, conifers, pentoxylales) followed by ferns. Apart from prolific occurrences of gymnospermous remains and subdued occurrences of ferns (both in macro and mioforms), the flora is also known for enigmatic occurrences of macro plant remains of angiospermous affinity from the early Cretaceous (Berriasian to Aptian) intertrappean beds of Rajmahal Formation (Sahni 1932a; Sharma 1997; Banerji 2000a); which have been challenged and subjected to reassessment by many workers over the time (e.g. Bose and Sah 1954; Srivastava and Krassilov 2012). However, definite palynological records for the angiospermous affinity are known from the uppermost lower Intertrappean beds which were radiometrically dated as 117 Ma (Aptian) (Tiwari and Tripathi 1995).

The present paper provides a detailed and updated synthesis of palaeobotanical studies pertaining to flora of Rajmahal Basin, with emphasis on upper Gondwana flora and an insight to its significance in understanding the evolution of angiosperms. Different floral groups recorded from the Permian and Mesozoic sequences of the Rajmahal Basin along with their biostratigraphic significance are discussed herein.

2 Geological Setting

Rajmahal Basin is a part of the north–south trending Rajmahal-Purnea Graben, which stretches across a large area of Jharkhand and West Bengal states, in the eastern Indian region (Fig. 1). The basin is bounded by the Ganga River in the north, trap hills in the east, metamorphic terrain in the west and laterite and alluvium in the south. A number of detached outcrops of Gondwana sediments occur along the western flank of Rajmahal Hills. Only the western margin of this master basin is exposed where the Gondwana sedimentaries abut against the Chhotanagpur Gneissic Complex along a prominent NNW-SSE trending fault. Other margins of the basin are concealed below thick piles of basic volcanics (Rajmahal Trap). In the Jharkhand State, four districts namely Sahibganj, Pakur, Godda, and Dumka covering an area of 2,600 km^2 encompass the 'Rajmahal Hills' (Raja Rao 1987). This hilly area is flanked by the river Ganga towards the northern and eastern parts and the undulating plains of Bhagalpur and West Bengal, towards the western and southern parts. The Rajmahal Hills comprise of a series of volcanic lava flows (Rajmahal Traps) and associated intertrappean sedimentary beds; the lava flows of the Rajmahal Formation have a characteristic topographic expression, entirely different from those of Precambrian or infratrappean formations (Sen Gupta 1988). The hill ranges formed by basalts of the Rajmahal Formation are generally flat-topped and show step-like terraces. Each terrace commonly represents an individual basaltic flow. The hills and plateau attain an abrupt height of 608 m, with a steep scarp face along the western flank of the

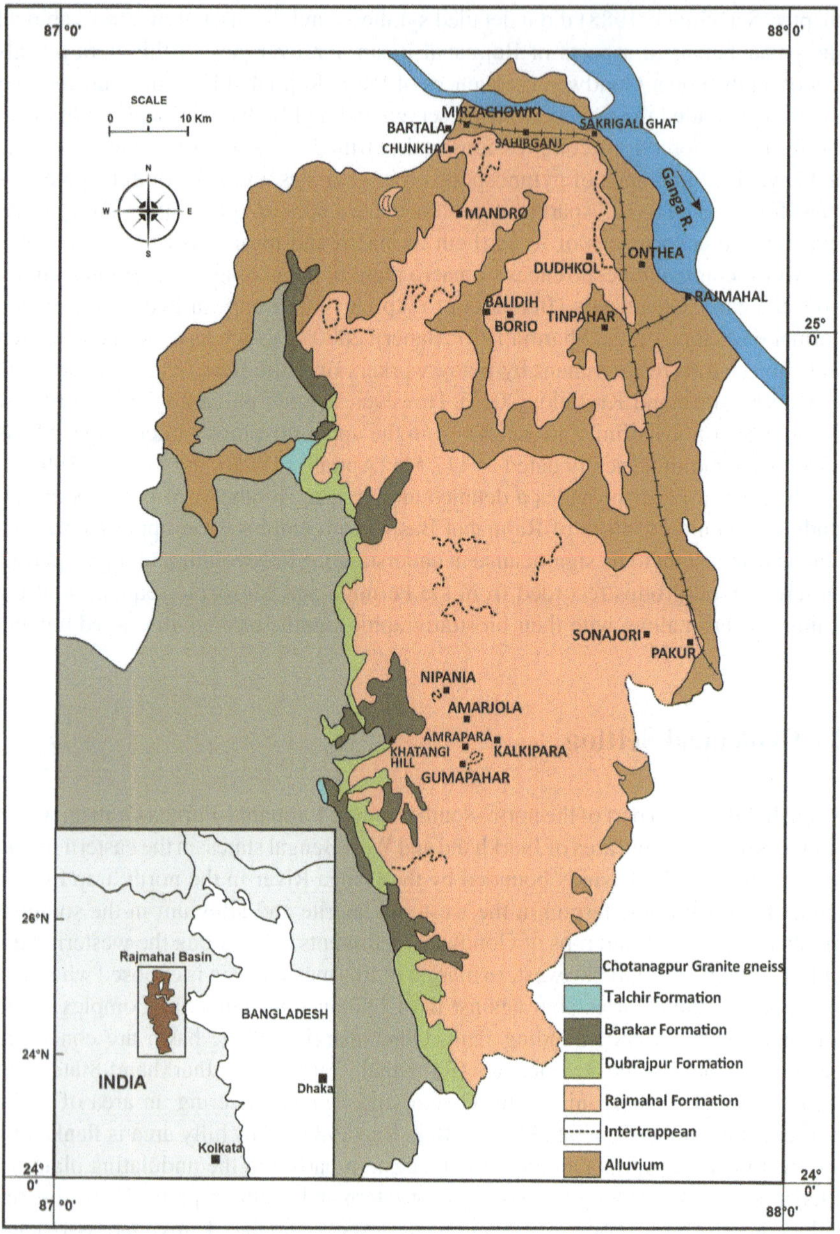

Fig. 1 Geological map of Rajmahal Basin depicting important Mesozoic plant fossil localities (modified after Sen Gupta 1988)

range. There is a rapid decline in altitude at the eastern and northern boundaries of the Rajmahal Hills.

Detailed geological mapping of the region was first undertaken by Ball (1877). The stratigraphy of the Rajmahal area was further elaborated by Raja Rao (1987) and Sen Gupta (1988). The general sequence of Gondwana strata present in the Rajmahal Basin is given in the Table 1. The complete succession of Gondwana is divided into lower and upper Gondwana; the lower Gondwana is represented by Talchir (Asselian-Sakmarian) and Barakar (Artinskian-Kungurian) formations while upper Gondwana is represented by Dubrajpur and Rajmahal formations. Dubrajpur rocks overlain Barakar Formation with a hiatus.

Lower Gondwana formations: Talchir Formation, the lowermost formation of Gondwana sediments is poorly exposed. This was deposited unconformably over the Precambrian basement complex, comprising Chhotanagpur Granite with gneisses, and migmatites (Sarkar 1972). This formation is succeeded by the Barakar Formation, comprising clayey sandstone, shale and carbonaceous shales with coal seams. It is well exposed laterally at the western flank of the Rajmahal basin in several NNW–SSE trending linear chains of detached outcrops known from north to south as

Table 1 Stratigraphical sequence depicting lava flows and intertrappean beds along with lithological characters (after Sen Gupta 1988)

Formation	Flows	Lithology		Age
Alluvium				
Laterite				
Rajmahal	Upper trap flows with or without intertrappean beds	Basalt flow		Cretaceous
		Basalt flow with unfossiliferous/scarcely fossiliferous intertrappean bed		
	Upper trap flows with fossiliferous intertrappean beds	Basalt flow		Lower Cretaceous
		8. Fossiliferous intertrappean bed with *Taeniopteris spatulata–Brachyphyllum rhombicum* Assemblage Zone (3)		
		Basalt flow		
	Lower trap flows with fossiliferous intertrappean beds	7. Fossiliferous intertrappean bed	With *Cladophlebis indica-Dictyozamites indicus* Assemblage Zone (2)	Middle to upper Jurassic
		Basalt flow		
		6. Fossiliferous intertrappean bed		
		Basalt flow		
		5. Fossiliferous intertrappean bed		
		Basalt flow		
		4. Fossiliferous intertrappean bed		

(continued)

Table 1 (continued)

Formation	Flows	Lithology	Age
		Basalt flow	
	Lower trap flows with unfossiliferous intertrappean beds	3. Intertrappean bed with rare vegetative remains like twig	
		Basalt flow	
		2. Unfossiliferous intertrappean	
		Basalt flow	
		1. Unfossiliferous intertrappean	
		Basalt flow	
Igneous contact			
Dubrajpur		Baked sandstone, siltstone, fine grained sandstone, ferruginous sandstone, coarse grained sandstone; with Ptilophyllum Flora at places	Lower Jurassic
		Fossiliferous siltstone with *Ptilophyllum acutifolium–Gleichenoids* Assemblage Zone (1)	Triassic
		Pebbly sandstone	
		Coarse grained sandstone, ferruginous sandstone, shale, sandstone with clay beds etc.	
Barakar Formation with Glossopteris flora			

Hura, Chuperbhita, Pachwara, Mahuagarhi, and Brahmani coalfields. In the outcrops, Barakar Formation is overlained by the Dubrajpur Formation representing a hiatus of late Permain and early-middle Triassic sediments. However, based on palynological analysis, late Permian sequences are recognized in the bore core sediments (Murthy et al. 2020a).

Upper Gondwana formations: The basal part of the upper Gondwana sequence—the Dubrajpur Formation occurs as a narrow, discontinuous swerving strip out skirting the Rajmahal Formation in the western margin of Rajmahal Hills, comprises bands of conglomerates, grits and sandstones with inclusions of fine-grained arenaceous beds/shales (Sah and Shah 1974) and is devoid of any coal seam. Dubrajpur Formation is overlain by the Rajmahal Formation, and the topmost part of the Dubrajpur Formation is often baked and charred by the heat of the molten basaltic lava flows of the Rajmahal Formation. In consequence, the contact between the formations is described as 'igneous contact' by Sen Gupta (1988). On the basis of mineralogical parameters Singh (1976) reconstructed palaeoclimate and suggested that subtropical climate existed during Mahadeva times, which was marked by prolonged periods of seasonal droughts, which gradually changed into tropical, humid climate during deposition of Rajmahal intertrappeans.

The Rajmahal Formation includes both the traps and the intertrappean beds. The intertrappean sediments crop out all over the Rajmahal Basin, except in the western part of the basin. This formation exhibits a gradual decrease in its thickness from west to east (Sen Gupta 1988). The thickness of the Rajmahal Formation is estimated to be around 457 m and the cumulative thickness of the intertrappean beds is estimated to be not more than 30 m (Pascoe 1968). The maximum exposed thickness of the formation is 263.4 m (Sen Gupta 1988). The intertrappean beds are very thin as compared to the thickness of individual lava flow in which thickness of intertrappean beds vary from a few centimetres to about few meters (upto 11 m), whereas in subsurface these are up to 100 m thick (Raja Rao 1987). It is considered that Rajmahal lava flows were poured out as fissure-type eruptions, although isolated volcanic vents have been recognised near Simra (Ball 1877), Tinpahar, Morang valley and Taljhari area (Sarbadhikari 1968). The lava flows comprise mainly tholeiitic basalt flows of a non-orogenic volcanic province (Sarbadhikari 1968) and intercalated freshwater intertrappean beds deposited in locally formed isolated lakes. Ramaswamy (1958) recognised fifteen lava flows with eleven intertrappean beds in the Rajmahal Hills. Lithologically, the intertrappean beds of the Rajmahal Formation are characterized by sandstone, shale, siltstone, and clay deposits.

The Rajmahal Formation is very well developed in the western part of the Rajmahal Hills, and in this part, the contact between Rajmahal Formation and underlying Dubrajpur Formation is exposed which is placed above the baked sandstone and shale of the topmost part of Dubrajpur Formation. The fossiliferous intertrappean beds of Rajmahal Hills are well exposed around Amarapara (lying in the southern part of the Rajmahal Basin), Borio (lying in the northern part of the Rajmahal Basin) and south of Mirzachawki (lying around the central part of the Rajmahal Basin). Their lateral correlation is difficult as these beds regularly pinch out within a short distance (Sen Gupta 1988). Furthermore, there is great degree of variation among the plant fossil content of the first intertrappean bed in the southern and northern parts of the Rajmahal Hills. Sen Gupta (1988) has dealt in great detail the stratigraphic positions of the intertrappean beds within the Rajmahal Formation and their exposures at different places, and for their correlation based on lithological characters, floral contents, and their elevation; he divided the Rajmahal area into four sectors (see Sen Gupta 1988 for detailed description for stratigraphy). He recognized eight interrappean beds sandwiched between the lava flows.

3 Age of Rajmahal Formation

The age of the Rajmahal Formation has been debatable since its initial studies; it has been dated as ranging between as old as latest Triassic and upper age limit as young as Early Cretaceous. It was dated as Liassic (earliest Jurassic) (Feistmantel 1877a), Lower Jurassic (Ball 1877), Middle Jurassic (Halle 1913; Bose 1953a, b; Ganju 1946; Rao 1972; Sahni 1932b, 1948; Sharma 1967, 1969b, 1971a, b, 1973b, 1974a, b), Upper Jurassic (Sahni and Rao 1933), and Early Cretaceous (Jacob 1937,

1938; Banerji and Jana 2003; Sen Gupta 1988; Banerji 1992, 1996, 1997, 2000a, b). Based on the palaeobotanical evidence, Seward (1917) considered these beds to be of Rhaetic to lower Jurassic. Du Toit (1927) also compared the Molteno flora of South Africa with the Rajmahal flora and suggested a Rhaetic age for the latter. Rao (1953) argued that many of the Rajmahal genera (like *Stachyotaxus* and *Phlebopteris*) have equivalents in other Rhaetic floras and based on this correlation he suggested a Rhaetic (latest Triassic) to middle Jurassic age for this flora and negated the age given by Jacob (1937) which was based on single occurrence of fern *Protocyathea*. Upper Jurassic is the widely accepted age for this formation. Palynological records also revealed long age bracketing for the beds of 'Rajmahal Series' especially that exposed in Nipania, Basko and Sakrigalighat localities, ranging between middle Jurassic to Cretaceous (Rao 1943a; Vishnu-Mittre 1954; Sah and Jain 1965). Palynological study carried out by Tiwari et al. (1984), based on bore-core sequence drilled near Kazigaon, in the northeastern part of Rajmahal Hills, encompassing complete Dubrajpur and Rajmahal formations, revealed 6 distinct palynoassemblages ranging from Triassic to early Cretaceous. This study suggested that Dubrajpur Formation is of Triassic to early Cretaceous and Rajmahal Formation is exclusively early Createceous in age. Palynological investigations based on the sediments of Dubrajpur Formation suggested that this formation is profoundly diachronous, as it includes strata of the Early–Late Triassic (Induan), late Early–early Mid Jurassic, Late Jurassic and Early Cretaceous age (Berriasian–Hauterivian) and the age of the overlying intertrappean sedimentary beds of the Rajmahal Formation has been established as Early Cretaceous (Tripathi and Ray 2006). Based on palynology, Tripathi (2008) suggested a Berriasian to Aptian age for the Rajmahal Formation. In a more recent study carried out by Murthy et al. (2020a), the palynological analysis of a 35 m deep borehole (BRS-7) intersected through sedimentary rock succession of Rajmahal Formation, Dubrajpur Formation and in its basal part coal bearing Barakar Formation of Brahmani Coalfield, Rajmahal Basin, have reveled three distinct palynoassemblages. The coal bearing sediments revealed *Densipollenites magnicorpus* Assemblage zone (Palynoassemblage-I) indicating late Permian (Lopingian) age corresponding to Raniganj Formation for the sediments lithologically designated as Barakar Formation. The sediments of overlying Dubrajpur Formation have yielded poor palynoassemblage and is devoid of precise age marker taxa of Dubrajpur Formation. However, the presence of *Callialasporites turbatus* and *Araucariacites australis* pollens in this palynoassemblage broadly reflects affinity with Jurassic- Early Cretaceous palynoflora. The overlying Rajmahal Formation has yielded a rich assemblage (Palynoassemblage-III) comprising of age marker taxa, viz. *Callialasporites* spp., *Podocarpidites* spp., *Januasporites spinulosus*, *Contignisporites* spp., *Podosporites tripakshi* and suggesting late Jurassic and earliest Cretaceous (Barriasian) age to this part of the Rajmahal Formation.

Thus in light of the recent studies, based on cumulative evidences of macrofloral (Sen Gupta 1988; Banerji 1994, 1995a, b, 1996; Banerji and Jana 2003), palynological (Tiwari et al. 1984; Tiwari and Tripathi 1995; Tripathi 2008; Murthy et al. 2020a, b, 2021) and geological (radiometric dates based on K–Ar and $^{40}Ar/^{39}Ar$, dating) data (McDougall and McElhinny 1970; Agarwal and Rama 1976; Baksi 1995;

Pandey and Dave 1998; Kent et al. 2002) an early Cretaceous age for the Rajmahal Formation is considered. Radiometric dates provided by Harland et al. (1990) dated ($^{40}Ar/^{39}Ar$), for the flows associated with uppermost lower intertrappean of Rajmahal deposits, is 117 Ma (Aptian). Kent et al. (2002) using the $^{40}Ar/^{39}Ar$ method, estimated the age of a 230 m thick lava pile in the Rajmahal Basin as 118 Ma.

4 Flora of Rajmahal Basin

Lower Gondwana flora is also known from the Rajmahal Basin, however, in contrast to Permian flora of the other peninsular Gondwana basins, Permian flora of Rajmahal is less abundant and less diversified. Lower Gondwana sequences of Rajmahal area were initially studied by Ball (1877), who established stratigraphy and also provided a brief note about the fossil plants of Talchir and Barakar sequences. Detailed palaeobotanical work of these sequences was further carried out by Feistmantel (1877a) and plant fossil assemblage included *Gangamopteris*, *Glossopteris*, equisetales (*Phyllotheca*), scale leaves, seeds and pteridophytic leaf remains. Later on, a rich and diversified plant assemblage was described by Maheshwari and Prakash (1964) comprising *Noeggerathiopsis*, *Gangamopteris*, *Glossopteris*, *Stellotheca* (*Phyllotheca*), *Schizoneura*, *Pecopteris*, *Alethopteris*, *Sphenopteris*, *Sphenophyllum* and *Vertebraria*. Bajpai and Maheshwari (1990) recorded *Neomariopteris* along with many *Glossopteris* species from Hura Coalfield. Peculiar Ginkgophyte leaves belonging to genera *Ginkgoites*, *Psygmophyllum*, *Rhipidopsis* and *Saportaea* were also reported from Lalmatia coalmine of the Hura coalfield (Maheshwari and Bajpai 1992). In addition, *Liknopetalon rajmahalensis* and *Lelstotheca harikrishnae* are also known from the Permian sequence of the Rajmahal Hills (Srivastava and Pant 2002). More recently, coal bearing sequences of Lalmatia coalmine, Hura Coalfield of Rajmahal are studied for its palaeofloristics and geochemical characteristics by Pillai et al. (2020) and Mathews et al. (2020).

The flora of Rajmahal Formation is known as 'Ptilophyllum Flora' which is world famous for its peculiarity of well-diversified flora preserved as impressions as well as silicified plant remains. It is one of the best formation for plant fossils of Mesozoic era. As mentioned earlier, the first systematic study of the rich Ptilophyllum Flora was carried out by Oldham (1854, 1860) and Oldham and Morris (1862), wherein they described 46 species belonging to different plant megafossil genera. Feistmantel (1877a) described 50 plant megafossils including 16 new species from the Upper Gondwana sequences of Rajmahal area and suggested its similarity with the Ptilophyllum Flora of Golapalli and Athgarh formations from the East Coast of India and assigned the Liassic (Jurassic) age to the Rajmahal Formation. After about four decades, the Rajmahal flora was reinvestigated by Seward and Sahni (1920) who consequently revised the classification of species based on morphological, cuticular and anatomical analyses and further worked on varied palaeobotanical aspects of this flora; the work was continued by Sahni and his associates over a considerable period of time. Later on, many workers have enriched the diversity and quantity of floral elements of Rajmahal Basin which are discussed in this chapter as and where needed.

Majority of the plant fossils of the Rajmahal Flora include bennettitales, cycads, pentoxylales, conifers (Podocarpaceae, Araucariaceae and Cupressaceae) and ferns, (Seward and Sahni 1920; Sahni and Rao 1933; Ganju 1947a, b; Sahni 1928, 1931, 1948; Gupta 1954; Rao 1972; Bose and Maheshwari 1974; Banerji 2000b; Banerji and Ghosh 2005, 2006, 2008; Ghosh et al. 2018 and references therein). In addition to these, there are a few lycopods (Vishnu-Mittre 1958), *Equisetites*, *Thinnfeldia*, *Ginkgoites* (Sah 1952, 1953; Sah and Sukh-Dev 1957; Bose and Sah 1967; Sharma et al. 1971), algae (Sharma and Harsh 1994; Sharma and Suthar 1989; Sharma and Tripathi 1997; Sharma et al. 2002; Banerji and Ghosh 2002), lichen (Sharma et al. 2001; Banerji and Ghosh 2002; Sharma and Harsh 2020), fungi (Sharma et al. 2017), bryophytes (Bose and Pal 1982; Sharma et al. 2015; Sharma 2017) and angiosperms (Sharma 1997; Sharma et al. 2001; Banerji 2000a, b) were also present in Rajmahal area during Jurassic-Early Cretaceous time period. Plenty of silicified plant remains have been recorded from different localities of the Rajmahal Basin, viz., Sonajori, Amarjola, Chligojuri, Hiranduba, and Nipania. Remarkably, the plant fossils in the northern portion of the Rajmahal Hills are found mostly as impressions, whereas those in the southern part as petrifactions (Gupta 1966). With the preponderance of conifers, bennettitales, cycadales and ferns, the most common genera of Rajmahal flora are *Ptilophyllum*, *Pterophyllum*, *Dictyozamites*, *Taeniopteris*, *Williamsonia*, *Brachyphyllum*, *Thinnfeldia*, *Nilssonia*, *Elatocladus* and *Cladophlebis*. The Rajmahal intertrappean beds are also unique in having many taxa that are either restricted to the Rajmahal Basin or very rare elsewhere, viz. *Podostrobus*, *Nipanioruha*, *Nipaniostrobus*, *Sitholeya*, *Mehtaia*, *Rajmahaliya*, *Mohriopsis*, *Murlipaharopteris*, *Ontheanthus*, *Hirandubia*, *Podocarpospermum*, *Bennetticarpus*, *Beaniopsis*, *Morrisia*, *Sahnia*, along with characteristic presence of *Pentoxylon* and many pteridophytic rhizomes and stems. Interestingly, except for a clupeiformes fish fossil (*Jhingrania roonwali*) from siliceous shales exposed at Sakrigalighat (Misra and Saxena 1964), there is no evidence for animal fossils from this basin. However, evidence of insect-plant interactions (feeding traces, ovipositional site, insect galls) have been reported from various localities of the Rajmahal Formation (Banerji 2004). Some significant plant taxa recorded from Dubrajpur and Rajmahal formations kept in the repository of BSIP are rephotographed by the authors and are shown in Plates 1, 2, 3 and 4. Besides this, Table 2 presents a compilation of macrofloral taxa documented in the Rajmahal Basin.

In comparison to enormous work done on the macrofloral aspects, palynological studies in the Rajmahal Basin began in the 1930s, and were confined to the intertrappean beds of the Rajmahal Formation (Rao 1936, 1943b; Vishnu-Mittre 1953a, 1954, 1955b, 1956b; Sah and Jain 1965). Subsequently, Maheshwari (1966), Srivastava and Maheshwari (1974), Pillai et al. (2020) analysed coal bearing Barakar sequences for palynocomposition and age correlation. Palynological assemblages of Talchir Formation were first reported by Banerjee and D' Rozario (1988, 1990). More recently a ?Late Pennsylvanian–Early Permian palynoflora is reported in a bore core sequence by Murthy et al. (2020b) from the (Salbadra-Gomarpahari sector

Plate 1 Plant macrofossils recorded from Rajmahal Formation. 1. *Ptilophyllum acutifolium* BSIP specimen no. 36554, 2. *Mohriopsis* sp. BSIP specimen no. 36548, 3. *Sphenopteris hislopi* BSIP specimen no. 36550, 4. *Marattiopsis macrocarpa* BSIP specimen no. 36544, 5. *Sphenopteris naukhoffiana* BSIP specimen no. 36551, 6. *Pachypteris indica* BSIP specimen no. 36552, 7. ?*Asplenites* sp. BSIP specimen no. 36549, 8. *Pterophyllum incisum* BSIP specimen no. 36557, 9. *Eborasia lobifolia* BSIP specimen no. 36546, 10. *Ctenis rajmahalensis* BSIP specimen no. 36559, 11. *Elatocladus confertus* sp. BSIP specimen no. 36560. (Scale bar 1 cm for all the specimens)

Plate 2 Plant macrofossils recorded from Rajmahal Formation. 1. *Cycadolepis pilosa* BSIP specimen no. 36555, 2. *Eborasia lobifolia* BSIP specimen no. 36547, 3. *Taeniopteris* sp. cf. *T. spatulata* BSIP specimen no. 36543, 4 and 5. *Marattiopsis macrocarpa.* BSIP specimen no. 36544, 6. *Equisetites rajmahalensis* BSIP specimen no. 36542, 7. *Selaginellites gracilis* BSIP specimen no. 36541. (Scale bar 1 cm for all the specimens)

of Brahmani Coalfield. Palynological investigations of subtrappean sediments of the Dubrajpur Formation have been undertaken by Tiwari et al. (1984), Tiwari and Tripathi (1995), Tripathi et al. (1988), Tripathi (2000, 2001), Tripathi and Ray (2005, 2006) and Murthy et al. (2020a, b). Palynological studies have so far successfully established the age relationship of the studied sediments of Rajmahal Basin and have

Plate 3 Plant macrofossils recorded from the Rajmahal Formation. 1. *Osmundopsis* sp. BSIP specimen no. 36475, 2. *Elatocladus conferta,* BSIP specimen no. 16799, 3. *Ptilophyllum acutifolium* BSIP specimen no. 36473, 4. *Ptilophyllum acutifolium* BSIP specimen no. 37394, 5. *Todites denticulatus* BSIP specimen no. 36473, 6. *Todites denticulatus* BSIP specimen no. 36473, 7. *Elatocladus jabalpurensis* BSIP specimen no. 37402A, 8. *Phyllopteroides laevis* BSIP specimen no. 36477, 9. *Murlipaharopteris indica* BSIP specimen no. 37387, 10 and 11. *Phyllopteroides laevis* BSIP specimen no. 36477, 12. *Desmiophyllum* sp. BSIP specimen no. 37402. (Scale bar 1 cm for all the specimens)

Plate 4 Plant macrofossils recorded from Dubrajpur and Rajmahal formations. 1. *Ctenis rajmahalensis* BSIP specimen no. 36260, 2. *Hirandubia cupressoides* BSIP specimen no. 39313, 3. *Hirandubia cupressoides* BSIP specimen no. 39317, 4. *Pterophyllum cf. P. footeanum* BSIP specimen no. 36252, 5. *Nilssonia* sp. BSIP specimen no. 36255, 6. *Pagiophyllum* sp. BSIP specimen no. 36258, 7. *Taeniopteris cf. T. oldhami* BSIP specimen no. 36253, 8. *Onychiopsis* sp. BSIP specimen no. 36261, 9. *Hirandubia cupressoides*, BSIP specimen no. 39314. (Scale bar 1 cm for all the specimens)

Table 2 List of macrofloral taxa recorded from the Lower (Talchir and Barakar formations) and Upper Gondwana (Dubrajpur and Rajmahal formations) sequences of Rajmahal Basin. (Localities abbreviations: Amarjola—Amj, Amrapara—Amp, Bagjori—Bgj, Balbhadri Hills—Balb-H., Balidih—Bald, Banchappa—Bnchp, Bartala—Bart, Basgo-Bedo—Basg-B, Bansoli Valley—Bans-V, Bindraban—Bndb, Borio—Bor, Buridhuba—Buri, Buskoghat—Busk, Butaha Pahar—But-P, Chilgojuri—Chilg, Chunakhal—Chnk, Daman-i-Koh—Dam-k, Dhokuti—Dhkt, Dhudkol—Dkol, Dubrajpur—Dub, Dumria—Dum, Gumaphahar—Gum-P, Ghutiari—Ghut, Gilamarhi—Gilam, Hiranduba—Hirnd, Jhenagaria—Jheng, Khatangi Hill—Kht-H, Kasamu—Kasm, Khairbani—Khrbn, Khutnashi—Khtn, Kulkipahar—Klkp, Lalmatia—Lalm, Maharajpur—Mahj, Mandro—Mndr, Mirzachowki—Mirzch, Mujhwa—Muj, Murcha—Mur, Murlipahar—Murl-P, Nipania—Nip, Onthea—Onth, Pakur—Pkr, Pathargama—Pathg, Segarbhana—Segb, Sakrigalighat—Skrg, Salempur—Salm, Sitalpur—Sitp, Sonajori—Sonj, Sripur—Srip, Sugadih—Sug, Tagutola—Tgt, Tetaria Basti—Tet-B)

Name of taxa	Talchir Formation	Barakar Formation	Dubrajpur Formation	Rajmahal Formation
Algae				
Algae cf. *Polysiphonia* Sharma et al.				Nip
Algal association with root of *Arauamyelon* Sharma and Suthar				Pkr
Characiosiphonites nipanica (Sharma et al.) Sharma et al.				Nip
Bryophytes				
Marchantiales thallus Sharma				Chilg
Leafy Jungermanniales Sharma				Nip
Hepaticites pantii Bose and Pal				Bndb
Fungi				
Saprophytic fungi on *Pentoxylon* bark				Nip
Saprophytic fungi on *Araucarites* integument				Sonj
Lichens				
Lichen A Sharma				Nip
Lichen B Sharma				Nip
Lichen C Sharma				Nip
Lichen D Sharma				Nip
Lichen E Sharma and Harsh				Nip
Lichen cf. *Parmelia nudecata* Sharma et al.				Nip
Lycopodiales				
Lycoxylon indicum Srivastava				Nip
Lycopodites rajmahalensis Banerji				Chnk
Lycoxylon sonajoriensis Banerji				Sonj

(continued)

Table 2 (continued)

Name of taxa	Talchir Formation	Barakar Formation	Dubrajpur Formation	Rajmahal Formation
Lycopodeaceous ligule Sharma				Chilg
Lycoxylon sp. (type 2) Vishnu-Mittre				Nip
Equisetales				
Equisetum rajmahalense (Oldham and Morris) Schimper				Dkol, Sonj, Dhkt, Balb-H, Bndb, Chnk, Bart, Bor, Skrg, Onth, Chilg, Nip, Sitp
Equisetites sp.				Gilam
Equisetaceous stem sheath				Chilg
Lelstotheca robusta (Feistmantel) Maheshwari		Bans-V		
Lelstotheca sp. Singh et al.		Lalm		
Stellotheca robusta Surange and Prakash		Bans-V		
Phyllotheca sp. Maheshwari and Prakash	Dam-k	Bans-V		
Schizoneura gondwanensis Feistmantel		Bans-V		
Sphenophyllales				
Sphenophyllum gondwanensis Singh et al.		Lalm		
Sphenophyllum speciosum (Royel) McClelland		Bans-V		
Selaginellales				
Selaginellites gracilis (Oldham and Morris) Pal				Bndb, Chnk, Dhkt
Selaginellites sp. A Sharma				Dhkt
Selaginellites sp. B Sharma				Dhkt

(continued)

Table 2 (continued)

Name of taxa	Talchir Formation	Barakar Formation	Dubrajpur Formation	Rajmahal Formation
Filicales				
?*Asplenites* sp. Banerji				Chnk
Cacumen sp. Banerji				Chnk
Cladophlebis denticulata Brongniart				Dhkt, Chnk
Cladophlebis denticulata (Brongniart) Fontain				Dhkt
Cladophlebis denticulata (Brongniart) Seward				Nip
Cladophlebis indica (Oldham and Morris) Sahni and Rao			Kht-H	Skrg, Dkol, Bart, Chnk, Murl-P, Dhkt, Sitp, Hirnd, Amj, Nip
Cladophlebis sahnii Vishnu-Mittre				Nip
Cladophlebis salicifolia (Morris) Pal and Arefin				Ghut, Sug, Busk, Bor, Pathg, Bart, Bald, Murl-P, Nip
Cladophlebis srivastvae Gupta				Bart, S of Mirzch, Bndb, Sitp
Cladophlebis sp. A-B Bose and Sah				Sug
Cladophlebis sp. Banerji				Murl-P
Cladophlebis sp. Banerji and Jana				Bart
Cladophlebis sp. Sah			Kht-H	
Cladophlebis sp. Sharma			Kht-H	

(continued)

Table 2 (continued)

Name of taxa	Talchir Formation	Barakar Formation	Dubrajpur Formation	Rajmahal Formation
Cladophlebis spp. Sahni and Rao				Dhkt
Clathropteris sp. Vishnu-Mittre				Nip
Coniopteris hymenophylloides (Brongniart) Seward				
Coniopteris sp. cf. *C. tatungensis* (Sze) Shuying				Murl-P
Coniopteris sp. Prakash				Sitp
Culcitites madagascariensis Appert			Kht-H	
Dennstaedia ramahalensis Sharma				Dhkt
Dichotomopteris ovate Maithy		Bans-V		
Dichotomopteris lyndleyii (Royle) Maithy		Bans-V		
Dicksonia speciosa Sharma				Dhkt
Dicksonia ramahalensis Sharma				Dhkt
Dryopteris cladophleboides Sharma				Dhkt
Dryopteris indica Sharma				Dhkt
Eboracia lobifolia (Phill) Thomas				Chnk
Fern rhizome Sharma				Nip
Gleichenia dhokutense Sharma				Dhkt
Gleichenites gleichenoides (Oldham and Morris) Seward and Sahni			Kht-H	Chnk, Dhkt Bndb, Onth, Dhkt
Gleichenia sonajoriensis Bohra and Sharma (rhizome)				Sonj
Gleichenites sp. Bose and Sah				Bndb
Gleichenites sp. Sharma				Dhkt

(continued)

Table 2 (continued)

Name of taxa	Talchir Formation	Barakar Formation	Dubrajpur Formation	Rajmahal Formation
Gleicheniomyelon diarcha Bohra and Sharma (root)				Sonj
Gleicheniorachis jurassica Sharma (petrified rachis)				
Gleicheniorachis mittrii Bohra and Sharma				Sonj
Gleicheniorachis sp. Sharma				Dhkt
Guptiorachis amarjolense Sharma (petrified petiole)				Amj
Hausmannia crenata (Nathorst) Möller				Chilg, Hirnd, Amj, Nip
Hausmannia sp. Banerji and Jana				Bart
Haydenia thyrsopteroides Seward				Dhkt, Dkol
Klukia rajmahalensis Sharma				Dhkt
Klukia sp. A Sharma				Dhkt
Klukia sp. B Sharma				Dhkt
Marattiopsis macrocarpa (Oldham and Morris) Seward and Sahni			Kht-H	Bnchp, Bndb, Onth, Ghut, Bart, Chnk, Dhkt, Basg-B
Marattiopsis reversa Sharma				Dhkt
?*Mohriopsis* sp. Banerji				Chnk
Morrisia mcclellandi (Oldham and Morris) Bose				Dhkt, Bndb, Khrbn, Skrg, ?Amj
Morrisia dentata (Rao and Jacob) Bose and Banerji				Mndr

(continued)

Table 2 (continued)

Name of taxa	Talchir Formation	Barakar Formation	Dubrajpur Formation	Rajmahal Formation
Morrisia rajmahalensis (Feistmantel) Bose and Banerji				?Bndb, Khrbn, Onth, ?Bor, ?Mur, ?Sitp
Murlipaharopteris indica Banerji				Murl-P
Neomariopteris hughesii (Feistmantel) Maithy		Lalm		
Onychiopsis psilotoides Stokes and Webb			Kht-H	Chnk
Onychiopsis sp. Banerji			Kht-H	
Osmundacaulis sahnii (Vishnu-Mittre) Miller				Bndb, Amj
Osmundites rajmahalense Gupta				Amj
Osmundopsis sp. Banerji				Chnk
Ophioglossaceous rhizome				Nip
Pecopteris lobata Oldham				Bndb
Pecopteris phegopteroids (Feistmantel) Arber		Bans-V		
Pecopteris sp. Maheshwari and Prakash		Bans-V		
Pecopteris sp. Srivastava			Kht-H	
Phlebopteris sp. Bose and Sah				
Phyllopteroides chunakhalensis Sah and Sukh-Dev				Chnk
Phyllopteroides laevis Cantrill and Webb				Gum-P, Bart, Bald, Chnk, Murl-P
Protocyathea cretaceae (Stenzel) Ogura				Skrg
Protocyathea cyatheoides (Unger) Feistmanlel				Skrg

(continued)

Table 2 (continued)

Name of taxa	Talchir Formation	Barakar Formation	Dubrajpur Formation	Rajmahal Formation
Protocyathea rajmahalensis (Feistmantel) stem				Skrg
Protocyathea rajmahalense Jacob				Skrg
Protocyathea tokunagae Ogura				Skrg
Protocyathea trichinopoliensis Feistmantel				Skrg
Petiol type A Sharma and Bohra				Pkr
Petiol type B Sharma and Bohra				Pkr
Santhalea bansloiensis Maithy		Bans-V		
Schizaeangium jurassica Bohra and Sharma				Sonj
Sporangia Type A Sharma and Bohra				Pkr
Tinpaharia sinuosa Jacob (rhizome)				Nip
Todites denticulatus (Brongniart) Krasser				Chnk
Todites indica (Oldham and Morris) Bose and Sah				Dhkt, Dkol, Sitp
Todites sp. cf. *T. williamsonii* (Brong.) Seward				Chnk
Insertae sedis				
Actinostelopteris pakurensis Sharma and Bohra				Sonj
Danaeopsis rajmahalensis Feistmantel				Onth
Dictyostelopteris fasciosteleoides Vishnu-Mittre				Nip
Dictyostelopteris jacobii Vishnu-Mittre				Nip

(continued)

Table 2 (continued)

Name of taxa	Talchir Formation	Barakar Formation	Dubrajpur Formation	Rajmahal Formation
Dictyostelopteris rajmahalense Vishnu-Mittre				Nip
Filicoamyelon actinostachyoide Bohra and Sharma				Sonj
Filicoamyelon cryptogrammoides Bohra and Sharma				Sonj
Solenostelopteris jurassica Sharma and Bohra				Sonj
Solenostelopteris nipanica Vishnu-Mittre				Nip
Solenostelopteris sahnii Vishnu-Mittre				Nip
Solenostelopteris sp. Sharma and Bohra				Pkr
Sphenopteris arguta Lindley and Hutton				Bor
Sphenopteris bindrabanensis (Feistmantel) Bose and Sah				Bndb, Mndr, Srip
Sphenopteris elaminata Sharma				Dhkt
Sphenopteris hislopii Oldham and Morris				Bndb, Balb-H pahar, Mirzch, Mahj, Khrbn, Chnk
Sphenopteris hughesii (Feistmantel) Zeiller		Bans-V		
Sphenopteris imbricata Sharma				Dhkt
Sphenopteris khairbaniensis Ganju				Khrbn, Chnk
Sphenopteris lobifolia Morris		Bans-V		
Sphenopteris sp. cf. *S. lobifolia* Sharma				Dhkt

(continued)

Table 2 (continued)

Name of taxa	Talchir Formation	Barakar Formation	Dubrajpur Formation	Rajmahal Formation
Sphenopteris membranosa Feistmantel				Muj pass near Bor
Sphenopteris naukhoffiana (Heer) Halle				Chnk
Sphenopteris patagonica Halle				Sug
Sphenopteris rajmahalensis Sahni and Rao				Balb-H, Khrbn, Bart, Bnchp, Onth
Sphenopteris sakrigaliensis Sah				Skrg
Sphenopteris sp. Sahni and Rao				Onth, Nip
Sphenopteris sp. Srivastava			Kht-H	
Sphenopteris sp. Sharma				Dhkt
Sphenopteris sp. Vishnu-Mittre				Nip
5 types of petioles and rachises Vishnu-Mittre				Nip
3 types of sori Vishnu-Mittre				Nip
3 types of isolated sporangia Vishnu-Mittre				Nip
Rhizomopteris ballii Feistmantel				Khrbn
Rhizomopteris rajmahalense Gupta				Khrbn
Rhizomopteris sahnii Gupta				Khrbn
Cordiatales				
Noeggerathiopsis hislopii Feistmantel		Lalm		
Noeggerathiopsis cf. *N. hislopii* Bunbury		Bans-V		
Pteridospermales				
Alethopteris sp. A Maheshwari and Prakash		Bans-V		

(continued)

Table 2 (continued)

Name of taxa	Talchir Formation	Barakar Formation	Dubrajpur Formation	Rajmahal Formation
Alethopteris sp. B Maheshwari and Prakash		Bans-V		
Alethopteris spp. A-B Sharma				Dhkt
Alethopteris sp. C Sharma				Dhkt
Arberiella-type sporangia Maheshwari and Prakash		Lalm		
Birbalsahnia divyadarshanii Bajpai and Maheshwari		Lalm		
Dicroidium sp. Sharma				Dhkt
Eretmonia-type leaf Maheshwari and Prakash		Lalm		
Gangmopteris clarkeana Feistmantel		Lalm		
Gangmopteris cyclopteroides Feistmantel	Dam-k			
Gangmopteris sp. cf. *cyclopteroides* Feistmantel		Bans-V		
Gangamopteris major Feistmantel		Lalm		
Gangmopteris sp. Maheshwari and Prakash		Bans-V		
Glossopteris ampla Dana		Lalm		
Glossopteris angustifolia Brongniart		Bans-V, Lalm, Dub and lower bed of Kht-H		
Glossopteris arberii Srivastava		Lalm		
Glossopteris barakarensis Kulkarni		Lalm		
Glossopteris browniana Brongniart		Lalm, Bans-V, Dub and lower bed of Kht-H		

(continued)

Table 2 (continued)

Name of taxa	Talchir Formation	Barakar Formation	Dubrajpur Formation	Rajmahal Formation
Glossopteris communis Feistmantel		Bans-V, Lalm, Dub and lower bed of Kht-H		
Glossopteris damudica Feistmantel		Bans-V, Kht-H		
Glossopteris emarginata Maheshwari and Prakash		Bans-V		
Glossopteris formosa Feistmantel		Bans-V		
Glossopteris indica Schimper		Lalm, Bans-V, Dub, lower bed of Kht-H		
Glossopteris intermittens Feistmantel		Lalm		
Glossopteris lanceolatus Pant and Singh		Lalm		
Glossopteris linearis McCoy		Lalm		
Glossopteris mohudaensis Feistmantel		Lalm		
Glossopteris nautiyalii Pant and Singh		Lalm		
Glossopteris parallela Feistmantel		Bans-V		
Glossopteris raniganjensis Chandra and Surange		Lalm		
Glossopteris retifera Feistmantel		Lalm, Bans-V		
Glossopteris retusa Mahcshwari		Lalm		
Glossopteris stenoneura Feistmantel		Lalm		
Glossopteris stricta Bunbury		Lalm		

(continued)

Table 2 (continued)

Name of taxa	Talchir Formation	Barakar Formation	Dubrajpur Formation	Rajmahal Formation
Glossopteris sp. cf. *G. cyclopteroides*		Bans-V		
Glossopteris sp. Feistmantel		Brahmini River		
Pachypteris indica (Oldham and Morris) Bose and Roy				Chnk
Pachypteris sp. cf. *P. indica* (Oldham and Morris) Bose and Roy				Baldih
Thinnfeldia amarjolense Sharma et al.				Amj
Thinnfeldia sp. cf. *T. amarjolense* Sharma				Dhkt
Thinnfeldia indica Feistmantel				Bart, Busk, Bor, Pathg
Thinnfeldia khatangiensis Sen Gupta			Kht-H	
Thinnfeldia nirmalensis Roy				Tet-B 2 km S of Mirzch RS
Thinnfeldia odontopteroides Brongniart				Nip
Thinnfeldia sp A Sharma et al.				Amj
Thinnfeldia sp B Sharma et al.				Amj
Thinnfeldia sp (A) Jacob				Skrg
Thinnfeldia sp Sharma et al. cf. *T. lancifolia* (Morris) Walkom				Amp
Thinnfeldia sp Sharma et al. cf. *T. feistmanteli*				Amp
Vertebraria indica Royle		Lalm		
Vertebraria sp. Maheshwari and Prakash		Bans-V		

(continued)

Table 2 (continued)

Name of taxa	Talchir Formation	Barakar Formation	Dubrajpur Formation	Rajmahal Formation
Veekaysinghia durgavatiae Bajpai and Maheshwari		Lalm		
Insertae sedis				
Rienitsia sp.				Gum-P
Cycadales				
Ctenis rajmahalensis Banerji			Kht-H	Chnk, Dhkt
Cycadeospermum indicum Gupta				Khrbn
Cycadites rajmahalensis Oldham				Bndb, Onth, Chnk, Dhkt
Cycadinorachis omegoides Sharma				
Fascisvarioxylon mehtae Jain				
Macrotaeniopteris crassinervis Feistmantel				Mndr, Busk, Dhkt
Macrozamiphyllum mucilagica Sharma et al.				Nip
Manomesarchioxylon heptaxylica Sharma et al.				Amj
Nilssonia distans (Morris) Sharma			Kht-H	Ghut, Bndb, But-P, Tet-B, Khtn
Nilssonia sp. (Srivastava; Banerji)			Kht-H	
Pseudoctenis sp. cf. *P. footeanum* (Feistmantel) Seward and Sahni			Kht-H	
Sewardioxylon sahnii Gupta				
Stangeriorachis heterospinulata Sharma et al.				Chilg

(continued)

Table 2 (continued)

Name of taxa	Talchir Formation	Barakar Formation	Dubrajpur Formation	Rajmahal Formation
Taeniopteris buskoghatensis Bose and Banerji			Kht-H	
Taenionteris crassinervis (Feistmantel) Walkom				Busk, Mndr, Bndb, Balb-H, Chnk, Mndr, Dhkt
Taenionteris crenata (McClelland) Bose and Banerji			Kht-H	Mndr
Taeniopteris lata Oldham				Bndb, Khrbn, Onth, Mndr, Skrg, Chnk, Dhkt, Mirzch
Taeniopteris oldhamii Bose and Banerji				Bndb
Taeniopteris sp. cf. *T. oldhami* Bose and Banerji			Kht-H	
Taeniopteris sarbadhikarii Sen Gupta				Buri
Taeniopteris spatulata McClelland				Mndr, Bor, Basg-B, Bndb, Mahj, Onth, Jheng, Murl-P, Sitp, Hirnd, Amj
Taeniopteris sp. cf. *Taeniopteris spatulata* McClelland				Chnk, Baldih
Taeniopteris sp. Banerji			Kht-H	
Taeniopteris sp. Banerji				Gilam
Taeniopteris spp. Feistmantel				Skrg
Taeniopteris sp. Sahni and Rao				But-P
Cycadean stem Specimen 1 Bose				Chilg
Cycadean stem Specimen 2 Bose				Chilg

(continued)

Table 2 (continued)

Name of taxa	Talchir Formation	Barakar Formation	Dubrajpur Formation	Rajmahal Formation
Cycadean stem Specimen 3 Bose				Chilg
Cycadean stem Specimen 4 Bose				Chilg
Cycadean stem Specimen 5 Bose				Chilg
Bennettitales				
Amarjolia dactylota Sharma				Amj
Anomozamites amarjolense Sharma, Surana and Singh				Balb-H, Bor, Amj, Bart
Anomozamites sp. Banerji				Khtn
Bucklandia dichotoma Sharma				Amj
Bucklandia guptai Sharma				Amj
Buckalandia indica Seward				Skrg
Buckalandia sahnii Bose				Nip
Bucklandia sp. Vishnu-Mittre				Hirnd
Cycadolepis indica Gupta				Khrbn, Bnchp, Bndb
Cycadolepis oldhamii (Feistmantel) Bose and Jain				Bndb, 2.5 miles S of Mirzch RS
Cycadolepis pilosa Feistmantel			Kht-H	Chnk
Cycadolepis (?) Sharma				Dhkt
Dictyozamites bagjoriensis Jacob				Bgj, Basg-B, Onth, ruins at Skrg
Dictiozamites falcatus (Morris) Medlicott and Blanford				Chilg, Amj, Kasm, Skrg, Sitp, Hirnd
Dictyozamites indicus Feistmantel				Mndr, Bndb, Basg-B, Mahj, Onth, Kasm, Chilg, Skrg, Sitp

(continued)

Table 2 (continued)

Name of taxa	Talchir Formation	Barakar Formation	Dubrajpur Formation	Rajmahal Formation
Dictyozamites sahnii Gupta and Sharma				Onth, Skrg Sitp
Dictiozamites sp. cf. *D. sahnii* Gupta and Sharma				Bart
Dictiozamites sp. Bose and Zeba-Bano				Basg-B
Dictiozamites sp. Bose and Kasat				Amj
Dictiozamites sp. Gururaja and Pant				Basg-B
Dictiozamites sp. Banerji				Gilam
Ontheostrobus sessilis Ganju				Onth
Otozamites gondwanensis Bose				Murl-P
Otozamites sp. cf. *O. walkomotaensis* Bose and Zeba-Bano				Hirnd
Otozamites sp. Srivastava			Kht-H	Khtn
Pterophyllum fissus Feistmantel			Kht-H	Bor, Basg-B, Bndb, Onth, Skrg, Gum-P, Bart, Bald, Hirnd, Amj
Pterophyllum sp. cf. *A. fissus* Feistmantel				Baldih
Pterophyllum guptai Bose and Banerji				Skrg
Pterophyllum incisum (Oldham) Sahni and Rao				But-P, due south of Tetaria basti, Chnk
Pterophyllum kingianum Feistmantel				Chnk

(continued)

Table 2 (continued)

Name of taxa	Talchir Formation	Barakar Formation	Dubrajpur Formation	Rajmahal Formation
Pterophyllum medlicottianum (Oldham and Morris) Feistmantel				Bndb, Ghut, Basg-B, Chnk, Dhkt, Mndr, Onth
Pterophyllum morrisianum Oldham				Bndb, Onth, Skrg, Chnk, Dhkt
Pterophyllum princeps Oldham and Morris				Bndb, Khrbn, Onth, Mndr, Skrg, Chnk, Dhkt, Mirzch
Pterophyllum rajmahalense Morris				Bndb, Onth, Bor
Pterophyllum valentinei Sen Gupta				Skrg
Pterophyllum sp. Sah			Kht-H	
Pterophyllum sp. Bose and Banerji			Kht-H	
Ptilophyllum acutifolium Morris			Kht-H	Skrg, Gum-P, Bald, Chnk, Murl-P, Nip, Amj, Gilam, Basg-B, Dum
Ptilophyllum amarjolense Bose				Amj
Ptilophyllum cutchense Morris			Kht-H	Bndb, Onth, Amp, Skrg, Dkol, Gum-P, Bart, Bald, Chnk, Murl-P, Nip, Hirnd, Amj, Salm
Ptilophyllum sp. cf. *cutchense* Morris				Nip, Onth

(continued)

Table 2 (continued)

Name of taxa	Talchir Formation	Barakar Formation	Dubrajpur Formation	Rajmahal Formation
Ptilophyllum nipanica Vishnu-Mittre				Nip
Ptilophyllum rarinervis (Feistmantel) Bose and Kasat				Onth
Ptilophyllum sahnii Gupta and Sharma				Amj
Ptilophyllum sakrigaliensis Sah				Skrg
Ptilophyllum tenerrimum Feistmantel				Onth, But-P, Skrg, Bart, Khtn
Ptilophyllum sp. A Ganju				Onth
Ptilophyllum spp. Feistmantel				?
Ptilophyllum sp. Banerji and Jana				Bart
Ptilophyllum rachis				Amj
Rajmahalia paradoxa Sahni and Rao				Balb-H, Bart, Bnchp, Khrbn, Chnk, Bndb, Mirzch, Onth, Basg-B
Sahnioxylon andrewsii Bose and Sah				Amp
Sahnioxylon rajmahalense (Sahni) Bose and Sah				Amp
Weltrichia polyandra (Ganju) Sitholey and Bose				Onth
Weltrichia santalensis (Sitholey and Bose) Bose				Skrg, Dhkt
Weltrichia singhii Bose				Skrg
Williamsonia amarjolense Sharma				Dhkt, Amj

(continued)

Table 2 (continued)

Name of taxa	Talchir Formation	Barakar Formation	Dubrajpur Formation	Rajmahal Formation
Williamsonia companulatiformis Sharma				Dhkt, Amj
Williamsonia gigas (Carruthers) Feistmantel				Bndb
Williamsonia guptai Sharma				Amj
Williamsonia harrisiana Bose				Amj
Williamsonia indica Seward				Bndb
Williamsonia sp. cf. *W. indica* Seward				Khrbn
Williamsonia microps Feistmantel				Busk
Williamsonia sahnii Gupta				Khrbn
Williamsonia santalensis Sitholey and Bose				Skrg
Williamsonia sp. cf. *W. scotica* Seward				Amj
Williamsonia sp. cf. *W. setosa* Nothorst				Nip
Williamsonia sewardiana Sahni				Amj
Williamsonia sp. cf. *W. sewardiana* Sahni				Bndb
Williamsonia sp. Seward and Sahni				Bndb
Williamsonia sp. Sharma				Amj, Dhkt
Williamsonia sp. Sah			Kht-H	
Williamsonia sp. Banerji			Kht-II	Baldih
Williamsonia sp Sharma				Hirnd, Dhkt
Williamsonia (flower) Gupta				Khrbn
Williamsonia (seed) Sharma				Amj
Zamites chunkhalensis Banerji				Chnk

(continued)

Table 2 (continued)

Name of taxa	Talchir Formation	Barakar Formation	Dubrajpur Formation	Rajmahal Formation
Pentoxylales				
Carnoconites compactus Srivastava				Nip, Sonj
Carnoconites rajmahalensis (Wieland) Bose et al.				Nip
Guptioxylon amarjolense Sharma				Amj
Guptioxylon endocentrica Sharma				Amj
Nipaniophyllum anomozamoides Sharma				Nip, Sonj
Nipaniophyllum hobsonii				Nip, Sonj
Nipaniophyllum raoi (Sahni) Vishnu-Mittre				Nip, Amj
Nipanioxylon guptai Srivastava				Amj
Pentoxylon sahnii Srivastava				Nip, Sonj, Amj
Purioxylon jurassica Sharma				
Sahnia nipaniaensis Vishnu-Mittre				Nip
Sahnixylon andrewsii Bose and Sah				Amj
Sonajorispermum heterotestoides Suthar et al. (seed)				Sonj
Pentoxyleae short shoots Sharma et al.				Sonj
Coniferales				
Allocladus rajmahalsnse (Feistmantel) Townrow				Bndb, Amp, Bor
Araucarites bindrabunensis Vishnu-Mittre				Bndb

(continued)

Table 2 (continued)

Name of taxa	Talchir Formation	Barakar Formation	Dubrajpur Formation	Rajmahal Formation
Araucarites cutchensis Feistmantel				Onth, Mirzch, Skrg, Baldih, Gilam, Sitp
Araucarites minutus Bose and Maheshwari				Sitp
Araucarites mittrii Bohra and Sharma				Sonj
Araucarites nipaniensis Singh				Nip
Araucarites sp. Sahni and Rao				Balb-H, Sitp
?*Araucarites* sp. Zeba-Bano et al.				Pathg
Araucarioxylon amraparense (Sah and Jain) Bose and Maheshwari				Amp
Araucarioxylon bindrabunense (Sah and Jain) Bose and Maheshwari				Bndb
Araucarioxylon mandroense (Sah and Jain) Bose and Maheshwari				Mndr
Araucarioxylon rajmahalense (Sahni) Bose and Maheshwari				
Araucarioxylon santalense (Sah and Jain) Bose and Maheshwari				Dhkt, Mndr
Arauamyelon diarcha Bohra and Sharma				Bald
Arauamyelon pakurense Bohra and Sharma				Sonj
Brachyphyllum expansum (Sternberg) Seward				Bart
Brachyphyllum florini Vishun-Mittre				Nip

(continued)

Table 2 (continued)

Name of taxa	Talchir Formation	Barakar Formation	Dubrajpur Formation	Rajmahal Formation
Brachyphyllum mamillare Brongniart			Kht-H	Nip
Brachyphyllum rhombicum (Feistmantel) Sahni				Gilam
Brachyphyllum spiroxylum Bose				Amj
Brachyphyllum sp. Banerji and Ghosh				Hirnd
Brachyphyllum sp. Prakash				Hirnd, Sitp
Cheirolepis sp. cf. *C. muneateri* Schimper. Feistmantel				Bndb
Circoporoxylon amarjolense Kräusel and Jain				Amj
Circoporoxylon sewardi (Sahni) Kräusel				Amj
Circoporoxylon shanense (Sahni) Kräusel				Amj
Coniferocaulon latisulcatum Sah				Amj
Coniferocaulon rajmahalense Gupta				Khrbn, Amj
Coniferocaulon sp. Banerji and Jana				Baldih
Coniferocaulon sp. Bancroft				Amp
Coniferocaulon sp. Banerji and Ghosh				Hiran
Conites rajmahalensis Sahni				Busk
Conites sessilis Sahni				Mndr, Dhkt
Cupressinoxylon (Taxodioxylon) rajmahalense Bhardwaj				Amj
Dadoxylon (*Araucarioxylon*) *jurassicum* Bhardwaj				Amj

(continued)

Table 2 (continued)

Name of taxa	Talchir Formation	Barakar Formation	Dubrajpur Formation	Rajmahal Formation
Desmiophyllum sp. Banerji				Murl-P
Araucarioxylon agathioides (Kräusel and Jain) Bose and Maheshwari				Mndr
Elatocladus conferta (Morris) Seward and Sahni				Onth, Mndr, Skrg, Bndb, Chnk, Bald, Murl-P, Hiran, Sitp, Nip
Elatocladus jabalpurensis (Feistmantel) Sahni				Nip, Murl-P
Elatocladus sahnii Vishnu-Mittre				Nip
Elatocladus tenerrima (Feistmantel) Sahni				Skrg, Hiran, Gilam, Khtn, Sitp
Hirandubia cupressoides Ghosh et al.				Hiran
Indophyllum nipanica Vishnu-Mittre				Nip
Indophyllum raoi Vishnu-Mittre				Nip
Indophyllum sahnii Vishnu-Mittre				Nip
Kaokoxylon jurassicum (Bharadwaj) Bose and Maheshwari				Amj
Mehtaia nipaniensis Vishnu-Mittre				Nip
Mehtaia rajmahalensis Vishnu-Mittre				Nip
Mehtaia santalensis Vishnu-Mittre				Nip

(continued)

Table 2 (continued)

Name of taxa	Talchir Formation	Barakar Formation	Dubrajpur Formation	Rajmahal Formation
Ontheodendron florinii (Sahni and Rao) Bose and Maheshwari				Onth
Podocarpoxylon indicum (Bharadwaj) Bose and Maheshwari				Amj
Podocarpoxylon rajmahalense (Jain) Bose and Maheshwari				Amj
Nipanioruha curvifolia Vishnu-Mittre				Nip
Nipanioruha granthia (Rao) Vishnu-Mittre				Bart
Nipanioruha lanceolata Vishnu-Mittre				Nip
Nipaniostrobus aciculifolia Vishnu-Mittre				Nip
Nipaniostrobus pagiophylloides Vishnu-Mittre				Nip
Nipaniostrobus sahnii (Rao) Vishnu-Mittre				Nip
Pagiophyllum araucaroites Vishnu-Mittre				Nip
Pagiophyllum sp. cf. *P. divaricatum* Bunb				Nip
Pagiophyllum peregrinum (Lindley and Hutton) Schimper				Skrg
Pagiophyllum sp. cf. *P. peregrinum* Lindley and Hutton				Nip, Bart
Pagiophyllum sp. Banerji				Khtn, Sitp
Pagiophyllum sp. Banerji			Kht-H	
Pakurispermum pachytestoides Suthar et al. (seed)				Sonj

(continued)

Table 2 (continued)

Name of taxa	Talchir Formation	Barakar Formation	Dubrajpur Formation	Rajmahal Formation
Paradoxospermum heterotestoides Suthar et al. (seed)				Sonj
Podocarpospermum rajmahalensis Banerji and Ghosh				Hiran
Podocarpospermum podocarpoides Ghosh et al.				Hiran
Podostrobus podocarpoides (Vishnu-Mittre) Rao and Bose				Nip
Podostrobus rajmahalensis (Rao) Rao and Bose				Nip
Podostrobus sahnii (Vishnu-Mittre) Rao and Bose				Nip
Podostrobus sp.				Nip
Sitholeya rajmahalensis Vishnu-Mittre				Nip
Scale leafe Banerji				Murl-P
Taxaceoxylon rajmahalense (Bharadwaj) Kräusel and Jain				Klkp, Amp
Taxaceoxylon sp. cf. *ramahalensis* (Bharadwaj) Kräusel and Jain				Dhkt
Taxaceoxylon cupressoides Sharma				Dhkt
Taxaliospermum arilloides Suthar et al. (seed)				Sonj
Torreyites sitholeyi Ganju				Onth
Torreyites sp. Sharma				Dhkt
Triangulospermum minimum Suthar et al. (seed)				Sonj

(continued)

Table 2 (continued)

Name of taxa	Talchir Formation	Barakar Formation	Dubrajpur Formation	Rajmahal Formation
Stachyotaxus rajmahalensis Bose and Maheshwari				Onth
Strobilites ontheaensis Ganju				
Strobilites sp. Zeba-Bano et al.				Pathg
Gymnospermous seed Sharma				Chilg
Araucarian young roots Sharma et al.				Sonj
Ginkgoales				
Ginkgoites like leaf		Hura coalfield		
Ginkgoalean leaf Specimens1 Mehta and Sud				Mahj
Ginkgoalean leaf Specimens1 Mehta and Sud				Skrg
Ginkgoalean leaf Specimens1 Mehta and Sud				Skrg
Ginkgo (Ginkgoites) digitata Brongniart				Tgt
Ginkgoites rajmahalensis Sah and Jain				Skrg, Bart
Ginkgo rajmahalensis (Sah and Jain) Zeba-Bano et al.				Skrg, Mahj, Basko, Segb, Pathg
Ginkgoites sp. Sharma				Dhkt
Seed attributed to *Ginkgo* Zeba-Bano et al.				Pathg
? Angiosperm (questionable occurrences)				
Lesqueria like fructification Sharma				Amj
Buteacarpon sonajoriensis Banerji				Sonj

(continued)

Table 2 (continued)

Name of taxa	Talchir Formation	Barakar Formation	Dubrajpur Formation	Rajmahal Formation
Sonajoripushpam pakurensis Banerji				Sonj
Monocot leaf cf. *Cortaderia selloana* and *Psammochloa villosa* (Metcalfe) Sharma et al.				Sonj
Others				
Mollusca				Hiran
Seed Type A (cf. medullosean seed) Sharma and Bohra				Pkr
Seed Type B (cf. *Carnoconites*) Sharma and Bohra				Pkr
Endogenus microorganism in coniferous plant root Banerji and Ghosh				Hiran
Incomplete megastrobilus belonging to the Araucariaceae (Bose and Jain)				Amj

helped to identify the presence of hitherto lithologically undesignated sequences of ?Late Pennsylvanian (Murthy et al. 2020b), Late Permian (Barren Measures and Raniganj Formation) as a condensed sequence (Tripathi 2001) and Late Triassic (Tripathi 2001).

Based on macroremains and palynological compositions, commonly occurring groups in the Upper Gondwana sequences of Rajmahal Basin are discussed in the followings.

(A) **Conifers**

In the Rajmahal flora, the majority of conifers were represented by the Podocarpaceae and Araucariaceae with rare occurrence of Cupressaceae (Rao 1963a). Some have also been recognised as belonging to the Taxaceae family. Conifers with Podocarpanean affinity are represented by sterile and fertile shoots, detached male and female cones, petrified woods, dispersed seeds and pollen grains. Genus *Nipaniostrobus* (Rao 1943a) is a seed-bearing cone with several single-seeded scales arranged spirally around a moderately thick axis. The strobili were lax or compact. The curved micropyles and the presence of epimatium-like tissue suggest a close comparison with the living genus *Dacrydium*, which also belongs to Podocarpaceae. However, in *Dacrydium* well-defined megastrobilus is not existed, the whole megastrobilus is reduced to a single-seeded scale. In the type specimen of *Nipaniostrobus sahnii*, the vegetative parts were unknown. The shoots of the two other species, viz. *N. aciculiifolia* and *N. pagiophylloides* described by Vishnu-Mittre (1957b) have short to linear, single-veind, decurrent, slightly spreading, or loosely overlapping leaves. The strobili are almost similar in all three species. *Nipanioruha* (Rao 1947) is a petrified coniferous shoot with picnoxylic wood, bearing spirally arranged needle-like leaves. In 1949, Rao described the megastrobilus of *Nipanioruha granthia* (Rao 1949). The seed-bearing scales are arranged spirally with a thick, possibly fleshy axis. The ovules have curved micropyles and are embedded in fleshy scales. *Nipanioruha lanceolata* and *Nipanioruha curvifolia* (Vishnu-Mittre 1957b) are two sterile shoots similar to *N. granthia*. Rao (1943a) recorded two lax and slender strobili which he believed might represent the young stage of *Nipaniostrobus sahnii*. It differs from *Nipaniostrobus* in that it has seed scales with erect seeds and no sterile parts. Vishnu-Mittre (1957b) described three species of similar cones: *Mehtaia nipaniensis*, *M. rajmahalensis* and *M. santalensis*. *M. nipaniensis* is fleshy and has loosely arranged scales whereas *M. rajmahalensis* resembles the former but has a different type of vascular supply to the cone scale. *M. santalensis* has compact seed cones and dimorphic foliage. *Sitholeya* is a fertile shoot terminally bearing a single, inverted ovule comprising a single species, *S. rajmahalensis* (Vishnu-Mittre 1957b). Despite the fact that their affinities are unclear, they can only be classified in Podocarpaceae. Recently, the specimens having affinity with Cupressaceae has been recorded from the Hirunduba locality by Ghosh et al. (2018) and are found preserved as silicified remains of branched shoots fragments with scale like leaves and seed cones decribed under the name *Hirandubia cupressoides*. This taxon exhibits close affinity with the extant cuprasseous genera *Platycladus* and *Chamaeocyparis*. Along with these shoots, a new dispersed seed taxon *Podocarpospermum podocarpoides* and an *Elatocladus*-type leaf twig have also been recorded from these specimens (Ghosh et al. 2018) (Plate 4).

Rajmahal cherts from Nipania locality have a lot of microspores in the matrix. One of them with three inflated wings grouped on the ventral side has been designated as *Podosporites tripakshi* (Rao 1943b) as they are quite comparable to the microspores of living Podocarpaceous genera namely *Microcachrys*, *Phaerosphaera* and *Podocarpus dacrydioides*. Similarly, some of the two-winged microspores found in the matrix are known as *Pityosporites nipanica* and *Pityosporites* sp. (Rao 1943b) which may also belong to the Podocarpaceae. Rao (1943a) described a male strobilus with two-winged pollens as *Masculostrobus rajmahalensis* from Nipania, Rajmahal Hills. Vishnu-Mittre (1956a) described another male cone with three-winged pollens as *Masculostrobus sahnii* from the same locality. Barnard (1968) on re-examination of the original specimen of the genus *Masculstrobus* studied by Seward (1911), found that the pollen sacs contained non-saccate pollens of the *Classopollis* type. Therefore, he proposed that male cones from the southern hemisphere superficially resembling *Masculostrobus* but having saccate pollen should be transferred to genus *Pityanthus* Nathorst ex Seward (1919). According to Rao and Bose (1970), the name *Pityanthus* was incorrect for this material since it suggested an affinity for abietineans, which is not clearly evident in the Indian specimens. In light of this, they, therefore, suggested the name *Podostrobus* for cones bearing microsporangia with two or three-winged pollen and with clear podocarpaceous affinities. Comparable isolated petrified pollen grains from the Nipania cherts have been recorded by Rao (1943b) and Vishnu-Mittre (1954). *Masculostrobus podocarpoides* Vishnu-Mittre (1958) and *Masculostrobus* sp. of Vishnu-Mittre (1958) are conspecific with *Podostrobus rajmahalensis*.

In addition, numerous petrified woods and shoots of suspected Podocarpaceous affinity have been found in the Rajmahal Hills, these are: *Podocarpoxylon indicum* (Bharadwaj 1953), *P. rajmahalense* (Jain 1964), *P. godavarianum* (Sahni 1931), *P. malerianum* (Sahni 1931), *P. parthasarthyi* (Sahni 1931), *P. sarmae* (Varma 1954), *P. schimidianum* (Schmid and Schleiden 1855), *P. tirumangalense* (Suryanarayana 1953), *P. trichinopaliense* (Varma), *Circoporoxylon amarjolense*, *Coniferocaulon latisulcatum* (Sah 1957) and *Retinosporites indica* (Oldham and Morris) Holden now known as *Pachypteris indica* (Bose and Roy 1968). A solitary impression of *Stachyotaxus sampathkumarani* Rao (1963b) with *Elatocladus conferta* like leaves found in the Rajmahal Hills has been referred to the Podocarpaceae. The megastrobilus bears two-seeded megasporophylls which are oppositely disposed. The fertile part of this specimen is indistinguishable from *Baeniopsis rajmahalensis* (Ganju 1947a). It is quite likely that the type described as *Beaniopsis rajmahalensis* by Ganju (1947a) is actually a *Stachyotaxus* which is essentially a Rhaetic genus. *Strobilites sewardii* associated with *Elatocladus jabalpurensis* type of foliage was also regarded by Sahni (1931) as Podocarpaceous. All these forms demonstrate emphatically that Podocarpaceous flora was quite abundant and flourished in India during the Early Cretaceous period. This Cretaceous age in India saw the emergence of a remarkable number of genera from the Podocarpaceae family. Currently, it is represented by the single living genus *Podocarpus*, which grows wild in some parts of peninsular India only.

Petrified woods with Araucarian type of pitting are also quite common in the Rajmahal Hills. These evidences put forward that the family Araucariaceae were co-existed with the Podocarpaceae in the Rajmahal flora although not quite so abundant. Araucariaceae though present in the Mesozoic flora of India are totally absent from the modern Indian flora except under cultivation. Banerji and Ghosh (2006) and Ghosh et al. (2018) established the presence of Podocarpaceae (*Podospermum rajmahalensis* from Sonajori) and Cupressaceae (*Hirandubia cupressoides* from Hiranduba) during the Early Cretaceous in the Rajmahal Basin of India respectively.

Characteristics peculiar to the order Taxales are exhibited by some petrified wood viz., *Taxaceoxylon* sp. cf. *rajamahalense* (Bharadwaj) Kräusel and Jain (1964), *Torreyites constricta* (Feistmantel) Seward and Sahni (1920) and *Torreyites sitholeyi* Ganju (1946), *Taxites lanceolata* and *Taxoxylon rajmahalense* Bharadwaj (1952) are described from Rajmahal Hills. This shows that the Taxales also coexisted with the Podocarpaceae and Araucariaceae in the Rajmahal flora. However, their unequivocal presence requires the discovery of definite Taxacean reproductive parts in the Rajmahal Basin.

(B) **Pentoxyleae**

Pentoxyleae were a very interesting and intriguing group of gymnosperms that appeared to be particularly noticeable in the Rajmahal flora. This group is instituted by Sahni (1948) to accommodate a variety of connected genera, some of which even demonstrated organic connections with one another. The reconstruction of *Pentoxylon sahnii* Srivastava (1946) reveals that it was a small tree with a thin branched stem that was covered by scale leaf scars. The stem was dimorphic, with long shoots that were thicker in diameter and had fewer leaf scars. These long shoots were pierced by five to six steles arranged in a ring each of which had enhanced secondary growth on the inner side—a rather unique condition in gymnosperms. Pentoxylaeae is a phytogeographically significant group, in addition to Rajmahal Basin, India (Sahni 1948; Vishnu-Mittre 1957a) it has also been recorded from the Gippsland Basin, Australia (Drinnan and Chambers 1986), Tasmania, New Zealand (Harris 1962) and the Byres Peninsula of the South Shetland Islands, Antarctica (Césari et al. 1998). Leaves of it are known as *Nipaniophyllum raoi* Sahni (1948) which were borne on short shoots and displayed a generally cycadean anatomy although the epidermal and stomatal structures were more bennettitalean. This plant bore cones known as *Carnoconites compactum*, which were long and bore seeds compactly on a long axis, with the micropylar part of each seed facing outwards, again a rather unusual feature in gymnosperms. *Carnoconites laxum* is another similar slightly thinner cone with more numerous seeds loosely arranged along an elongated axis. The male reproductive parts are known as *Sahnia nipaniensis* Vishnu-Mittre (1953b). According to him these were filiform, branched microsporophylls surrounded by deciduous bracts, which united at the base into a disc surrounding the broad conical receptacle. However, Sharma (1996) suggested that the microsporophylls did not unite at the base. They were free and were disposed of in a spiral phyllotaxis on the base of the receptacle. Associated with *Pentoxylon sahnii* occurs the *Nipanioxylon guptai* Srivastava (Sahni 1948), another petrified stem which in cross-section shows a wide cortex and pith and a ring of eight vascular bundles, each with equally well-developed secondary growth all around. Ample efforts of Sharma (1969a, d, e, 1971b, 1972a, b, 1973a, 1974a, b, 1979, 1989, 1996) helped to better figure out the actual structures of this interesting group. This group is of great interest for tracing the evolutionary history of angiosperms and their precursors.

(C) **Bennettitales**

Bennettitales are another group of gymnosperms having a strong presence in the Rajmahal flora. Essentially, the Bennettitales must have had contributed a large proportion of the flora but morphologically similar Bennettitales and Cycads are not easily distinguishable from each other in the Cycadophytean complex of stems and leaves. Several species of Bennettitalean fronds are described under the names *Ptilophyllum, Dictyozamites, Otozamites* etc., which constitute a big portion of the flora. The most important one among them is *Ptilophyllum*, which is an index upper Gondwana form-genus. A large number of species of this genus have been recorded from the Rajmahal itself of which *P. acutifolium, P. cutchense* are preserved as impressions. They are also preserved as petrified specimens e.g., Vishnu-Mittre (1956b) described the anatomy of *P. nipanica.* Bose (1953a) studied the anatomy of *P. amarjolense* while Rao and Achuthan (1967) studied the morphology and anatomy of a species of *Ptilophyllum* comparable to *P. cutchense. Ptilophyllum guptai* Sharma and *P. sparsifolium* Sharma are two species of petrified specimens studied by Sharma (1967). Associated with these *Ptilophyllum* leaves occur stems known as *Bucklandia* and flowers known as *Williamsonia.* One of the most important contributions of Sahni (1932b) is the reconstruction of the bennettitalean plant *Williamsonia sewardiana.* Sahni (1932b) established *Bucklandia indica* as the stem, *Ptilophyllum cutchense* as the foliage and *Williamsonia* as the female flower of the same plant under a single name *Williamsonia sewardiana* Sahni. It is also supposed that the foliage known as *Ptilophyllum amarjolense* Bose (1953a) was borne on stems known as *Bucklandia sahnii* (Bose 1953b). A dichotomously branched specimen of *Bucklandia* has been described by Sharma (1969b) under the name *Bucklandia dichotoma.* It has mixed characters of Bennettitales, Cycadeoidales and vessel-less Magnoliales. Several new species of *Williamsonia* flower have been described from the Rajmahal Hills like *W. gigas, W. indica, Williamsonia* sp. (Gupta 1955) and *Williamsonia harrisiana* Bose (1968). *Weltrichia santalensis* Sitholey and Bose (1953), *Weltrichia singhii* Bose (1967), *Ontheanthus polyandra* Ganju (1947b), *Williamsonia companulatiformis* Sharma (1969c) are the male flowers from Rajmahal Basin. Bennettitalean wood *Sahnioxylon andrewsi* was discovered by Bose and Sah (1954) and *Sahnioxylon rajmahalense* was previously described as *Homoxylon rajmahalense* by Sahni (1932a), suggesting its angiosperm affinity as it is devoid of vessels. It is compared with the modern homoxylous family Magnoliaceae. But subsequent investigations of more specimens of the same kind by Hsü and Bose (1952) showed that it was bennettitalean. Similarly, the impression fossils of Stem – *Fascivarioxylon mehtae* Jain (1962), strobilus of *Ontheodendron florin* (Sahni and Rao 1933) and scale leaves of *Cycadolepis indica* have uncertain bennettitalean and cycadean affinities.

(D) **Caytoniales**

A group of Mesozoic plants which are rather noticeable by their absence in the Rajmahal flora are the Caytoniales although, leaves like *Sagenopteris bhambhani* Jacob and pollens like *Alisporites jurassicus, A. auriculiformis* (Rao 1943b) from the Rajmahal Hills may belong to the Caytoniales. It is possible that Caytonialean remains are there, but are not recovered yet. *Thinnfeldia indica* Feistmantel and *T. chunakhalensis* Sah and Sukh-Dev (1957) represent other seed ferns foliage in the Rajmahal flora.

(E) **Ginkgoales**

Ginkgoales were meagerly present in the Rajmahal flora. Mehta and Sud (1953) reported genera like *Ginkgoites* and *Baiera* from two localities in the Rajmahal Hills. Sah (1953) also reported *Ginkgoites* like impressions from Sakrigalighat in the Rajmahals. The leaves of *Ginkigoites rajmahalensis* Sah and Jain (1964) with much divided lamina and forked once or twice are found in the Rajmahal Basin. These records substantiate that Ginkgoales had worldwide distribution during the Mesozoic period.

(F) **Pteridophytes**

Pteridophytes are quite abundant in the Rajmahal Basin. They were represented by lycopods, sphenopods and ferns. Of these, ferns were very much diversified and allied of modern ferns. Equisetales are represented by impressions of a slender species,

Equisetites rajmahalensis with many nodal diaphragms. Lycopodiales include shoot impressions like *Lycopodites gracilis* and petrified specimen *Lycoxylon indicum* (Srivastava 1946); internal anatomy of *L. indicum* is very similar to that of the modern *Lycopodium clavatum.* No reproductive parts have been hitherto found. Earlier, Rao (1943b) described some pteridophytic sporangia and spores like *Sporites maghadenses, S. naviculae* and others from the Nipania cherts. Many of them could be compared to the spores of *Lycopodium* and *Selaginella* while a few could not be determined. Further, Vishnu-Mittre (1954), Bose and Sah (1967) and a few others have described several spores from the silicified sediments of the Rajmahal Basin. A general perusal of Rajmahal flora indicates that the fern flora of the Rajmahal basin was as modern as present-day flora. It is comprised of many families like Osmundaceae: *Todites indicus* (Oldham and Morris) Bose and Sah (1967), *Cladophlebis sahnii* Vishnu-Mittre and *Osmundites sahnii* Vishnu-Mittre (1955a); Schizaeaceae: *Azonomonolete* Type I (spore) and stem *Solenostelopteris nipanica* Vishnu-Mittre (1958); Cyatheaceae: *Liratosporites* Type I (spores) Vishnu-Mittre (1958); Dicksoniaceae: rhizome—*Tinpaharia sinuosa* Jacob (1943, 1950); Marattiaceae: *Marattiopsis macrocarpa* (Oldham and Morris) Seward and Sahni (1920); Gleicheniaceae: *Gleichenia gleichenoides, Gleichenia*-like petioles (Vishnu-Mittre 1958); Matoniaceae: *Phlebopteris* (*Lacopteris*) sp. Rao (1950), *Phlebopteris* sp. Bose and Sah (1967); Dipteridaceae: *Hausmannia crenata* (Nathorst) Möller (Bose and Sah 1967); Thyrsopterideae(?): *Coniopteris hymenophylloides* (Brongniart) Seward. Besides this *Culcitites madagascarensis* Appert and *Phyllopteroides laveae* Cantril and Webb were also reported by Banerji (1987, 1992) respectively which are significant as age-marker taxa. Apart from these taxa there are a number of petrified petioles, rachii and rizhomes reported from Rajmahal Basin whose exact affinities could not be traced.

(G) **Algae, Fungi and Bryophytes**

Sharma (2017) reported a fossil leafy Jungermaniales plant in a thin section of silicified chert of Nipania, Rajmahal Hills. The leafy thalii were either simple or bifurcated and had two lateral rows and a ventral row of leaves, resembling some extant genus *Porella* sp. Horn af Rantzien (1957) and Vishnu-Mittre (1952) described charophytic oospores probably belonging to the Nitelleae from the Rajmahal Hills. Sharma et al. (2001), Sharma and Harsh (1994, 2020) also reported fossil algae and lichen in a thin section through a piece of silicified chert of Nipania. Algae are in the form of dark coloured filaments that resemble non-flagellate spores of Rhodophyceae and lichen looks like a small propagule in which algal cells are surrounded by fungal filaments as in *Parmelia nudecata.* Sharma and Tripathi (1997) reported a petrified coenobial alga from the Rajmahal Hills. On the basis of study of a number of sporangia like specimens in thin sections of Nipania chert which resembled with sporogonium of *Notothylas,* Sharma et al. (1984) established a new taxon *Sporangioceros nipanica,* but after further study Sharma et al. (2002) suggested its similarity with an extant fresh water alga *Characiosiphon* and gave a new name *Characiosiphonites nipanica.* Fungal hyphae inhabiting partly decayed seeds of *Nipaniostrobus sahnii* have been reported by Rao (1943a). Sharma et al. (2017) reported two types of saprophytic fossil fungi in the forms of non-septate (coenocytic) mycelia and hyphae with numerous conidia (nucleated on the bark of *Pentoxylon* stem) and septate mycelia and hyphae with only few conidia (in the integument of an ovule of *Araucarites mittrii*) in thin section of silicified cherts of Nipania and Sonajori localities in the Rajmahal Hills.

(H) **Angiosperms**

Macrofloral angiospermous remains recorded from the Rajmahal area have been dubious and were under constant revaluation for their authenticity. Fossil wood genus *Homoxylon* with two species (*H. rajmahalense* and *H. andrusii*) from Rajmahal, recorded and described by Sahni (1932a) was considered to have angiospermous affinity (Sahni 1932a). However, later Bose and Sah (1954) re-examined it and discussed its putative bennettitalean affinity and transferred it to the new genus *Sahnioxylon*. There are three alleged fructification records assigned to angiosperms. Sharma (1997) described a fructification from Amarjola locality of Rajmahal Basin which is known for rich bennetitalean fossils along with *Williamsonia*, showing similarity with *Lesqueria*, a mid-Cretaceous infructescence from the Kansas (Crane and Dilcher 1984). This fructification was re-investigated by Srivastava and Krassilov (2012), and they argued that the specimen described as *Lesqueria*-like fructification is a hemispherical structure on a short pedicel, covered with scale-like appendages. These appendages were considered as multi-ovulate condition of follicles by Sharma (1997) which were regarded as stalked ovules adhering to interseminal scales typical of bennettitalean fruiting structures (Srivastava and Krassilov 2012). Banerji (2000a) described impressions of an unnamed flower-like structure and a fruit-like body and assigned it to a new genus *Sonajoricarpon* from the Sonajori locality, Rajmahal Formation. *Sonajoricarpon rajmahalensis* Banerji (2000a) is based on a rough cast of a globose body in an irregularly split chert block. A linear knobby structure extends out of the globose body and forks at a short distance from it. A well-marked series of ovate scars surrounded by concentric ridges of fibrous tissue over the periphery of the cast were interpreted as seeds. Banerji (2000a) compared *Sonajoricarpon* with the fruits of the extant fabaceous genus *Butea* Roxb. ex Willdenow (1802). The unnamed flower-like structure has an elliptical body with vaguely marked structures that appears to be radiating from it as perianth lobes and also depicts stamen-like appendages attached between the lobes. This specimen was also scrutinized by Srivastava and Krassilov (2012) and they regarded it as tuberose stem base of *Pentoxylon* wrapped in swollen leaf bases (see Srivastava and Krassilov 2012, pp. 69–70 for detailed discussion). Sharma et al. (2001) reported anatomical structure of leaf from a thin section of Nipania chert resembling with a monocot leaf, however they also suggested further investigation of that material.

Vishnu-Mittre (1955b) reported a sporomorph of angiospermous affinity from Nipania chert, under the name *Sporojuglandoidites jurassicus* distinguished by the presence of foraminoid apertures, a character not met within any plant group except in the angiosperms. Long after, Tiwari and Tripathi (1995) reported palynomorphs viz., *Retimonocolpites peroreticulatus*, *Retimonocolpites* sp., Aff. *Clavatipollenites* sp. 1, Cf. *Stephanocolpites* sp., *Liliacidites* sp., Cf. *Stellatopollis* sp., from the uppermost lower intertrappean of Rajmahal deposits, from the borehole section (RJNE-32), near the village Kirtanya, in the northern part of Rajmahal Basin. Harland et al. (1990) dated ($^{40}Ar/^{39}Ar$) flows associated with this intertrappean as 117 Ma (Aptian).

5 Palaeofloral Zonation of Rajmahal Basin

Sen Gupta (1988) carried out in depth study of the stratigraphy and palaeobotany of Rajmahal Hills. He identified and systematically described 35 plant species belonging to Sphenophyta, Pteridophyta, Pteridospermophyta, Cycadeodophyta and Coniferophyta. Three new species, namely *Thinnfeldia khatangiensis*, *Taeniopteris sarbadhikarii* and *Pterophyllum valentinei* were erected by him.

In an attempt to resolve inter-sector correlation of intertrappean beds, Sen Gupta (1988) provided a biostratigraphic zonation based on the distribution of megaplant species occurring in intertrappean beds. He established three biozones, namely *Ptilophyllum acutifolium–Gleichenites gleichenoides* assemblage zone (Zone 1; lower Jurassic age), *Cladophlebis indica–Dictyozamites indica* assemblage zone (Zone 2; middle-late Jurassic age) and *Taeniopteris spatulata–Brachyphyllum rhombicum* assemblage zone (Zone 3; early Cretaceous age). Thereafter, Banerji (1989, 1992, 1993, 1995a, b, 1996, 1997, 2000a), Prakash (2000), Pal et al. (2009), Banerji and Jana (1998, 2000, 2003) and Banerji and Ghosh (2005, 2006, 2008) added more plant megafossils in the Rajmahal flora. Based on the reappraisal of the flora, Banerji (2000b) classified it into four assemblages and retained the three assemblages of Sen Gupta (1988). Salient features of each assemblage/zone as mentioned by Sen Gupta (1988) and Banerji (2000b) are briefly provided and discussed in the followings.

Assemblage 1: (*Ptilophyllum acutifolium–Gleichenites gleichenoides* assemblage zone)

This assemblage of Sen Gupta (1988), is the oldest assemblage and includes the floral assemblage retrieved from the upper part of the Dubrajpur Formation and is less diversified than overlying assemblages. This assemblage/zone is exposed only in the western sector of the basin. The fragmentary plant remains make their first appearance in the third intertrappean bed. First and second intertrappean beds are unfossiliferous (Sen Gupta 1988). The type section of the Dubrajpur Formation i.e. Sarwan Pahar locality has not yielded any plant fossil (Sen Gupta 1985). Both Banerji (1989) and Sen Gupta (1988) have procured plant remains from the section exposed at Khatangi Hill. Upper shale bed of this section has yielded abundant plant megafossils.

Assemblage composition of Dubrajpur flora comprises of *Pterophyllum*, *Ptilophyllum*, *Cladophlebis*, *Culcitites*, *Elatocladus*, *Gleichenites*, *Nilssonia*, *Onychiopsis*, *Otozamites*, *Ctenis*, *Pagiophyllum*, *Pecopteris*, *Sphenopteris*, *Taeniopteris* and *Thinnfeldia* (Sah 1964; Sastry et al., 1977; Sen Gupta 1984, 1988; Banerji 1989). The assemblage is characterised by the frequent occurrence of *Culcitites*, *Onychiopsis*, *Ctenis* and *Nilssonia* along with the preponderance of *Ptilophyllum* and *Gleichenites*.

Frequent occurrence of *Onychiopsis* along with the abundance of *Gleichenites* indicates a Lower Cretaceous affinity. The Dubrajpur flora shows some resemblance to the Upper Jurassic flora of the Manamana Basin of Madagascar (Appert 1973) with which it has certain genera in common (viz. *Equisetites*, *Gleichenites*, *Cladophlebis*, *Phlebopteris*, *Mohriopsis*, *Culcitites*, *Eboracia*, *Haydenia* and *Onychiopsis*). Sen Gupta (1988) designated the *Ptilophyllum acutifolium–Gleichenites glecichenoides* Zone for this assemblage. An early and Late Jurassic–Early Cretaceous age was assigned by Sah (1964), Sah and Shah (1974), Sen Gupta (1988) and Banerji (1989). Based on the occurrence of some dinoflagellate cysts viz., *Batiacasphaera laevigata*, *Batiacasphaera* sp., *Dissiliodinium* sp. Feist-Burkhardt and Monteil (1997); suggested an Early Cretaceous age to these intertrappean beds. This assemblage is dominated by cycads–bennettitales followed by pteridophytes. Conifers are rare.

The palynological investigations of subtrappean sediments of the Dubrajpur Formation revealed long age ranges for this formation. In the palynological analysis based on bore core sediments, Tripathi and Ray (2006) concluded that Dubrajpur Formation includes strata of Early–Late Triassic, late Early–early Mid Jurassic, Late Jurassic and Early Cretaceous, thereby indicating that this unit is profoundly diachronous. The presence of marker palynomorphs suggests that the lower age limit of the Dubrajpur Formation is Early Triassic. Identification of Early Cretaceous palynomorph markers from the uppermost Dubrajpur Formation suggests that the upper age limit is Early Cretaceous (Tiwari et al. 1984; Tripathi and Ray 2006). Similarly, the presence of *Callialasporites turbatus* and *Araucariacites australis* pollens in the bore core sediments (BRS-7, Salbadra-Gompapahari section) belonging to Dubrajpur Formation reflects affinity with Jurassic–Early Cretaceous palynoflora (Murthy et al. 2020a).

Assemblage 2: (*Cladophlebis indica–Dictyozamites indica* assemblage zone)

Strata yielding this assemblage are exposed in the northern, central and southern sectors of the Rajmahal Hills. It includes exposed first to fourth intertrappean beds in the northern and central sectors (Sen Gupta 1988). Whereas, in the southern sector of the Rajmahal Hills, it is exposed in the lower three intertrappean beds. In this assemblage, well preserved fossils are found in baked siltstone and fine grained sandstone units of intertrappeans. Significant localities of this assemblage are Sakrigalighat, Amarjola and Chunakhal. This assemblage is characterized by the dominance of cycads, bennettitales and followed by pteridophytes and conifers. Filicales are well diversified. Pteridospermales, ginkgoales, and pentoxylales are relatively uncommon. Sen Gupta (1988) described that this zone is marked by the characteristic presence of *Ptilophyllum acutifolium*, *P. cutchense*, *Pterophyllum fissum*, *Cladophlebis indica*, *Dictyozamites indicus* and *D. falcatus*. The first appearance of *P. cutchense*, *Dictyozamites indicus* and *Marattiopsis macrocarpa* mark the lower boundary of the zone whereas upper boundary is marked by appearance of *Brachyphyllum rhombicum*, *Taeniopteris sarbadhikarii* and proliferation of *T. spatulata* in the overlying assemblage zone. Besides the above mentioned taxa, some other important taxa recorded in this zone are *Equisetites rajmahalense, Cladophlebis srivastave, Todites* sp. cf. *Todites williamsoni, Gleichenites gleichenoides, Pecopteris lobata, Marattiopsis macrocarpa, Sphenopteris sakrigalensis, Thinnfeldia indica, T. chunakhalensis, Cycadites rajmahalensis, Taeniopteris crenata, T. lata, T. crassinervis, Ptilophyllum tenerrimum, P. sakrigalensis, Pterophyllum morrisianum, P. medlicottianum P. princeps, P. velentinei, Otozamites gondwanensis, Dictyozamites bagjoriensis, D. sahnii, D. hallei, D. indicus, Brachyphyllum mamillare* and *Elatocladus sahnii, E. tenerrima.* Sen Gupta (1988) also recognized a subzone *Pterophyllum princeps- Taeniopteris lata* within this zone based on the persistent and restricted occurrence of these two taxa in this subzone; he suggested middle to upper Jurassic age for Assemblage 2 zone.

Banerji (2000b) reported *Phyllopteroides laevis*, a significant Neocomian (earliest Cretaceous) taxa of eastern Australia, that has been recorded from five localities exposed at Gumapahar, Bartala, Balidih, Chunakhal and Murlipahar. With this discovery, she considered Neocomian age for this assemblage. Many taxa (i.e. *Eboracia, Murlipaharopteris, Coniopteris, Anomozamites* and *Dictyozamites*) are restricted to this assemblage only.

A diverse palynological assemblage having the typical elements of early Cretaceous time are reported by Sah and Jain (1965) from the Basko and Sakrigalighat carbonaceous shale exposures correlatable with the intertrappean beds of Assemblage 2. Significant palynotaxa are *Araucariacites, Cyathidites, Podocarpidites, Glecheniidites* and *Callialasporites.* Recovered palynotaxa have the affinities with pteridophytes, cycadophytes, conifers and some unidentified plant groups.

Assemblage 3: (*Taeniopteris spatulata–Brachyphyllum rhombicum* assemblage zone)

This assemblage/zone is exposed only in the southern sector of the Rajmahal Hills, where it corresponds with the exposed fourth intertrappean bed exposed around the village of Nipania. This biostratigraphic assemblage zone is correlatable with 'Nipania beds'. The assemblage is characterized with the appearance of *Taeniopteris sarbadhikarii* and *Brachyphyllum rhombicum* and also increased abundance of *T. spatulata* along with disappearance of many elements occurring in the previous zone e.g. *Dictyozamites, Ptilophyllum acutifolium, P. cutchense, Pterophyllum fissum* and *Cladophlebis indica.* The other significant taxa present in this zone are *Pentoxylon, Nipaniophyllum, Sahnia, Carnoconites, Indophyllum, Nipaniostrobus, Podostrobus, Mehtaia* and *Nipanioruha* etc. Based on the correlation of the "floral assemblage of Raghavapuram Shale" and "Jabalpur Stage", Sen Gupta (1988) assigned a lower Cretaceous age to this zone. This assemblage shows the abundant co-occurrence of Pentoxylales and Coniferales.

Corresponding to these beds, a palynoassemblage was described by Vishnu-Mittre (1954) who reported forty-six types of spores and pollen grains in the thin sections made from the intertrappeans of Nipania chert; indicating the presence of pteridophytic spores, cycadophytic pollen grains and the coniferous saccate pollen grains which were almost equally represented in the assemblage. The important constituents of this assemblage are *Gleicheniidites*, *Cicatricosisporites*, *Cycadopites*, *Podosporites*, *Pitysporites*, *Alisporites*, *Podocarpidites* and *Callialasporites*.

Assemblage 4

Based on the fossil find of angiospermous macroremains as described by Banerji (2000a, b), she established one more assemblage to the assemblage scheme given by Sen Gupta, and suggested this new assemblage as the youngest assemblage from the Sonajori Intertrappean Bed intercalated with trap rocks exposed at the Sonajori locality. The zone is characterized by the preponderance of coniferales, pentoxylales and filicales and absence of cycads and bennettitales, and the appearance of angiospermous remains. Numerous silicified remains of plants belonging to the filicales, pentoxylales and coniferales have been reported by Sharma and Bohra (1976a, b, 1987) and Bohra and Sharma (1978, 1979, 1980a, b) from these beds. Among the conifers, araucarian elements were dominant as evidenced by the occurrence of araucarian seeds from this locality viz., *Pakurispermum*, *Paradoxospermum*, *Taxaliospermum*, *Sonajorispermum*, *Triangulospermum* (Suthar et al. 1988). Hundreds of *Acmopyle*-like ovule/seed namely *Podospermum rajmahalensis* of podocarpean affinity are also described suggesting the presence of Podocarpaceae during the Early Cretaceous (Banerji and Ghosh 2006). Pteridophytic remains are also found frequently preserved. Based on the occurrence of angiospermous remains i.e. *Buteacarpon* and *Sonajoripushpam* by Banerji (as mentioned in table, 2000b, p. 138), this youngest assemblage is considered as the late early Cretaceous age (Barremian-Aptian). However, later these proclaimed remains of angiospermous affinity were questioned by Srivastava and Krassilov (2012) and *Sonajoripushpam* was deciphered to have Pentoxylalean affinity as discussed earlier. Moreover, sediments from this locality did not yield spores and pollen (Tiwari et al. 2013).

6 Patterns of Floral Change in Rajmahal Basin Through Time

The detailed analysis of the palaeobotanical evidence indicates a remarkable change in floral patterns in the Lower and Upper Gondwana flora. Typical lower Gondwanan elements i.e. Glossopterids are not encountered in the sediments younger than Permian in Rajmahal Basin, as is the case with other contemporaneous Gondwana basins (Saxena et al. 2014, 2019; Goswami et al. 2018). From the beginning of the Dubrajpur Formation (late Triassic to Jurassic) to the end of the Rajmahal Formation (early Cretaceous) significant transformation in the upper Gondwana flora is observed that began with the first appearance of *Ptilophyllum* and culminated in the evolution of early angiosperms that paved way for the subsequent proliferation of angiosperms.

The basal beds of Khatangi hills are characterised by the presence of gymnosperms which are represented by 4 species of *Glossopteris* (*G. ampla*, *G. communis*, *G.*

retifera, G. parallela) and one species of *Vertebraria* (*V. indica*), based on these floral remains, these beds are considered as belonging to the Barakar Formation of Permian age (Sah and Shah 1974, Sen Gupta 1984, 1985). The upper zone at this section has yielded typical elements of upper Gondwana and are represented by *Marattiopsis macrocarpa, Gleichenites gleichenoides, Ptilophyllum acutifolium, Nilssonia princeps, Macrotaeniopteris lata* and *Williamsonia* sp. (reported by Srivastava 1958, Sah 1964, Sen Gupta 1988, Banerji 1989) and belong to the Dubrajpur Formation of Early Cretaceous age Sukh-Dev (1987). These findings clearly put forward that the place of Permian pteridosperms dominating flora (Glossopteris flora) was taken over by Upper Gondwana flora (Ptilophyllum Flora) during Permian–Triassic and Jurassic transitions. Moreover, the first instance of the occurrence of a Ptilophyllum Flora in Khatangi Plant Bed which is infratrappean in position to the Rajmahal volcanic series indicates that floral change already started before the initiation of the Rajmahal volcanic episodes. Another striking observation is the absence of Corystospermales (particularly genus *Dicroidium*—an abundant element of Triassic) in the Rajmahal Basin, although it is quite frequent in adjacent Damodar and Son-Mahanadi Basins (Saxena et al. 2019 and reference therein), and is often considered as the peculiar flora existed between the extinction of Glossopteris Flora and appearance of Ptilophyllum Flora. Their absence can be linked with period of non-deposition/erosion between the deposition of Barakar and Durbrajpur sediments.

During Jurassic, the landscape of Rajmahal Basin was dominated by lush green forests of Bennititales, Cycadales followed by Pteridophytes. Bennettitalean shrubs of medium height, were represented by *Ptilophyllum, Pterophyllum* and *Williamsonia.* The former one is the most dominant element whereas the later two are meagerly represented. Large leaved cycadales were also dominant, of which *Nilssonia* and *Ctenis* were quite frequent, whereas *Taeniopteris* was rare. Ferns also contribute large portion of the flora after bennettitales and cycadales, represented by *Gleichenites, Culcitites, Onychiopsis* and *Cladophlebis.* Lycopsids and sphenopsids are absent in this assemblage. Conifers are extremely rare and represented by a fragment of *Pagiophyllum* and *Elatocladus conferta.* Thus, the palaeovegetation is dominated by short hight gymnosperm plants i.e. Benntitales and Cycadales (Srivastava 1958, Sah 1964, Sen Gupta 1988, Banerji 1989).

The preponderance of Bennettitals and Cycadales point towards the warm and humid climate. The dominance of ferns also confirms the existence of humid, subtropical climatic conditions in the certain places. Broad-leaved taxa e.g., *Pterophyllum, Nilssonia* and *Taeniopteris* and reticulate veined leaves e.g., *Dictyozamites* and *Ctenis* also indicate warm, humid climatic conditions prior to the initiation of the volcanic episode in the Rajmahal basin (Banerji 1989, 2000b, Jacob and Shukla 1955).

The assemblage is characterised by the frequent occurrence of *Culcitites, Onychiopsis, Ctenis* and *Nilssonia* along with the preponderance of *Ptilophyllum* and *Gleichenites.* The frequent occurrence of *Onychiopsis* along with the abundance of *Gleichenites* indicates a Lower Cretaceous affinity (Banerji 1989). However, unlike other Lower Cretaceous assemblages, the conifers are relatively infrequent. The flora is dominated by ferns, bennettitales and cycadales. An Upper Jurassic affinity cannot be ruled out as the flora contains an Upper Jurassic marker genus *Culcitites*. Thus, before the start of the Rajmahal Formation, during the later phase of the Dubrajpur Formation, the flora is dominated by ferns, bennettitales and cycadales however, the conifers are present, but they are very rare, it is depicted that a tropical to subtropical climate was prevailing at the time of deposition (Banerji 1989).

The next floral change is evident after the second and third volcanic eruptions, reflected by the palaeoassemblage of lower first to third fossiliferous intertrappean beds at Bartala, Balidih, Chunakhal, Gumapahar and Murlipahar. Like the earlier assemblage, the palaeovegetation is dominated by Bennettitales (*Pterophyllum*, *Ptilophyllum*, *Dictyozamites*, *Otozamites*, *Anomozamites*), followed by ferns (*Marattiopsis*, *Cladophlebis*, *Todites*, *Osmundopsis* sp., *Phyllopteroides*, *Cacumen*, *Gleichenia*, *Eborasia*, *Mohriopsis*, *Asplenites*, *Onychiopsis*, *Muralipaharopteris*, *Coniopteris*, *Hausmania*) (Banerji 1992, 1993, Banerji and Jana 1998). However, floral change is conspicuous as evident by the more diversified forms of Bennettitales as well as ferns. Contrary to the earlier scenario, conifers which are meager in the infratrappean palaeofloral assemblage of Dubrajpur Formation became diversified and increased in numbers and variety (*Brachyphyllum*, *Pagiophyllum*, *Araucarites*, *Coniferocaulon*, *Elatocladus*, *Desmiophyllum*). A stratigraphically significant Neocomian fern of eastern Australia, *Phyllopteroides laevis* has been found preserved in five localities. Pteridosperms (*Thinnfeldia*, *Pachypteris*) and ginkgoales (*Ginkgoites rajmahalensis*) have also been recovered from some localities. This transformation (species richness and diversity) in the flora indicates that before the volcanic episodes, the climate was warm and humid enough with ample rainfall which favours the better growth of Bennettitales and moisture-loving ferns. However, after the second and third volcanic events, the climate might have changed to arid with xerophytic conditions, as reflected by the increasing number and variety of conifers.

Contrastingly, *Culcitites madagascarensis* which is quite common in Dubrajpur flora, is absent from the flora of Rajmahal Formation (Banerji 1987). Interestingly, in a few localities, some specific taxa dominate, viz., *Equisetites* (Sphenopsid) in the Dudhkol assemblage (Prakash 2000) and *Thinnfeldia* (Pteridosperm) in the Balidih locality (Banerji and Jana 1998) and *Anomozamites* (Bennettitales) in the Bartala locality (Banerji and Jana 2000). Probably it is due to the local environmental conditions which favour the progressive growth of these specific taxa.

During the deposition of Rajmahal Formation, after the fourth volcanic outburst, there was a significant change in the palaeovegetation, which can be recognised by the flora from the intertrappean beds associated with the fourth and fifth lava flows at Nipania, Sonajori, and Pakur localities. These localities are monopolised by an interesting and peculiar group of plants, the Pentoxyleae which dominate the assemblage together with conifers (Srivastava 1946; Sahni 1948; Bose et al. 1985; Srivastava and Krassilov 2012; Ghosh et al. 2018) and pteridophytes whereas cycadophytes are relatively rare (Vishnu-Mittre 1956b, 1958).

The conifers are represented by three families, the Podocarpaceae, the Araucariaceae and Cupressaceae with a dominant representation of the Podocarps than those of the Araucarians and rare occurrence of Cupressians. Both the famous localities namely, Nipania and Sonajori, have yielded abundance of Pentoxylales and conifers followed by pteridophytic remains. Although several undoubted records of the Araucarians from the upper Gondwana of India and other countries are known, the *Podocarps* of Rajmahal are distinct. The characteristic forms of this assemblage are *Pentoxylon, Nipaniophyllum, Sahnia, Carnoconites, Indophyllum, Nipanioruha, Podostrobus, Mehtaia, Sitholeya* and *Nipaniostrobus.* Another significant evidence of prevalence of conifers comes form the Hiranduba locality in the form of fragmentary branched shoots and seed cones (*Hirandubia cupressoides*) of cupressian affinity and Podocarpean seed (*Podocarpospermum podocarpoides*) (Ghosh et al. 2018) further strengthens the fact that conifers were dominant during Eraly Cretaceous in Rajmhal. This co-dominance of Pentoxyleae and Conifers, with subdued occurrences of pteridophytes and rare Bennettitales along with absence of cycadaes in the upper part of Rajmahal Formation is interpreted as to indicate a more xerophytic community. The questionable occurrence of angiospermous fruit *Sonajoricarpon* which was later interpreted as underground storage organ attached to branching rhizome with Pentoxylean affinity has been interpreted as evidence to illustrate the incipient geophytic habit in Mesozoic plants, which has been also seen in early angiosperms of far East (Srivastva and Krassilov 2012 and references therein). The Pentoxyleae assumes importance not only for their unique characters but also because attempts have been made to derive information about the evolution of angiosperms (Meeuse 1961). On the basis of leaf cuticular features, Meeuse (1961) suggested that the Pandanales (monocot clade of angiosperms) might have been derived from the Pentoxyleae.

The pattern of microfloral change in the Dubrajpur and Rajmahal formations also indicates distinct changes in the flora which helped in the demarcation of late Triassic, Late Jurassic/Early Cretaceous ages to these sediments. A detailed study of the Dubrajpur Formation by Tiwari and Tripathi (1995) revealed dominance of gymnospermous pollen in the sequence, pteridophytic spores are consistently present but in low and fluctuating frequency. Similarly, bryophytic spores are recorded throughout the sequence, but they are rare. The qualitative pattern of spore-pollen occurrences suggests that the early Cretaceous infratrappean flora (recovered just below the first trap) of Dubrajpur Formation continues in the first intertrappean beds of Rajmahal Formation. Tiwari and Tripathi (1995) also suggested that due to the limited impact of the first flow, the effect on the vegetation was less and the flora of the infratrappean was not wiped out completely from the region hence the flora of the infratrappean is represented in the intertrappeans, after this radical decline in the floral diversity in the second intertrappean is observed followed by gradual flourishing and diversification during the deposition of the third and fourth traps.

Miofloral evidence for angiosperms affinity recorded by Vishnu-Mittre (1955b) as pollen taxon *Sporojugiandoidites jurassicus* from Nipania chert and later, palynological investigations carried out by Tripathi and Tiwari (1990), Tripathi (1993), Tiwari and Tripathi (1995) from outcrops and borehole section near Mirzachowki also revealed appearance of angiospermous pollen (*Retimonocolpites* and *Clavatipollenites*) in the uppermost part of the lowest intertrappean deposits (the first intertrappean bed) which is radiometrically dated as Aptian (117 Ma, $^{40}Ar/^{39}Ar$). Thus, the presence of distinct angiospermous taxa from the Rajmahal traps suggests that angiosperm evolved in the Early Cretaceous. Although definite macrofloral evidence are still lacking. This scarcity of records of angiosperm flora from the Rajmahal Basin could be attributed to various factors, such as less exploration of the Rajmahal Basin or less quantitative diversity in angiosperms, or growth of angiosperms distinctly far away from the basin. Banerji (2000b) envisaged two distinct floral communities in the Rajmahal during Jurassic to early Cretaceous, one as open land mesophytic type and another a xerophytic community. Both of these communities had smaller sub-communities represented by different compositional assemblages at different sites. Open land mesophytic community comprised of lycophytes, sphenophytes and marattiales ferns growing near lakes or other water bodies and cycadyophytes with other pteridophytes growing across the open and sloping lands (Banerji 2000b). The xerophytic community comprised by dominance of conifers and pentoxylales with fewer pteridophytes and cycades.

Correlations: The flora of the Rajmahal Formation can be categorize into two broad groups. The older group (Group-I) of Rajmahal floral assemblage (including Dubrajpur) is characterized by the abundance of large leaved Cycads, Bennettitales and Filicales together with remains of Equisetales, Lycopodiales, Coniferales and a few Ginkgoales. The younger group (Group-II) of Rajmahal floral assemblage belonging to Sonajori, Nipania and contemporaneous beds are dominated by Conifers and Pteridophyte together with Pentoxylales but rare Bennettitalean remains.

The flora of Dubrajpur Formation shows similarity with the upper Jurassic flora of Manamana Basin, Madagascar (Appert 1973) in having common occurrences of *Cladophlebis*, *Culcitites*, *Eboracia*, *Equisetites*, *Gleichenites*, *Haydenia*, *Onychiopsis*, and *Phlebopteris*. The flora of Rajmahal Formation can be compared with the early Cretaceous megaflora of Victora, Australia as the latter is also *Ptilophyllum* dominant flora. Sukh Dev (1987) recognized 10 megafloristic zones in the Mesozoic formations of India. On the basis of the overall dominance of Bennettitales and Cycadales over Pteridophytes and Conifer in Rajmahal flora including flora of Dubrajpur Formation (Group-I) categorized under *Dictyozamites-Pterophyllum-Anomozamites* Zone (8) and showed its resemblance with Shivaganga Formation, Tamil Nadu; Sarnu Hill Formation and Pariwar Formation, Jaisalmer Basin, Rajasthan and Bhuj Formation, Kuchchh, Gujarat.

A comparative study of the floristic composition of all the localities in the Rajmahal Hills shows that some localities, viz. Amarjola, Chilgojori, Malipara, Kulkipara, Bindrabun (Brindabun) near Tinpahar, Sakrigali Ghat (softer bed), Kunjbana, etc., have not yielded any large-leaved cycadophytes but are in the abundance of other cycadophytes, ferns and a negligible presence of conifers, the fossil flora of these localities resembles the older group of the Rajmahal Formation. Moreover, the dominance of *Thinnfeldia* in Balidih and Pathargama and *Anomozamites* in Bartala localities perhaps shows local variations within the Group-I flora of Rajmahal Formation. But megafloral assemblages of these localities have certain common elements viz., *Equsetites*, *Thinnfeldia*, *Ptilophyllum*, *Anomozamites*, *Taeniopteris*, *Pterophyllum* and *Elatocladus*. The overall composition of the Bartala assemblage indicates affinity with the Utatur Plant Bed of Tiruchirapalli District, Tamil Nadu described by Feistmantel (1879), Gopal et al. (1957), Mamgain et al. (1973) and Maheshwari (1986). The common elements are—*Thinnfeldia indica, Taeniopteris indica, Taeniopteris spatulata, Elatocladus, Brachyphyllum* and *Ginkgoites* sp. In generic comparison it shows close similarity with early Cretaceous flora of Koonwarra, Victoria, Australia (Drinnan and Chambers 1986) and seems to represent a transitional flora between the zone B and C of Victoria's early Cretaceous flora (Douglas 1969). The common taxa are *Ptilophyllum, Taeniopteris, Thinnfeldia, Hausmannia* and *Ginkgoites.*

In overall composition, the Chunakhal flora seems to be contemporaneous to the Dhokuti assemblage in having dominance of filicales and cycadophytes. Chunakhal composition is somewhat similar to the Neocomian flora of the Surat Basin (Day 1964) in having *Phyllopteroides laevis, Taeniopteris spatulata* (? = *T. daintreei*) and species of *Ptilophyllum* (Cantrill and Webb 1987) The absence of Bennettitales and a possible indication of angiosperms in the Koonwarra mega-and microflora of the Gippsland Basin, Victoria, Australia suggests a younger age for this sequence than the Chunakhal megaflora.

The younger Flora (Group-II) of Rajmahal Formation i.e. Nipania, Sonajori, Dumurchir flora from the intertrappean beds associated with lava flows 4–5, consists of an abundance of ferns, conifers and pentoxyleae. Nilssoniales are completely absent and Bennettitales are rare. Floristic comparisons made with the other Gondwanan basins show that in having the abundance of ferns and conifers, the Nipania fossil flora compares with the Floras of Athgarh, Jabalpur and Kota Formation (Wadia 1953). In the absence of the large leaved cycadophytes, the fossil flora of Nipania further exhibits closer comparison with that of the Jabalpur group (Feistmantel 1877b). The silicified plant remains known from the Cretaceous rocks of Japan (Stopes and Fuji 1910) show common elements with the Nipania fossils, viz., *Schizaeopteris mesozoica, Solenostelopteris loxosomoides, Fasciostelopteris japanica, Niponophyllum cordaitiforme*, etc.

Early Cretaceous flora of the Rajmahal Formation can be compared with the Early Cretaceous flora of East coast of India, namely Atgarh (Mahanadi Basin), Krishna-Godavari, Pranhita Godavari, Palar Basin and Cauvery Basin (Chinnappa et al. 2021). Some of the common elements are Cycadales (*Taeniopteris*), Coniferales (*Elatocladus, Desmiophyllum, Pagiophyllum, Torreyites*), Bennititales (*Ptilophyllum, Pterophyllum, Dictyozamites, Otozamites, Cycadites, Bucklandia, Williamsonia*). Pteridospermales (*Thinnfeldia, Pachypteris*), Ginkogales (*Ginkgoites*) and Pteridophytes (*Equisetites, Cladoplebis, Marattiopsis, Todites, Gleichenia, Phlebopteris, Onychiopsis*).

Osmundites are known from Nipania chert. There are only two well-defined species of petrified *Osmundites* known so far from the Jurassic horizon, viz. *O. dunlopi* and *O. gibbiana* (Kidston and Gwynne-Vaughan) from the Jurassic of New Zealand, Central Australia and Queensland (Kidston and Gwynne-Vaughan 1907; Sinnott 1914; Edwards 1933; Sahni 1920). *Osmundites aucklandicus* (Marshal 1926) described from the upper Jurassic of New Zealand, according to Edwards (1933), is identical with *O. dunlopi.*

The peculiar occurrences of Pentoxylae in the Rajmahal have wide phytogeographic significance as it has also been recorded from Tasmania and New Zealand (Harris 1962), the Gippsland Basin, Australia (Drinnan and Chamber 1986) and Antarctica (Césari et al. 1998). Besides extensive occurrence of pentoxylales in Nipania beds of Rajmahal Hills, India, Harris (1962) described *Carnoconites* (*C. cranwelli*) from the *Taeniopteris* bed at Port Waikato, New Zealand which has been dated as Tithonian. Later, White (1981) reported male and female cones from Jurassic of Talbragar Fish Bed, New South Wales, Australia. The main trunk of Pentoxylon-plant was recorded by Rozefelds (1982) from south east Queensland of Australia. Drinnan and Chamber (1985) published a comprehensive work of pentoxylopsids (*C. cranwelli*, *Sahnia laxiphora* and *Taeniopteris daintreei*) from Early Cretaceous of Koonwarra Whitelaw beds of Victoria, Australia. Recent finding of *Taeniopteris* sp. and *Carnoconites llambiasii* by Césari et al. (1998) indicate the occurrence of Pentoxylales from Early Cretaceous bed of the Byers Peninsula, South Shetland Island, Antarctica.

7 Conclusions

In light of the palaeobotanical data derived from the upper Gondwana sequences of the Rajmahal Basin so far, and the absence of definite macro-remains of angiosperms; it can be argued that the Assemblage 4 of Banerji (2000b) (based on the assemblage of Sonajori locality) is more or less similar to the assemblage recorded from Nipania locality as the latter also has abundant Pentoxylales and conifers and seems to the continuation of the preceding zone. The only difference is the absence of cycadophytes at Sonajori. Abundance of Pentoxylales together with conifers is indicative of prevalence of warm and arid climate conducive for the development of xerophytic plants. On the basis of palaeofloristics of the flora of Rajmahal Basin it is envisaged that, during the deposition of Lower Gondwana (Permian) sequences, climate was warm and humid as evident by the deposition of coal bearing sequences with abundant glossopteridalean remains, that became arid and dry near the transition from Permian to Triassic. In this basin, Triassic and early Jurassic period is the time of denudation/non-deposition. During the deposition of Dubrajpur Formation, climate was warm and humid as manifested by the dominance of broad leaved cycadophytes (*Nilssonia*, *Pterophyllum*) and pteridophytes; such conditions were continued till the deposition of early part of Rajmahal Formation which gradually change to warm and arid in the later part owing to the volcanic eruptions as marked by the dominance of conifers. The climate change in the Rajmahal Basin during Mesozoic is reflected in floral constituents. It can be summarized that in the beginning (Jurassic), the dominance of bennettitales and cycadophytes is witnessed and is followed by gradual decline of cycads, bennettitales, and pteridophytes along with the dominance of conifers and pentoxylalean during late Jurassic and early Cretaceous times. Acme of pentoxylales, and disappearance of bennettitales, along with characteristic appearance of angiosperms (pollen records) mark the culmination of the Rajmahal Flora during the early Cretaceous.

Despite the commendable work that has been carried out by various workers, the early Cretaceous flora of Rajmahal still poses many challenges for palaeobotanists, the most important one is to trace the definite meso/macro remains of early angiosperms and their precursors. The volcano-sedimentary combination of the preserved sequences of Rajmahal Formation makes it most promising horizon in India to look for the angiospermic evidences. In addition to this, well preserved continental sequences of Rajmahal Basin ranging from Permian to Early Cretaceous provide an invaluable window to understand the evolution, radiation and subsequent proliferation and demise of many plant groups/communities across the late Palaeozoic and Mesozoic time.

Acknowledgements Authors are thankful to Director, BSIP for providing facilities and support and granting permission (RDCC/58/2023-24). We also acknowledge the help renderd by Dr. Suyash Gupta in preparation of figures and plates.

References

Agarwal JK, Rama FA (1976) Chronology of Mesozoic volcanics of India. Proceeding Indian Acad Sci-Sect A 84(4):157–179

Appert O (1973) Die Pteridophyten aus dem Oberen Jura des Manamana in Südwest-Madagaskar. Schweiz Paläontol Abh 94:1–62

Bajpai U, Maheshwari HK (1990) On two enigmatic infructescences from Permian Gondwana of the Rajmahal Basin. Palaeobotanist 39(1–3):9–19. https://doi.org/10.54991/jop.1990.1678

Baksi AK (1995) Petrogenesis and timing of volcanism in the Rajmahal flood basalt province, northeastern India. Chem Geol 121(1–4):73–90

Ball V (1877) Geology of Rajmahal Hills. Mem Geol Surv India 13(2):1–94

Banerjee M, D'Rozario A (1988) Biostratigraphy and environment of deposition in the Lower Gondwana sediments of Chuperbhita Coalfield, Rajmahal Hills. J Palaeontol Soc India 33:73–90

Banerjee M, D'Rozario A (1990) Palynostratigraphic correlation of Lower Gondwana sediments in the Chuperbhita and Hura basins, Rajmahal Hills, Eastern India. Rev Palaeobot Palynol 65:239–255

Banerji J (1987) First record of *Culcitites* Appert from India and its significance. Palaeobotanist 36:201–204. https://doi.org/10.54991/jop.1987.1578

Banerji J (1989) Plant fossils from Dubrajpur Formation, Bihar and their significance in stratigraphy. In: Jain KP, Tiwari RS (eds) Proceedings of a symposium on vistas in Indian palaeobotany. Palaeobotanist 38:122–130. https://doi.org/10.54991/jop.1989.1646

Banerji J (1992) Osmundaceous fronds in Lower Cretaceous beds at Chunkhal, Rajmahal, Hills, Bihar, India. Alcheringa 16(1):1–13

Banerji J (1993) Plant fossils from Chunakhal, Rajmahal Hills, Bihar. Geophytology 23(1):71–80

Banerji J (1994) Upper Gondwana megafloral succession of Rajmahal Basin, India. In: Mitra ND, Acharya SK, Chandra PR, Ghosh A, Ghosh S, Guha PKS (eds) Proceedings of the 9th international Gondwana symposium, Hyderabad, India, pp 269–281

Banerji J (1995a) Megafloral assemblages from two new localities of Rajmahal Formation. Geophytology 24(2):205–208

Banerji J (1995b) Recent records of Mesozoic ferns from Rajmahal Basin. Indian Fern J 12(1–2):70–78

Banerji J (1996) Early Cretaceous megaflora from Muralipahar, Rajmahal Basin, India. Geophytology 25:41–46

Banerji J (1997) Significant occurrence of angiospermous remains in the Early Cretaceous flora of Sonajori, Rajmahal Basin, Bihar. In: Abstracts of the national symposium on biodiversity, conservations and evolution of plants, Allahabad, India

Banerji J (2000a) Occurrence of angiosperm remains in an Early Cretaceous intertrarappean bed, Rajmahal Basin, India. Cretac Res 21:781–784. https://doi.org/10.1006/cres.2000.0225

Banerji J (2000b) Megafloral diversity of the Upper Gondwana sequence of the Rajmahal Basin, India. J Afr Earth Sci 31(1):133–144. https://doi.org/10.1016/S0899-5362(00)00078-6

Banerji J (2004) Evidence of insect–plant interactions from the Upper Gondwana Sequence (Lower Cretaceous) in the Rajmahal Basin, India. Gondwana Res 7(1):205–210. https://doi.org/10.1016/S1342-937X(05)70320-8

Banerji J, Ghosh AK (2002) Mutualism/symbiosis from the Early Cretaceous (intertrappeans) of Rajmahal Basin, Jharkhand, India. Curr Sci 83(9):1073–1074

Banerji J, Ghosh AK (2005) Diversity of early Cretaceous mega flora from Hiranduba locality of Rajmahal Basin, Jharkhand. Geophytology 35:7–13

Banerji J, Ghosh AK (2006) *Podospermum* gen. et sp. nov., an *Acmopyle*-like dispersed silicified ovule/seed from Lower Cretaceous intertrappean beds of the Rajmahal Basin, India. Cretac Res 27(5):707–711. https://doi.org/10.1016/j.cretres.2006.03.003

Banerji J, Ghosh AK (2008) *Podocarpospermum*, a replacement name for the *Acmopyle*-like ovule/seed *Podospermum* Banerji and Ghosh from Lower Cretaceous intertrappean beds of the Rajmahal Basin, India. Cretac Res 29(4):561–562. https://doi.org/10.1016/j.cretres.2008.01.003

Banerji J, Jana BN (1998) Early Cretaceous megafossils from Balidih Rajmahal Basin, India. Geophytology 27(1–2):33–36

Banerji J, Jana BN (2000) Early Cretaceous megaflora from Bartala Hill, Rajmahal Basin India. Palaeobotanist 49(1):51–56. https://doi.org/10.54991/jop.2000.131

Banerji J, Jana BN (2003) Petrified araucarian remains from Sonajori, Rajmahal Basin, India. Palaeobotanist 52(1–3):55–62. https://doi.org/10.54991/jop.2003.1752

Barnard PDW (1968) A new species of *Masculostrobus* Seward producing *Classopolllis* pollen from the Jurassic of Iran. Bot J Linn Soc 61(384):167–176

Bharadwaj DC (1952) On a new species of *Taxoxoylon* Unger from the Jurassic of Rajmahal Hills, Bihar, India. Lloydia 15(4):234–240

Bharadwaj DC (1953) Jurassic woods of the Rajmahal Hills, Bihar. Palaeobotanist 2:59–70. https://doi.org/10.54991/jop.1953.433

Bohra DR, Sharma BD (1978) Petrified schizaeaceous sporangia from the Jurassic of Rajmahal Hills, India. Ameghiniana 15(3–4):321–326

Bohra DR, Sharma BD (1979) Jurassic petrified filician plants from the Rajmahal Hills, India. Ann Bot, Lond 44(6):749–756. https://doi.org/10.1093/oxfordjournals.aob.a085789

Bohra DR, Sharma BD (1980a) *Arauamyelon pakurense* gen. et sp. nov. from the Jurassic of Rajmahal Hills, India. Bull Natil Sci Mus Tokyo Ser C (geol Paleontol) 6:93–96

Bohra DR, Sharma BD (1980b) *Araucarites mittrii* sp. nov., a petrified megastrobillus from the Rajmahal Hills, India. Ameghiniana 17:3–9

Bose MN (1953a) *Ptilophyllum amarjolense* sp. nov. from the Rajmahal Hills, India. Proc Natl Inst Sci India 19(5):605–612

Bose MN (1953b) *Bucklandia sahnii* sp. nov. from the Jurassic of the Rajmahal Hills, Bihar. Palaeobotanist 2:41–50. https://doi.org/10.54991/jop.1953.430

Bose MN (1967) *Weltrichia singhii* n. sp. from the Rajmahal Hills, Bihar. Curr Sci 36(2):48–49

Bose MN (1968) A new species of *Williamsonia* from the Rajmahal Hills, India. Bot J Linn Soc 61(384):121–127. https://doi.org/10.1111/j.1095-8339.1968.tb00109.x

Bose MN, Maheshwari HK (1974) Mesozoic conifers. In: Surange KR, Lakahnpal RN, Bhardwaj DC (eds) Aspects and appraisal of Indian palaeobotany. Birbal Sahni Institute of Palaeobotany, Lucknow, pp 212–223

Bose MN, Pal PK (1982) *Hepaticites pantii* sp. nov. from the Rajmahal Hills, Bihar. In: Nautiya DD (ed) Studies on living and fossil plants (Pant Commemorations Volume). Phyta 1:31–33

Bose MN, Pal PK, Harris TM (1985) The *Pentoxylon* plant. Philos Trans R Soc Lond Ser 310(1142):77–108. https://doi.org/10.1098/rstb.1985.0100

Bose MN, Roy SK (1968) On the occurrence of *Pachypteris* in Jabalpur Series of India. Palaeobotanist 16(1):1–8

Bose MN, Sah SCD (1954) On *Sahnioxylon rajmahalense*, a new name for *Homoxylon rajmahalense* Sahni and *S. andrewsii*, a new species of *Sahnioxylon* from Amrapara in the Rajmahal Hills, Bihar. Palaeobotanist 3:1–8. https://doi.org/10.54991/jop.1954.443

Bose MN, Sah SCD (1967) Some pteridophytic remains from the Rajmahal Hills, Bihar. Palaeobotanist 16(1–3):12–28

Cantril DJ, Webb JA (1987) A reappraisal of *Phyllopteroides* Medwell (Osmundaceae) and its stratigraphic significance in the Lower Cretaceous of eastern Australia. Alcheringa 11(1):59–85. https://doi.org/10.1080/03115518708618980

Césari SN, Parica CA, Remesal MB et al (1998) First evidence of Pentoxylales in Antarctica. Cretac Res 19(6):733–743

Chinnappa C, Kavali PS, Rajanikanth A et al (2021) Early Cretaceous flora from the east coast sedimentary basins of India: their chronostratigraphic and palaeobiogeographic significance. In: Mesozoic stratigraphy of India: a multi-proxy approach, pp 469–528

Crane PR, Dilcher DL (1984) *Lesqueria*: an early angiosperm fruiting axis from the mid-Cretaceous. Ann Missouri Bot Gard 71(2):384–402. https://doi.org/10.2307/2399031

Day RW (1964) Stratigraphy of the Roma – Wallumbilla area. Geol Surv Queensland Publ 318:1–23

Douglas JG (1969) The Mesozoic floras of Victoria Part 1&2. Mem Geol Surv Victoria 28:1–310.

Drinnan AN, Chambers TC (1985) A reassessment of *Taeniopteris daintreei* from the Victorian Early Cretaceous: a member of the Pentoxylales and a significant Gondwanaland plant. Austr J Bot 33:89–100

Drinnan AN, Chambers TC (1986) Flora of the Lower Cretaceous Koonwarra fossil beds (Korumburra Group), South Gippsland, Victoria. Mem Assoc Australas Palaeontol 3:79–110

Du Toit AL (1927) The fossil flora of the Upper Karroo beds. Ann South African Mus 22(2):289–420

Edwards WN (1933) *Osmundites* from Central Australia. Ann Mag Nat Hist Ser 10, 11:661–663

Feist-Burkhardt S, Monteil E (1997) Dinoflagellate cysts from the Bajocian stratotype (Calvados, Normandy, Western France). Kystes de dinoflagellés du stratotype du Bajocien (Calvados, Normandie, France). Bull Cent de rech Exp Prod Elf Aquint 21(1):31–105

Feistmantel O (1876a) On some fossil plants from the Damuda Series in the Raniganj Coalfield, collected by J Wood Mason. J Asiat Soc Bengal 45:329–380

Feistmantel O (1876b) Notes on the age of fossil floras of India. Rec Geol Surv India 9(2):28–42

Feistmantel O (1877a) The fossil flora of the Upper Gonswanas. Jurassic (Liassic) flora of the Rajmahal Group, in the Rajmahal Hills. Mem Geol Soc India, Palaeont Indica, Ser 2 1(2):53–162

Feistmantel O (1877b) Flora of the Jabalpur group (Upper Gondwanas) in the Son Narbada region. Mem Geol Surv India. Pal Ind Ser 11 2(2):81–105

Feistmantel O (1879) The fossil flora of the Upper Gondwana Outliers on the Madras Coast. Mem Geolo Surv India, Palaeontol indica Ser. II 1:191–224

Feistmantel O (1881) Notes on some Rajmahal Plants. Rec Geol Surv India 14(1):148–152

Ganju PN (1946) On a collection of Jurassic plants from the Rajmahal Hills, Behar. J Indian Bot Soc (M.O.P. Iyengar Commem vol), pp 51–58

Ganju PN (1947a) On *Beaniopsis rajmahalensis* gen et. sp. nov., a new type of gymnosperm female fructification from the Jurassic of Behar. Proc Indian Acad Sci 25B(5):95–104

Ganju PN (1947b) *Ontheanthus polyandra* gen et. sp. nov., a new type of fossil gymnosperm male fructification from the Ramahal Hills. Proc Indian Acad Sci 25B(5):105–118

Ghosh AK, Kar R, Chatterjee R et al (2018) Two new conifers from the Early Cretaceous of the Ramahal Basin, India: implications on palaeoecology and palaeogeography. Ameghiniana 55(4):437–450. https://doi.org/10.5710/AMGH.17.02.2018.3124

Gopal V, Jacob C, Jacob K (1957). Stratigraphy and palaeontology of the upper Gondwana of Ramnad District on the East Coast. Rec Geo Sur India 84:477–496

Goswami S, Saxena A, Singh KJ et al (2018) An appraisal of the Permian palaeobiodiversity and geology of the Ib River Basin, eastern coastal area, India. J Asian Earth Sci 157:283–301

Gupta KM (1954) Notes on some Jurassic plants from the Rajmahal Hills, Bihar, India. Palaeobotanist 3:18–25. https://doi.org/10.54991/jop.1954.445

Gupta KM (1955) *Hausmannia indica* sp. nov. Gupta a Dipteridaceous leaf from the Jurassic of Rajmahal Hills, Bihar (India). Proc Natl Acad Sci India 21B(3):147–148

Gupta KM (1966) Significance of the study of cycadean fronds from the Upper Gondwanas of India (Rajmahal Hills). In: Proceedings of the symposium on floristics and stratigraphy of Gondwanaland, Lucknow, pp 137–142

Halle TG (1913) The Mesozoic flora of Grahamland. Wiss Ergeb Schwed Südpolar - Exped 3(14):123

Harland WB, Armstrong RL, Cox AV et al (1990) A geological time scale 1989, vol 16. Cambridge University Press, Cambridge, p 263

Harris TM (1962) The occurrence of the fructification *Carnoconites* in New Zealand. Trans R Soc N z, Geol 1(4):17–27

Hobson GV (1928) Progress Report (unpbs), Geological Survey of India, partly published in general report. Rec Geol Surv India 62:145–147

Horn af Rantzien H (1957) Nitellaceous Charophyte gyrogonites in the Rajmahal series (Upper Gondwana) of India with notes on the flora and stratigraphy. Stock Contrib Geol 1:1–29

Hsü J, Bose MN (1952) Further information on *Homoxylon rajmahalense* Sahni. J Indian Bot Soc 32(1, 2):1–12

Jacob K (1937) On *Protocyathe arajmahalense* sp. nov., a Cyatheaceae tree-fern, with notes on the geological distribution of the Cyatheaceae. Proc Indian Acad Sci 6B(2):73–90

Jacob K (1938) Fossil plants from Sakrigalighat in the Rajmahal Hills, with remarks on the age of the beds. In: Proceedings of 25th Indian Science Congress, Calcutta, vol 3, pp 152–153

Jacob K (1943) *Tinpaharia* gen. nov. J Ind Bot Soc 22:175–177

Jacob K (1950) The stele in *Tinpaharia*, a genus of petrified Mesozoic ferns from the Rajamahl Hills, Bihar, India. In: Proc 7th Intern Bot Cong, Stockholm, pp 564–566

Jacob K, Shukla BN (1955) Jurassic plants from the Saighan series of northern Afghanistan and their palaeo-climatological and palaeo-geographical significance. Mem Geol Surv India Palaeont Indica (ns) 33(2):1–64

Jain KP (1962) *Fascivarioxylonmehtae* gen.et sp. nov., a new petrified cycadean wood from the Rajmahal Hills, Bihar, India. Palaeobotanist 11(3):138–143. https://doi.org/10.54991/jop.1962.633

Jain KP (1964) A new species of *Mesembrioxylon, M. rajmahalense* from the Rajmahal Hills, Bihar, India. J Palaeosci 13(2):153–154. https://doi.org/10.54991/jop.1964.691

Kent RW, Pringle MS, Müller RD et al (2002) ^{40}Ar/^{39}Ar geochronology of the Rajmahal Basalts, India, and their relationship to the Kerguelen Plateau. J Petrol 43(7):1141–1153. https://doi.org/10.1093/petrology/43.7.1141

Kidston R, Gwynne-Vaughan DT (1907) On the fossil Osmundaceae. Phil Trans Roy Soc Edinburgh 45(27):759–780

Kräusel R, Jain KP (1964) New fossil coniferous woods from the Rajmahal Hills, Bihar, India. Palaeobotanist 12(1–3):59–67. https://doi.org/10.54991/jop.1963.643

Maheshwari HK (1966) Studies in the Glossopteris flora of India-29 Miospore assemblage from the Lower Gondwana exposures along Bansloi River in Rajmahal Hills, Bihar. Palaeobotanist 15(3):258–280. https://doi.org/10.54991/jop.1966.770

Maheshwari HK (1986) *Thinnfeldia indica* Feistmantel and associated plant fossils from the Tiruchirapalli District, Talmil Nadu. Palaeobotanist 35:13–21

Maheshwari HK, Bajpai U (1992) Ginkgophyte leaves from Permian Gondwana of Rajmahal Basin, India. Palaeontogr Abt B 224:131–149

Maheshwari HK, Prakash G (1964) Studies in the Glossopteris flora of India – 21. Plant megafossils from the Lower Gondwana exposures along Bansloi River in Rajmahal Hills, Bihar. Palaeobotanist 13(1–3):115–128. https://doi.org/10.54991/jop.1964.687

Mamgain VD, Sastry MVA, Subhraraman IV (1973) Report of ammonoides from Gondwana plant beds at Terani, Tiruchirapalli District, Tamil Nadu. J Geol Soc India 14:198–200

Marshal P (1926) A new species of *Osmundites* from Kawhai, New Zealand. Trans Proc New Zealand Inst 56:210–213

Mathews RP, Pillai SSK, Manoj MC et al (2020) Palaeoenvironmental reconstruction and evidence of marine influence in Permian coal-bearing sequence from Lalmatia Coalmine (Rajmahal Basin), Jharkhand, India: a multi-proxy approach. Int J Coal Geol 224:103485. https://doi.org/10.1016/j.coal.2020.103485

McClelland J (1850) General remark II. Geognosy-III description of plates and collections. In: Report of the Geological Survey of India, for the Season 1848–1849. Military Orphan Press, Calcutta

McDougall I, McElhinny MW (1970) The Rajmahal Traps of India: K-Ar ages and palaeomagnetism. Earth Planet Sci Lett 9(4):371–378

Meeuse ADJ (1961) The Pentoxyleae and the origin of the Monocotyledons. Proc Koninkl Akad Wetensch Amsterdam Ser C 64:543–559

Mehta KR, Sud JD (1953) On some Ginkgoalean leaf impressions from the Rajmahal Hills. Palaeobotanist 2:51–54. https://doi.org/10.54991/jop.1953.432

Misra KS, Saxena RS (1964) A new fossil fish, *Jhingrania roonwali*, from the Rajmahal Hills, India. J Palaeontol Soc India 4:30–32

Murthy S, Saxena A, Chakraborti B (2020a) Palynostratigraphy of Permian and Mesozoic subsurface sediments of Brahmani coalfield, Rajmahal Basin, India. J Palaeontol Soc India 65(2):149–161

Murthy S, Kavali PS, Di Pasquo M (2020b) Late Pennsylvanian and Early Cisuralian palynofloras from the Rajmahal Basin, eastern India, and their chronological significance. Hist Biol 32(2):143–159

Murthy S, Mendhe VA, Dieter U et al (2021) Palaeobotanical and biomarker evidence for Early Permian (Artinskian) wildfire in the Rajmahal Basin, India. J Palaeogeogr 10(1):1–21. https://doi.org/10.1186/s42501-021-00084-2

Oldham T (1854) A brief summary of the results of an examination of the Rajmahal Hills made during the cold season of 1852–53. J Asiat Soc Bengal 23:263–276

Oldham T (1860) On the geological relations and probably geological age of the several systems of rocks in Central India and Bengal. Mem Geol Surv India 2:299–335

Oldham T, Morris J (1862) Fossil flora of the Rajmahal Series in the Rajmahal Hills. Mem Geol Surv India, Palaeontol Indica Ser II 1:1–52

Pal PK, Arefin MB, Basu M (2009) Addition to the knowledge of megefloral diversity from the Rajmahal Formation. Palaeobotanist 58(1–3):83–91. https://doi.org/10.54991/jop.2009.85

Pandey J, Dave A (1998) Stratigraphy of Indian petroliferous basins. In: Presidential address-XVI Indian Colloq Micropalaeontol and Strat. National Institute of Oceanography, Goa, India, pp 15–16

Pascoe Edwin H (1968) A manual of the Geology of India and Burma, vol 2, 3rd edn. Geological Survey India, Gov India Press, Calcutta, pp 485–1338 (reprinted)

Pillai SSK, Mathews RP, Murthy S et al (2020) Palaeofloral investigation based on morphotaxonomy, palynomorphs, biomarkers and stable isotope from Lalmatia Coal Mine of Rajmahal Lower Gondwana Basin, Godda District, Jharkhand: an inclusive empirical study. J Geol Soc India 96:43–57

Prakash N (2000) Floral diversity of two fossil sites (Dudhkol and Sitalpur) of Rajmahal Formation, Bihar, India. Palaeobotanist 49(1–3):57–64. https://doi.org/10.54991/jop.2000.601

Raja Rao CS (1987) Rajmahal Coalfields: coalfields of India—coal resources of Bihar (excluding Dhanbad District). Bull Geol Surv India Ser A 45:300–322

Ramaswamy SK (1958) Progress report of Geological Survey of India for field season 1954 on Rajmahal Hills, Santhal Paragana, Bihar. Rec Geol Surv India 88(1):20–21

Rao AR (1936) Winged pollen from the Jurassic of India. In: Proc 23rd Indian Sci Cong

Rao AR (1943a) *Nipaniostrobus*, a new genus of *Dacrydium*-like seed bearing cones, and other silicified plants from the Rajmahal Series. Proc Natl Acad Sci 13:113–132

Rao AR (1943b) Jurassic spores and sporangia from the Rajmahal Hills, Bihar. Proc Natl Acad Sci, India 13:181–197

Rao AR (1947) *Nipanioruha granthia* gen. et sp. nov.—a new petrified coniferous shoot from the Rajmahal Hills, Bihar. J Indian Bot Soc, M.O.P. Iyengar Commem vol, pp 389–397

Rao AR (1949) The megastrobilus of *Nipanioruha granthia* Rao. Curr Sci 18:447–448

Rao AR (1950) Two hitherto unreported plant fossils from the Rajmahal Hills, Bihar. Curr Sci 19(12):378–380

Rao AR (1953) Some observations on the Rajmahal flora. Palaeobotanist 2:25–28. https://doi.org/10.54991/jop.1953.427

Rao AR (1963a) The Podocarpaceae in India. Mem Indian Bot Soc 4:150–157

Rao AR (1963b) *Stachyotaxus sampathkumarani* sp. nov. from Onthea in the Rajmahal Hills, Bihar. Palaeobotanist 12(1–3):217–219

Rao AR (1972) The Jurassic flora of the Rajmahal Hills. In: 18th Sir Albert Charles Seward memorial lecture, 1970. Birbal Sahni Institute of Palaeobotany, Lucknow, pp 1–14

Rao AR, Achuthan V (1967) Further contribution to our knowledge of the anatomy of *Ptilophyllum*. Palaeobotanist 16(3):249–257. https://doi.org/10.54991/jop.1967.525

Rao AR, Bose MN (1970) *Podostrobus* gen. nov. a petrified podocarpaceous male cone from the Rajmahal Hills, India. Palaeobotanist 19(1–3):83–85. https://doi.org/10.54991/jop.1970.856

Rozefelds AC (1982) *Pentoxylon*: An unusual fossil tree. In: Australian Gem and Treasure Hunter, pp 10–11

Sah SCD (1952) Occurrence of Ginkgoales in the Rajmahal Series of Behar. Curr Sci 21(5):129–130

Sah SCD (1953) On some species of *Ginkgoites* from the Jurassic of the Rajmahal Hills, Bihar. Palaeobotanist 2:55–58. https://doi.org/10.54991/jop.1953.431

Sah SCD (1957) *Coniferocaulon latisulcatum* sp. nov. from the Rajmahal Hills, Bihar with remarks on the affinities of the genus. Palaeobotanist 6(1–2):71–76. https://doi.org/10.54991/jop.1957.558

Sah SCD (1964) Palaeobotanical evidence on the age of the Khatangi beds (? Dubrajpur), in the Rajmahal Hills, Bihar. Palaeobotanist 13(1–3):218–221. https://doi.org/10.54991/jop.1964.697

Sah SCD, Jain KP (1964) *Ginkgoites rajmahalensis* sp. nov. from the Rajmahal Hills, Bihar. Palaeobotanist 13(1–3):155–157. https://doi.org/10.54991/jop.1964.692

Sah SCD, Jain KP (1965) Jurassic spores and pollen grains from the Rajmahal Hills, Bihar, India: with a discussion on the age of the Rajmahal intertrappean beds. Palaeobotanist 13(1–3):264–290. https://doi.org/10.54991/jop.1964.702

Sah SCD, Shah SC (1974) Dubrajpur Formation and its biozones. In: Surange KR, Lakhanpal RN, Bharadwaj DC (eds) Aspect and appraisal of Indian palaeobotany. Birbal Sahni Institute of Palaeobotany, Lucknow, pp 447–451

Sah SCD, Sukh-Dev (1957) *Thinnfeldia chunakhalensis* sp. nov. from the Jurassic of the Rajmahal Hills, Bihar. Palaeobotanist 6(1–2):22–24. https://doi.org/10.54991/jop.1957.550

Sahni B (1920) Petrified plant remains from the Queensland and Tertiary Mesozoic Formations. Queensland Geol Surv Publ No 267

Sahni B (1928) Revisions of Indian fossil plants. Pt. 1. Coniferales (a. Impression and incrustations). Mem Geol Surv India, Palaeontol Indica, N Ser 11:1–49

Sahni B (1931) Revisions of Indian fossil plants. Part II. Coniferales (b. Petrifactions). Mem Geol Surv India, Palaeontol Indica, N Ser 11:53–124

Sahni B (1932a) *Homoxylon rajmahalense* gen. et sp. nov., a fossil angiospermous wood, devoid of vessels, from the Rajmahal Hills, Behar. Mem Geol Surv India, Palaeontol Indica, N Ser 20(2):1–19

Sahni B (1932b) A petrified *Williamsonia* (*W. sewardiana* sp. nov.) from the Rajmahal Hills, India. Mem Geol Surv India, Palaeontol Indica, N Ser 20(3):1–19

Sahni B (1948) The Pentoxyleae—a new group of Jurassic gymnosperms from the Rajmahal Hills of India. Bot Gaz 110(1):47–80

Sahni B, Rao AR (1933) On some Jurassic plants from the Rajmahal Hills, Bihar. J Proc Asiat Soc Bengal, (N Ser) 27(2):183–208

Sarbadhikari TR (1968) Petrology of the north-eastern portion of the Rajmahal Traps. Quat J Geol, Miner Metallurg Soc India 40:151–171

Sarkar SN (1972) Present status of Precambrian Geochronology of Peninsular India. Int Geol Cong, Sec 3 (1972):262

Sastry MVA, Acharyya SK, Shah SC et al (1977) Stratigraphic Lexicon of Gondwana Formation of India. Geol Surv India, Misc Publ 36:1–170

Saxena A, Singh KJ, Goswami S (2014) Advent and decline of the genus *Glossopteris* Brongniart in the Talcher Coalfield, Mahanadi Basin, Odisha, India. Palaeobotanist 63(1–2):157–168. https://doi.org/10.54991/jop.2014.329

Saxena A, Singh KJ, Cleal CJ et al (2019) Development of the Glossopteris flora and its end Permian demise in the Tatapani-Ramkola Coalfield, Son-Mahanadi Basin, India. Geol J 54(4):2472–2494

Schmid EE, Schleiden MJ (1855) Über de Natur der Kieselhölzer. Jena 36:1–4

Sen Gupta S (1984) On the stratigraphy of Khatangi Hill Section, Rajmallal Hills, Bihar, India. Rec Geol Surv India 114(2):63–67

Sen Gupta S (1985) Dubraipur Formation and its type section from Rajmahal Hills, India. Rec Geol Surv India 113(3):99–105

Sen Gupta S (1988) Upper Gondwana stratigraphy and palaeobotany of Rajmahal Hills, Bihar (India). Geol Surv India, Palaeontol Indica Monogr, N Ser 48:1–182

Seward AC (1911) The Jurassic flora of Sutherland. Trans R Soc Edinburgh 47(4):643–709

Seward AC (1917) Fossil plants, III. Cambridge Univ Press, Cambridge, p 656

Seward AC (1919) Fossil plants: a text-book for students of botany and geology. Vol. 4. Ginkgoales, Coniferales, Gnetales. Cambridge Biol Ser. Cambridge Univ Press, Cambridge, p 543

Seward AC, Sahni A (1920) Indian Gondwana plants: a revision. Mem Geol Surv India, Palaeontol Indica, N Ser 7(1):1–54

Sharma BD (1967) Investigations on the Jurassic flora of Rajmahal Hills, India, 3. A review of the genus *Ptilophyllum* of Morris, with description of two new species from Amarjola in the Rajmahal Hills. Palaeontogr B 120:139–150

Sharma BD (1969a) On the Pentoxyleae remains from Amarjola in the Rajmahal Hills, India. Ameghiniana 6(1):50–56

Sharma BD (1969b) *Bucklandia dichotoma* sp. nov. from the Middle Jurassic of Rajmahal Hills, India. Ameghiniana 6(4):303–308

Sharma BD (1969c) Further observation on *Williamsonia santalensis* Sitholey and Bose with description of a new species. Palaeontogr B 125(4–6):93–103

Sharma BD (1969d) *Guptioxylon amarjolense* gen. et sp. nov. from Amarjola in the Rajmahal Hills, India. Palaeontogr B 126:145–153

Sharma BD (1969e) On a collection of fossil ferns from Dhokuti in the Rajmahal Hills, India. Palaeontogr B 128:56–63

Sharma BD (1971a) Further studies on fossil pteridophytic fronds collected from the middle Jurassic of Dhokuti in the Rajmahal Hills, India. Palaeontogr B 133:61–71

Sharma BD (1971b) Further observations on *Pentoxylon sahnii* Srivastava from the Jurassic of Amarjola in the Rajmahal Hills, India. Palaeobotanist 20(1–3):216–220. https://doi.org/10.54991/jop.1971.899

Sharma BD (1972a) *Guptioxylon endocentrica* sp. nov. du Jurassique D' Amarjola daus les Rajmahal Hills (Inde). Extr, Bull Mens Soci Linn Lyon 21:114–120

Sharma BD (1972b) *Purioxylon jurassica* gen. et sp. nov. from Amarjola in the Rajmahal Hills, India. In: Advances in plant morphology. Puri Commem vol, pp 233–242
Sharma BD (1973a) On the anatomy of dwarf shoots of *Pentoxylon sahnii* Srivastava collected from Amarjola in the Rajmahal Hills, India. Acta Palaeobot 14(3):195–204
Sharma BD (1973b) Anatomy of Osmundaceous rhizomes collected from the middle Jurassic of Amarjola in the Rajmahal Hills, India. Palaeontogr B 140:151–160
Sharma BD (1974a) *Pentoxylon* and allied fossil woods from the Jurassic of Amarjola in the Rajmahal Hills, India. Bull Natl Sci Mus, Tokyo 17:75–85
Sharma BD (1974b) Observations on branching in *Pentoxylon sahnii* Srivastava. Bull Natl Sci Mus, Tokyo 17:315–324
Sharma BD (1979) Further observations on the dwarf shoots of *Pentoxylon sahnii* Srivastava collected from the Jurassic of Rajmahal Hills, India. Acta Palaeobot 20:129–136
Sharma BD (1989) Possible occurrence of polyembryony in Pentoxyleae. Phytomorphology 39:199–201
Sharma BD (1996) The pentoxyleae: an overview. Palaeobotanist 45:50–56. https://doi.org/10.54991/jop.1996.1218
Sharma BD (1997) An early angiosperm fructification resembling *Lesqueria* Crane and Dilcher from the Rajmahal Hills, India. Phytomorphology 47(3):305–310
Sharma BD (2017) Occurrence of a leafy Jungermanniales in the Mesozoic of the Rajmahal Hills, India. Palaeobotanist 66(1):81–83. https://doi.org/10.54991/jop.2017.281
Sharma BD, Bohra DR (1976a) A new assemblage of fossil plants from the Jurassic of Rajmahal Hills, India. Geobios, France 9(2):111–123
Sharma BD, Bohra DR (1976b) A petrified matoniaceous rhizome from Amarjola in the Rajmahal Hills, India. Palaeobotanist 25(1–3):457–460. https://doi.org/10.54991/jop.1976.1031
Sharma BD, Bohra DR (1987) The fossil flora of the Rajmahal Hills, recent investigations with critical remarks. Geol Surv India Spec Publ 11(1):253–260
Sharma BD, Bohra DR, Suthar OP (1984) Two isolated petrified sporangia from the Rajmahal Hills, India. J Earth Sci 11(1):87–91
Sharma BD, Bohra DR, Suthar OP (2001) Some interesting plant fossils from the Mesozoic rocks of the Rajmahal Hills, India. Palaeobotanist 50(1–3):207–212. https://doi.org/10.54991/jop.2001.1823
Sharma BD, Bohra DR, Suthar OP (2002) Reinterpretation of an extinct taxon *Sporangiocerousnipanica* Sharma et al. from the Rajmahal Hills, India. Palaeobotanist 51:31–36. https://doi.org/10.54991/jop.2002.1731
Sharma BD, Bohra DR, Suthar OP et al (2015) Some permineralized plant fossils from the Early Cretaceous sediments of Rajmahal Hills, Jharkhand India. Geophytology 45(1):41–46
Sharma BD, Bohra DR, Suthar OP et al (2017) Saprophytic fossils fungi from the Early Cretaceous sediments of the Rajmahal Hills, Jharkhand, India. Palaeobotanist 66(2):217–222. https://doi.org/10.54991/jop.2017.287
Sharma BD, Harsh R (1994) Polysiphonous algae from Mesozoic non-marine deposits of the Ramahal Hills, India. Phytomorphology 44:261–264
Sharma BD, Harsh R (2020) An extinct lichen from the Mesozoic of the Rajmahal Hills, Jharkhand, India. Nelumbo 62(2):264–265. https://doi.org/10.20324/nelumbo/v62/2020/155471
Sharma BD, Surana AC, Singh AP (1971) Jurassic plants from Amarjola in the Rajmahal Hills. J Palaentological Soc India 16:27–34
Sharma BD, Suthar OP (1989) Algal symbiotic gymnospermous roots from the Jurassic of Rajmahal Hills, India. Phytomorphology 39:161–163
Sharma BD, Tripathi RP (1997) A petrified coenobial alga from the Mesozoic rocks of the Rajmahal Hills, Bihar (India). Phytomorphology 47(4):371–374
Singh IB (1976) Mineralogical evidence for climatic vicissitudes in India during Gondwana times. Geophytology 6(2):174–185
Sinnott EW (1914) On some Jurassic Osmundaceae from New Zealand. Ann Bot 28(109): 471–79

Sitholey RV, Bose MN (1953) *Williamsonia santalensis* sp. nov.—a male fructification from the Rajmahal Series, with remarks on the structure of *Ontheanthus polyandra* Ganju. Palaeobotanist 2:29–39. https://doi.org/10.54991/jop.1953.429

Srivastava BP (1946) Silicified plant remains from the Rajmahal Series of India. Proc Natl Acad Sci, India 15(6):185–211

Srivastava RN (1958) In: Krishnan MS (ed) General report on Geological Survey of India for year 1954. Rec Geol Surv India 88(1):21–23

Srivastava PC, Pant DD (2002) Some new pteridophytic remains from the Permian strata of Rajmahal Hills, India. Palaeobotanist 51(1–3):45–55. https://doi.org/10.54991/jop.2002.1733

Srivastava R, Krassilov VA (2012) Revision of Early Cretaceous angiosperm remains from the Rajmahal Basin, India, with notes on the palaeoecology of the Pentoxylon plant. Cretac Res 33:66–71

Srivastava SC, Maheshwari HK (1974) Palynostratigraphy of the Damuda Group in the Brahmini Coalfield, Rajmahal Hills, Bihar. Geophytology 4(1):35–45

Stopes MC, Fujii K (1910) Studies on the structure and affinities of Cretaceous plants. Phil Trans Roy Soc Edinburgh B 201:1–90

Sukh Dev (1987) Floristic zones in the Mesozoic formations and their relative age. Palaeobotanist 36:161–167. https://doi.org/10.54991/jop.1987.1572

Suryanarayana K (1953) *Mesembrioxylon tirumangalense*, a new species from the Sripermatur Group near Madras. J Indian Bot Soc 32(4):159–164

Suthar OP, Bohra DR, Sharma BD (1988) Petrified isolated gymnospermous seeds from the Jurassic of the Rajmahal Hills, India. Acta Palaeobot 28(1, 2):15–20

Tiwari RS, Kumar P, Tripathi A (1984) Palynodating of Dubrajpur and Intertrappean beds in subsurface strata of northeastern Rajmahal Basin. In: Tiwari RS, Awasthi N, Singh HP, Sharma BB (eds) Proc 5th Indian geophytology conf, Lucknow 1983. Palaeobot Soc, Lucknow, Spec Pub, pp 207–225

Tiwari RS, Tripathi A (1995) Palynological assemblages and absolute age relationship of intertrappean beds in the Rajmahal Basin, India. Cretac Res 16(1):53–72. https://doi.org/10.1006/cres.1995.1004

Tripathi A (1993) Advent of angiospermous pollen in India and its spatial relationship in Gondwanaland. In: Balkema AA (ed) Proc 8th Gondwana symp, Hobart, Tasmania, Australia, 1991, pp 277–289

Tripathi A (2000) Palynological events during Late Triassic–Early Jurassic time in India. Palaeobotanist 49(1–3):399–408. https://doi.org/10.54991/jop.2000.157

Tripathi A (2001) Permian, Jurassic and Early Cretaceous palynofloral assemblages from subsurface sedimentary rocks in Chuperbhita Coalfield, Rajmahal Basin, India. Rev Palaeobot Palynol 113(4):237–259. https://doi.org/10.1016/S0034-6667(00)00062-2

Tripathi A (2008) Palynochronology of Lower Cretaceous volcano-sedimentary succession of the Rajmahal Formation in the Rajmahal Basin, India. Cretac Res 29(5–6):913–924. https://doi.org/10.1016/j.cretres.2008.05.008

Tripathi A, Jana BN, Verma O et al (2013) Early Cretaceous palynomorphs, dinoflagellates and plant megafossils from the Rajmahal Basin, Jharkhand, India. J Palaeontol Soc India 58:125–134

Tripathi A, Ray A (2005) Palynodating of coal bearing sediments in Pachwara Coalfield, Rajmahal Basin, Santhal Pargana, Bihar. In: Majumdar S, Maiti RP, Roy UK, Ghatak G (eds) Proceedings of the workshop on solid fossil fuel: an introspection and perception in Indian context, vol 81. Geol Surv India Spec Pub, pp 135–142

Tripathi A, Ray A (2006) Palynostratigraphy of the Dubrajpur Formation (Early Triassic to Early Cretaceous) of the Rajmahal Basin, India. Palynol 30:133–149

Tripathi A, Tiwari RS (1990) Early Cretaceous angiospermous pollen from the intertrappean beds of Rajmahal Basin, Bihar. Palaeobotanist 39(1–3):50–56. https://doi.org/10.54991/jop.1990.1682

Tripathi A, Tiwari RS, Kumar P (1988) Palynology of the subsurface Mesozoic sediments in Rajmahal Basin, Bihar. Palaeobotanist 37(1–3):367–388. https://doi.org/10.54991/jop.1988.1631

Varma CP (1954) On two new species of *Mesembrioxylon* from the Cretaceous rocks of Trichinopoly District, Madras. Palaeobotanist 3:97–102. https://doi.org/10.54991/jop.1954.458

Vishnu-Mittre (1952) Charophyte remains from the Jurassic of the Rajmahal Hills, Bihar. Curr Sci 21:41–42

Vishnu–Mittre (1953a) Microfossils from the Jurassic of Rajmahal Hills, Bihar. In: Proc 40th Indian Sci Cong Assoc, Lucknow, p 112

Vishnu–Mittre (1953b) A male flower of the Pentoxyleae with remarks on the structure of the female cones of the group. Palaeobotanist 2:75–84. https://doi.org/10.54991/jop.1953.4354

Vishnu-Mittre (1954) Petrified spores and pollen grains from the Jurassic rocks of the Rajmahal Hills, Bihar. Palaeobotanist 3:117–127. https://doi.org/10.54991/jop.1954.461

Vishnu–Mittre (1955a) *Osmundites sahnii* sp. nov., a new species of petrified osmundaceous rhizomes from India. Palaeobotanist 4:113–118. https://doi.org/10.54991/jop.1955.479

Vishnu-Mittre (1955b) *Sporojuglandoidites jurassicus* gen. et sp. nov., a sporomorph from the Jurassic of the Rajmahal Hills, Bihar. Palaeobotanist 4:151–152. https://doi.org/10.54991/jop.1955.481

Vishnu-Mittre (1956a) *Masculostrobus sahnii* sp. nov., a petrified conifer male cone producing three-winged and one- and four-winged abnormal pollen grains from the Jurassic of the Rajmahal Hills, Bihar. Grana palynol 1(2):99–107

Vishnu-Mittre (1956b) Studies on the fossil flora of Nipania, Rajmahal Series, India–Bennettitales. Palaeobotanist 5(2):95–99. https://doi.org/10.54991/jop.1956.493

Vishnu-Mittre (1957a) Studies on the fossil flora of Nipania (Rajmahal Series), India–Pentoxyleae. Palaeobotanist 6(1–2):31–46. https://doi.org/10.54991/jop.1957.553

Vishnu–Mittre (1957b) Studies on the fossil flora of Nipania (Rajmahal Series), Bihar-Coniferales. Palaeobotanist 6(1–2):82–112. https://doi.org/10.54991/jop.1957.560

Vishnu-Mittre (1958) Studies on the fossil flora of Nipania, Rajmahal Series, India—pteridophyta, and general observations on Nipania fossil flora. Palaeobotanist 7(1–2):47–66. https://doi.org/10.54991/jop.1958.570

Wadia DN (1953) Geology of India. London

Willdenow CL (1802) Caroli a Linne Species Plantarum Editioquarta. Berlin 3:1

Palynology of Upper Cretaceous-Early Paleocene Deccan Volcanic Associated Sediments: Implication in Understanding Age, Climate and Depositional Environments

Bandana Samant, Deepali Thakre, and D. M. Mohabey

Abstract The Upper Cretaceous (Maastrichtian) infratrappean (Lameta Formation) and the Upper Cretaceous-early Paleocene intertrappean sedimentary beds associated with Deccan volcanic flows are windows to understand floral biodiversity, age, and paleoclimate during Deccan volcanic activity. Palynological studies of these geographically separated sedimentary beds at different stratigraphic levels show that the initiation of Deccan volcanism in the area brought climatic changes that consequently affected the biota and the depositional environments. The fluvio-lacustrine infratrappean (Lameta Formation) sediments which were deposited in a semiarid-arid climate, were having a mixture of gymnosperm and angiosperm plants. The onset of Deccan volcanic activity changed the climate and depositional environments. During the repose in Deccan volcanic activity, many small closed lakes were formed over the lava surface between the two volcanic flows. The sediments associated with these ponds have a pteridophyte-angiosperm rich flora. This vegetation is distinctly different from that of Lameta sediments. The Late Cretaceous intertrappean sediments shows dominance of palynotaxa such as *Azolla cretacea, Gabonisporis vigourouxii, Aquilapollenites bengalensis, Ariadnaesporites, Farabeipollis* spp. *Jiangsupollis* spp., *Scollardia conferta,* and *Triporoletes reticulatus.* The presence of epiphyllous fungi, algal remains (*Ovoidites*), pteridophytes (*Azolla, Crybelosporites, Gabonisporis*) and mega-plant fossils, collectively suggest the prevalence of warm, humid tropical climatic conditions. Intermittent dry spells are also indicated at many places. In the Deccan volcanic province, the Paleocene is marked by disappearance/scarcity of Maastrichtian taxa. The Paleocene vegetation is distinctly different and it shows some similarity with the Paleocene-Eocene palynoflora of western India.

Present Address:
B. Samant (✉)
Department of Geology, School of Environment and Earth Sciences, Central University of Punjab, Bathinda 151401, Punjab, India
e-mail: bandanabhu@gmail.com

B. Samant · D. Thakre · D. M. Mohabey
Postgraduate Department of Geology, RTM Nagpur University, Nagpur 440 001, Maharashtra, India

B. Samant and D. Thakre (eds.), *Applications of Palynology in Stratigraphy and Climate Studies*, Society of Earth Scientists Series, https://doi.org/10.1007/978-3-031-51877-5_5

Keywords Deccan volcanic province · Lameta Formation · Intertrappean beds · Late Cretaceous · Early Paleocene · Palynofloral change

1 Introduction

The Cretaceous-Paleogene mass extinction event (66 My) is marked by the demise of non-avian dinosaurs that ruled the earth for more than 150 My. Their decline paved the way for the proliferation of mammals. The probable cause of this extinction event is believed to be the voluminous Deccan lava eruptions (3×10^6 km^3, Jay and Widdowson 2008) or the Chicxulub asteroid impact (Schulte et al. 2010) or a combination of both (Renne et al. 2015). Earlier studies suggested that the voluminous outpouring of lava flows was believed to have erupted in three pulses. The first phase erupted in 67–68 Ma, the main volcanic phase (poured 80%) in C29R, and the last spanned across the early Paleocene (Chenet et al. 2007, 2008). However, recent studies suggest that the major pulse of volcanism erupted in < 1 million years, with ~75% magma pouring out post K-Pg boundary (Renne et al. 2015; Schoene et al. 2015).

The impact of these voluminous lava flows is expected to be more pronounced on the Indian subcontinent, as it was closer to the volcanic activity. Here, we review the significance of palynology in understanding the age, paleoclimate, and depositional environment of the Deccan volcanic associated sediments, as well as our understanding of the pattern of floral changes across the Late Cretaceous-early Paleocene.

2 Mega and Microflora from Deccan Volcanic Associated Sediments

Sediments associated with Deccan volcanic sequences are known as "infratrappean (Lameta Formation)", sediments that occur just below the volcanic flows, and "intertrappean", sediments in between the two lava flows. Numerous megaflora and microflora are recorded from the Deccan volcanic associated infra and intertrappean sediments.

2.1 *Flora from the Lameta Sediments*

The fluvio-lacustrine Lameta sediments were deposited as detached outcrops in Maharashtra, Madhya Pradesh, and Gujarat, in six inland basins, namely, (I) Nand-Dongargaon Basin, (II) Jabalpur Basin, (III) Balasinor-Jhabua Basin, (IV) Sagar Basin, (V) Amarkantak-Ambikapur Basin and, (VI) Salbardi-Belkhar Basin (Mohabey 1996; Mankar and Srivastava 2015).

Of these, most of the megaflora is known from the Nand-Dongargaon (N-D) Basin of Maharashtra, and palynoflora from the N-D and Jabalpur basins. The megaflora and microflora rich Lameta sediments of the Nand-Dongargaon basin occupy an area of > 700 km^2 and the thickest sediments are about 20 m thick. These fluvio-lacustrine sediments were deposited in semiarid-arid climates with seasonality (Mohabey and Udhoji 1993; Mohabey et al. 1993; Mohabey and Samant 2005).

Megaflora from the N-D Basin is represented by fossil wood of the family Sapindaceae and Euphorbiaceae (Kar et al. 2004), palm and dicot leaves, *Bracheophyllum,* conifer seeds, *Araucaria,* (Mohabey and Udhoji 1996), Capparidaceae fruit (Dutta and Ambwani 2007), and impressions of the monocotyledon fruit *Viracarpon* (Birader and Bonde 1979; Bonde and Biradar 1981). In addition, permineralized seeds of *Araucaria*, angiosperm flowers, ovules, and seeds have also been recorded from the sauropod coprolites (Mohabey 1996, 2005; Mohabey and Samant 2003; Ghosh et al. 2003; Prasad et al. 2005, 2011; Khosla and Verma 2015). Apart from the N-D Basin, till date there has been a single record of a seed of Boraginaceae from the Lameta sediments of Kheda area (Mathur and Mathur 1985).

In comparison to megaflora, the palynofloral record from the Lameta sediments is distinctly diverse. The palynoflora is recorded in two basins, namely the Jabalpur Basin and the Nand-Dongargaon Basin. The Lameta sediments of Jabalpur Basin comprise of green sandstone, lower limestone, mottled nodular marls, upper limestone, and upper sandstone (Huene and Matley 1933). Palynoflora from the Lameta sediments of Jabalpur is recorded from the lowermost green sandstone/coaly shales and the uppermost upper sandstone with bands of carbonaceous clays (Dogra et al. 1994). Overall, palynoflora from the Jabalpur Basin is divided into two palynozones. The lower one is the *Aquilapollenites indicus* zone, and the upper one is the *Proxapertites crassimurus* zone. The *Aquilapollenites indicus* zone, has dominance of gymnosperms (*Cycadopites* and *Callialasporites*), angiosperms (*Aquilapollenites indicus* and *Palmidites maximus),* and many dinocysts, whereas the *Proxapertites crassimurus* zone has a dominance of angiosperms (*Proxapertites crassimurus, Palmidites plicatus,* and *Polyadopollenites mesozoicus*) and dinocysts (Dogra et al. 2010).

The palynoflora from the N-D Basin (both from the sediments and sauropod coprolites) is represented by pteridophytes (*Azolla, Cyathidites, Gabonisporis*; Mohabey and Samant 2003; Ghosh et al. 2003; Sonkusare et al. 2017), gymnosperms (*Araucariacites, Cycadopites, Classopollis, Podocarpidites, Calliasporites, Trilobosporites,* and *Belmiopsis* (Mohabey and Udhoji 1996; Samant and Mohabey 2005) and angiosperms (*Palmaepollenites, Graminidites, Retimonosulcites, Sapotaceoidaepollenites, Cretacaeiporites,* C*ompositoipollenites*). In addition, algal remains (*Oedogonium, Lecaniella* and *Botryococcus*), fungal spores (*Alternaris, Meliola, Helminthosporum, Aspergillus, Frasnacritetrus*), mycorrhizal fungi (*Collestrochum, Erysiphe* and *Uncinula)* (Sharma et al. 2004; Samant and Mohabey 2009), grass phytoliths, silicified cuticles and tracheids (Prasad et al. 2011; Sonkusare et al. 2017), trichomes, sponge spicules, centric (*Aulacoseira*) and pinnate diatoms (Mohabey 2001; Ambwani et al. 2003), remains of insects and thecamoebians (testate amoebae) (Sonkusare et al. 2017) have also been recorded from the Lameta Formation. Overall, the palynoflora of Lameta sediments of both Jabalpur and N-D basins shows dominance of angiosperm and gymnosperm flora (Dogra et al. 1994; Samant and Mohabey 2009) (Table 1).

Table 1 Palynoflora recorded from terrestrial infra- and intertrappean localities

Lameta Formation		Upper Cretaceous (Maastrichtian) intertrappean beds			
Nand-Dongargaon Basin (30N-29R)	**Jabalpur Basin (29R)**	**Daiwal, Khandala and Ashta Anandvan (29R); Sindhi IT**	**Singpur IT**	**Mohgaon Kalan IT (MKWS-30N + MKFF-30N/29R)**	**Ranipur IT (29R Maastrichtian)**
Samant and Mohabey (2005, 2009, 2014)	Dogra and Singh (1996)	Samant et al. (2005); Samant and Mohabey (2005, 2009, 2014)	Samant et al. (2008)	Kumaran et al. (1997); Kar and Shrinivasan (1998); Thakre et al. (2017)	Mathur and Sharma (1990)
	Megaflora not known	Megaflora present in Sindhi IT	Megaflora present	Megaflora present in MKFF	Dinosaur remains
Dinosaur remains, diatoms & mycorrhizal fungi	Dinosaur remains & dinoflagellates	Diatoms, dinoflagellates, algal & fungal remains			
Pteridophytes	**Pteridophytes**	**Pteridophytes**	**Pteridophytes**	**Pteridophytes**	**Pteridophytes**
Azolla	*Todisporites*	*Azolla cretacea*	*Azolla cretacea*	*Azolla cretacea*	*Azolla cretacea*
	Ariadnaesporites	*Ariadnaesporites* sp.	*Ariadnaesporites* sp.	*Ariadnaesporites* sp.	*Cicatricosisporites dorogensis*
	Cyathidites	*Gabonisporis vigourouxii*	*Gabonisporis vigourouxii*	*Cicatricosisporites dorogensis*	*Gabonisporis vigourouxii*
	Baculatisporites	*G. bacricumulus*	*G. bacricumulus*	*Cicatricosisporites* sp.	*Gabonisporis* sp.
	Dictyophyllidites	*Triporoletes reticulatus*		*Contignisporites* sp.	*Lycopodiumsporites* sp.
	Cicatricosisporites	*Lycopodiumsporites* sp.		*Cyathidites australis*	*Todisporites major*
	Leptolepidites verrucatus	*Lygodiumsporites sp.*		*Cyathidites minor*	
	Contignisporites	*Osmundacidites sp.*		*Gabonisporis vigourouxii*	

(continued)

Table 1 (continued)

Lameta Formation		**Upper Cretaceous (Maastrichtian) intertrappean beds**			
	Lametatriletes	*Triplanosporites sp.*		*Gabonisporis labyrinthus*	
	Lycopodiumsporites umstewensis	*Todisporites* sp.		*Lygodiumsporites lakiensis*	
	Polypodiisporites	*Pediastrum*		*Schizaeoisporites* sp.	
	Arecellites			*Striatriletes* sp.	
	Coptospora			*Todisporites major*	
Gymnosperms	**Gymnosperms**	**Gymnosperms**	**Gymnosperms**	**Gymnosperms**	**Gymnosperms**
Araucariacites sp.	*Araucariacites limbatus*	*Araucariacites* sp.	*Equisetosporites ovatus*	*Cycadopites* sp.	*Equisetosporites* sp.
Callialasporites sp.	*Podocarpidites elipticus*	*Cycadopites* sp.	*Equisetosporites* sp.	*Ephedripites multicostatus*	
Cycadopites sp.	*Baculatisporites*	*Equisetosporites* sp.		*Ephedripites ovatus*	
Classopollis sp.	*Alisporites*	*Podocarpidites* sp.		*Ephedripites rugularis*	
Podocarpidites sp.	*Cedripites*			*Ephedripites* sp.	
	Callialasporites trilobatus				
	C. segmentatus				
	Ginkgopites				
	Laricoidites				
	Cycadopites couperi				
	Ginkgocycadophytus				

(continued)

Table 1 (continued)

Lameta Formation		**Upper Cretaceous (Maastrichtian) intertrappean beds**			
	Ginkgoretectina vastus				
	Psilospora				
	Schizosporis				
	Equisetosporites				
	Narbadapollis indicus				
Angiosperms	**Angiosperms**	**Angiosperms**	**Angiosperms**	**Angiosperms**	**Angiosperms**
Palmaepollenites sp.	*Aquilapollenites indicus*	*Aquilapollenites bengalensis*	*Aquilapollenites bengalensis*	*Aquilapollenites bengalensis*	*Aquilapollenites bengalensis*
Longapertites sp.	*A. andamanensis*	*Proxapertites operculatus*	*Pulcheripollenites cauveriana*	*Crototricolpites reticlavatus*	*Cricotriporites guianensis*
Phytolith of *Graminidites*	*Proxapertites crassimurus*	*Palmidites* sp.	*Racemonocolpites maximus*	*Cranwellia* sp.	*Echitricolporites maristellae*
Retimonosulcites sp.	*P. assamicus*	*Neocouperipollis* sp.	*Scabrastephanocolpites* spp.	*Cricotriporites guianensis*	*Foveotriletes* sp.
Compositoipollenites sp.	*P. granulatus*	*Racemonocolpites maximus*	*spinizonocolpites echinatus*	*Cricotriporites minutiporus*	*Foveotricolporites* sp.
Cretacaeiporites sp.	*Palmaepollenites eocenicus*	*Racemonocolpites* sp.	*Tetracolporites* sp.	*Dipterocarpuspollenites retipilatus*	*Kindopollis* sp.
Periporopollenites sp.	*Palmidites maximus*	*Echitricolpites* sp.	Normapolles group pollen	*Echitricolpites* sp.	*Pulcheripollenites cauveriana*
Sapotaceoidaepollenites sp.	*P. plicatus*	*Tricolpites* sp.		*Echitricolpites longicolpatus*	*Rutihesporites trochuensis*

(continued)

Table 1 (continued)

Lameta Formation		**Upper Cretaceous (Maastrichtian) intertrappean beds**			
	Proteacidites amolosexinous	*Scollardia conferta*		*Echitricolpites* sp.	*Stellatopollis araripensis*
	Liliacidites microreticulatus	*Scabrastephanocolpites* spp.		*Ericipites sahnii*	*Triangulipollis turonicus*
	L. giganticus	Normapolles group pollen		*Farabeipollis minutus*	*Triporoletes reticulatus*
	Lakiapollis			*Inaperturopollenites* sp.	*Triporotetradites* sp.
	Eucommidiites troedssonii			*Intrareticulites brevis*	
	Scollardia			*Jiangsupollis* spp.	
	Polyadopollenits mesozoicus			*Paleosantalaceaepites miocenicus*	
				Proteacidites sp.	
				Proxapertites cursus	
				Proxapertites operculatus	
				Proxapertites sulcatus	
				Psilodiporites sp.	
				Spinizonocolpites spp.	
				Tribrevicolporites eocenicus	
				Tricolpites sp. A	
				Tricolpites levis	
				Tricolpites reticulatus	

(continued)

Table 1 (continued)

Lameta Formation	Upper Cretaceous (Maastrichtian) intertrappean beds				
				Triporotetradites sp.	
Padwar IT (29R)	**Anjar IT (29R)**	**Lakshmipur IT**	**Meghalaya IT**	**Shimbala IT**	**Shankar Lodhi IT**
Prakash et al. (1990); Kar and Shrinivasan (1998)	Dogra et al. (2004)	Samant and Bajpai (2005)	Nandi (1990); Nandi and Chattopadhyay (2002)	Samant et al. (2019)	Samant et al. (2019)
Pteridophytes	**Pteridophytes**	**Pteridophytes**	**Pteridophytes**	**Pteridophytes**	**Pteridophytes**
Azolla cretacea	*Gabonisporis vigourouxii*	*Araucariacites australis*	*Azolla cretacea*	*Azolla cretacea*	*Azolla cretacea*
Foveosporites sp.	*Gabonisporis bacricumulus*	*Callialasporites* sp.	*Araucariacites* sp.	*Concavissimisporites sp.*	*Cyathidites australis*
Gabonisporis vigourouxii	*Gabonisporis* sp.	*Concavissimisporites sp.*	*Cingulatisporites*	*Cyathidites australis*	*Crybelosporites intertrappea*
*Lygodiumsporites*sp.		*Gabonisporis vigourouxii*	*Cyathidites cutchensis*	*Crybelosporites intertrappea*	*Gabonisporis vigourouxii*
Lycopodiumsporites podocarpidites			*Selaginellites* sp.	*Gabonisporis vigourouxii*	*Lygodiumsporites lakiensis*
*Osmundacidites*sp.					
Todisporites sp.					
Triporoletes reticulatus					
Gymnosperms	**Gymnosperms**	**Gymnosperms**	**Gymnosperms**	**Gymnosperms**	**Gymnosperms**
Ephedripites sp.		*Cycadopites* sp.	*Appendicisporites* sp.	*Ephedripites* sp.	*Podocarpidites* sp.
			Cicatricosisporites sp.		*Triplanosporites sinusus*
			Cycadopites sp.		*Aquilapollenites intertrappeus*

(continued)

Table 1 (continued)

Padwar IT (29R)	Anjar IT (29R)	Lakshmipur IT	Meghalaya IT	Shimbala IT	Shankar Lodhi IT
			Ephedripites sp.		*Jiangsupollis striatus*
					Proteacidites reticulatus
					Proxapertites sulcatus
					Spinizonocolpites sp.
					Tricolpites
Angiosperms	**Angiosperms**	**Angiosperms**	**Angiosperms**	**Angiosperms**	**Angiosperms**
Aquilapollenites bengalensis	*Aquilapollenites bengalensis*	*Aquilapollenites bengalensis*	*Aquilapollenites bengalensis*	*Aquilapollenites bengalensis*	
Diporoconia sp.	*Aquilapollenites indicus*	*Liliacidites* sp.	*A.* (*Integricorpus*) *brevocolpatum*	*Crototricolpites karii*	
Palmaepollenites sp.	*Palmidites* sp.	*Longapertites* sp.	*A. indicus*	*Echitricolpites* sp.	
Proxapertites cursus	*Proteacidites reticulatus*	*Proxapertites microreticulatus*	*A. nemuroensis*	*Incrotonipollis neyvelii*	
Proxapertites operculatus	*Proteacidites* sp.	*Proxapertites operculatus*	*A. truncatus*	*Jiangsupollis* sp.	
Spinizonocolpites spp.	*Proxapertites* sp.	*Racemonocolpites* sp.	*Andreisporis*	*Neocouperipollis* sp.	
Tricolpites	*Proxapertites granulatus*	*Retitricolpites vulgaris*	*Araliaceoipollenites*	*Proxapertites operculatus*	
Triporolites sp.	*Racemonocolpites maximus*	*Tricolpites* sp.	*Araucariacites*	*Sparganiaceaepollenites* sp.	
	Tricolpites sp.	*Tricolpites reticulatus*	*Callialasporites*	*Stricolporites striatus*	
	Triorites sp.		*Clavatipollenites*		

(continued)

Table 1 (continued)

Padwar IT (29R)	Anjar IT (29R)	Lakshmipur IT	Meghalaya IT	Shimbala IT	Shankar Lodhi IT
			Complexiopollis		
			Constantinisporis		
			Engelhardtioidites		
			Farabeipollis simplex		
			Farabeipollis longicolpus		
			Farabeipollis longiprocessus		
			Farabeipollis longicolpus		
			Interporopollenites		
			Lakiapollis		
			Lycopodiumsporites		
			Microfoveolatosporites		
			Palmepollenites		
			Polypodiidites		
			Proxapertites sp.		
			Pulcheripollenites		
			Retimonocolpites		
			Retitricolpites		
			Spinizonocolpites sp.		
			Tricolpopollenites		

(continued)

Table 1 (continued)

Padwar IT (29R)	Anjar IT (29R)	Lakshmipur IT	Meghalaya IT	Shimbala IT	Shankar Lodhi IT
			Tricolpites		
			Tricolporopollenites		
			Vacuopollis		
Mahurzhari IT	**Kanhobagholi IT**	**Gaumukh IT (29R)**	**Jamsavli IT**	**Bharudpura IT (30N)**	**Naskal IT**
Samant et al. (2020a, b)	Sonkusare et al. (2021)	Mishra et al. (2021)	Samant et al. (2022)	Mohabey et al. (2018); Manchester et al. (2023)	Singh et al. (2006)
Megaflora present	–	–	Algae & megaflora present	Dinosaur remains & megaflora present	Megaflora present
Pteridophytes	**Pteridophytes**	**Pteridophytes**	**Pteridophytes**	**Pteridophytes**	**Pteridophytes**
Azolla cretacea	*Ariadnaesporites*sp.	*Cicatricosisporites* sp.	*Azolla cretacea*	*Azolla cretacea*	*Azolla cretacea*
Cyathidites australis	*Cyathidites australis*	*Cyathidites australis*	*Cyathidites australis*	*Gabonisporis vigourouxii*	*Gabonisporis vigourouxii*
Gabonisporis vigourouxii	*Gabonisporis vigourouxii*	*Cyathidites minor*	*Crybelosporites intertrappea*		*Ariadnaesporites intermedius*
Lygodiumsporites lakiensis		*Gabonisporis vigourouxii*	*Gabonisporis vigourouxii*		*Triporoletes reticulatus*
*Triplanosporites*sp.		*Gabonisporis bacricumulus*			*Cyathidites* sp.
		Leiotriletes sinuosoides			*Cyathidites minor*
		Polypodiisporites constrictus			*Todisporites major*
		Todidporites major			*Polypodiisporites* sp.
					Maculatospora sp.

(continued)

Table 1 (continued)

Mahurzhari IT	**Kanhobagholi IT**	**Gaumukh IT (29R)**	**Jamsavli IT**	**Bharudpura IT (30N)**	**Naskal IT**
					Minerisporites triradiatus
Gymnosperms	**Gymnosperms**	**Gymnosperms**	**Gymnosperms**	**Gymnosperms**	**Gymnosperms**
	Ephedripites ovatus	*Araucariacites australis*	*Ephedripites*sp.		*Cyacadopites* sp.
	Ephedripites sp.	*Cycadopites follicularis*	*Pteridacidites* sp.		
		Ephedripites lanceolatus			
		Ephedripites multicostatus			
		Ephedripires ovalis			
Angiosperms	**Angiosperms**	**Angiosperms**	**Angiosperms**	**Angiosperms**	**Angiosperms**
Aquilapollenites bengalensis	*Aquilapollenites bengalensis*	*Acanthotricolpites bulbospinosus*	*Aquilapollenites bengalensis*	*Aquilapollenites bengalensis*	*Mulleripollis bolpurensis*
Farabeipollis sp.	*Aquilapollenites intertrappeus*	*Arecipites bellus*	*Jiangsupollis* sp.	*Psilodiporites* sp.	*Mulleripollis* sp.
Foveotriletes sp.	*Echitricolpites* sp.	*Aquilapollenites bengalensis*	*Liliacidites* sp.	*Echitricolpites* sp.	*Lithoparaphysis* sp.
Spinizonocolpites echinatus	*Farabeipollis minutus*	*Aquilapollenites intertrappeus*	*Sparganiaceaepollenites reticulatus*	*Jiangsupollis* sp.	*Proxapertites cursus*
	Farabeipollis deccanensis	*Aquilapollenites indicus*	*Tricolpites magnus*	*Proxapertites operculatus*	*Spinizonocolpites echinatus*
	Incrotonipollis neyvelii	*Cranwellia striata*		*Scollardia conferta*	
	Jiangsupollis minutes	*Crototricolpites protoannemariae*		*Spinizonocolpites* spp.	

(continued)

Table 1 (continued)

Mahurzhari IT	Kanhobagholi IT	Gaumukh IT (29R)	Jamsavli IT	Bharudpura IT (30N)	Naskal IT
	Proxapertites operculatus	*Dipterocarpuspollenites cretacea*		*Triporoletes reticulatus* *Malvacipolloides intertrappea*	
	Spinizonocolpites sp.	*Echitricolpites microechinatus*			
	Stricolporites ovatus	*Euphorbiacites reticulatus*			
	Stricolporites sp.	*Graminidites annulaus*			
		Graminidites gramineoides			
		Icacinoipollenites spinulosus			
		Kielmeyerapollenites eocenicus			
		Longapertites sp.			
		Monocolpites sp.			
		Myrtaceidites eucalyptoids			
		Palaeomalvaceaepollis mamilatus			
		Palmaepollenites kutchensis			
		Polyporina chenopodiaceoides			
		Periporopollenites spp.			

(continued)

Table 1 (continued)

Mahurzhari IT	Kanhobagholi IT	Gaumukh IT (29R)	Jamsavli IT	Bharudpura IT (30N)	Naskal IT
		Proxapertites sp.			
		Proxapertites assamicus			
		Proxapertites operculatus			
		Psilaperiporites robustus			
		Retitribrevicolporites rubra			
		Retipollenites confusus			
		Retitriporites federicii			
		Rhoipites waimumuensis			
		Spinizonocolpites baculatus			
		Spinizonocolpites indicus			
		Spinizonocolpatus prominatus			
		Sapotaceoidaepollenites neyvelli			
		Sapotaceoidaepollenites obscurus *Tricolporopollis globosus*			
		Tricolporopollenites wackersdorfensis			

(continued)

Table 1 (continued)

Palaeocene intertrappean beds		
Ninama Samant et al. (2013)	**Lalitpur** Singh and Kar (2002)	**Surli (29N)** Thakre et al. (2016)
Pteridophytes	**Pteridophytes**	**Pteridophytes**
Araucariacites sp.	*Cyathidites minor*	*Cyathidites australis*
Pteridacidites sp.	*Dandotiaspora dilate*	*Gabonisporis vigourouxii*
	Dandotiaspora plicata	*Gabonisporis labyrinthus*
	Dandotiaspora pseudoauriculata	*Gabonisporis* sp.
	Inapertusporites kedvesii	*Gleicheniidites senonicus*
	*Todisporites*sp.	*Lygodiumsporites lakiensis*
		Striacolporites striatus
Angiosperms	**Angiosperms**	**Angiosperms**
Aquilapollenites bengalensis	*Lakiapollis ovatus*	*Aquilapollenites bengalensis*
Aquilapollenites ovatus	*Spinizonocolpites echinatus*	*Ericipites sahnii*
Arecipites communis		*Haloragacidites amolosus*
Crototricolpitesdensus		*Incrotonipollis neyvelii*
Graminidites sp.		*Longapertites vaneendenburgii*
Intrareticulites brevis		*Mulleripollis bolpurensis*
Liliacidites sp.		*Palmaepollenites* sp.
Neocouperipollis sp.		*Palmaepollenites elongates*
Proteacidites sp.		*Palmaepollenites eocenicus*
Retitricolpites crassimarginatus		*Palmaepollenites nadhamunii*
Retitricolpites sp.		*Polybravicolporites* sp. A

(continued)

Table 1 (continued)

Palaeocene intertrappean beds		
Ninama Samant et al. (2013)	**Lalitpur** Singh and Kar (2002)	**Surli (29N)** Thakre et al. (2016)
Striacolporites striatus		*Proxapertites operculatus*
Triporopollenites sp. A		*Retidiporites megdalensis*
Typha-like pollen		*Rhoipites* sp.
		Sparganiaceaepollenites spp.
		Spinizonocolpites echinatus
		Striacolporites striatus

***Possible Cretaceous Palaeocene boundary intertrappean beds**	
K-Pg sections *Naskal (29R)	*Rangapur (29R)
Wilson Mantilla et al. (2022)	Wilson Mantilla et al. (2022)
Pteridophytes	**Pteridophytes**
Cingulatisporites sp.	*Cyathidites australis*
Crybelosporites intertrappea	
Cyathidites australis	
Gabonisporis vigourouxii	
Gabonisporis sp.	
Angiosperms	**Angiosperms**
Aesculipollis sp.	*Mulleripollis bolpurensis*
Echimonocolpites sp.	*Incrotonipollis neyvelii*
Echistephanocolpites meghalyensis	*Sparganiaceaepollenites* sp.
Margocolporites sp.	
Mulleripollis bolpurensis	
Palmaepollenites neyvelii	

(continued)

Table 1 (continued)

***Possible Cretaceous Palaeocene boundary intertrappean beds**	
K-Pg sections *Naskal (29R)	*Rangapur (29R)
Sparganiaceaepollenites sp.	
Striacolporites striatus	
Tricolpites reticulatus	
Malvacipolloides deccanensis	

2.2 Flora from the Intertrappean Sediments

The Geological Survey of India, based on systematic mapping of lava flows, petrogenetic studies, magnetostratigraphy, and geochemical studies (Deshmukh 1984; Deshmukh et al. 1996; Cox and Hawkesworth 1984; Beane et al. 1986; Courtillot et al. 1986; Nair et al. 1996; Jay and Widdowson 2008), established the litho-stratigraphy of the Deccan lava flows. The flows in Deccan volcanic province are classified into the Sahyadri, Amarkantak, Malwa and Satpura groups (Geological Survey of India 2001a, b, c, d, 2003; Nair and Bhusari 2001). However, the flows of Kutch and Saurashtra, though systematically mapped, are still unclassified and not formally designated (Fig. 1).

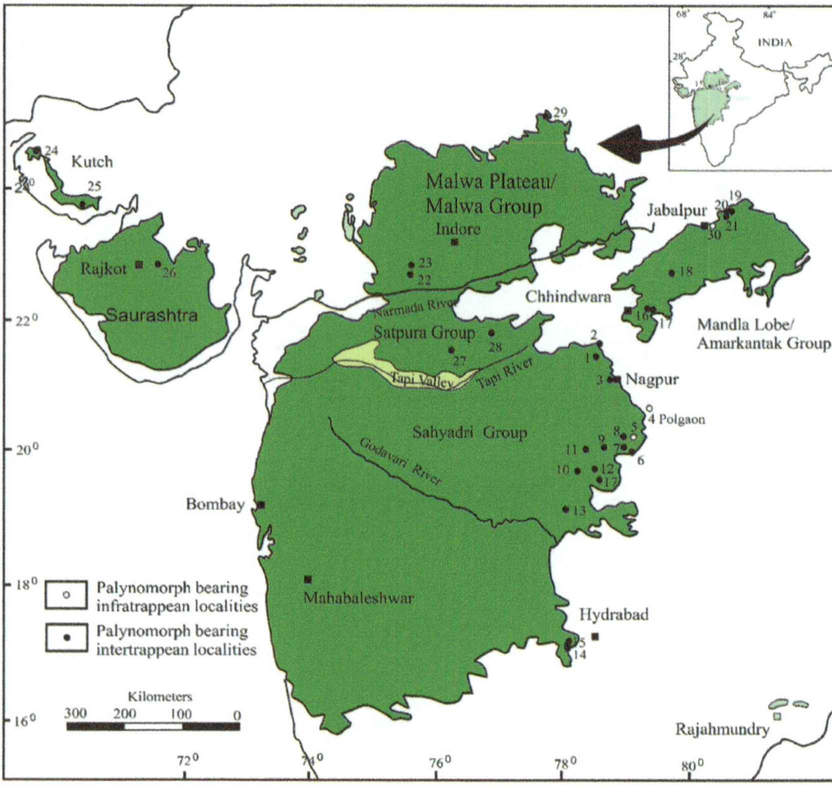

Fig. 1 Palynomorph bearing infra and intertrappean localities Mega and microflora bearing localities: 1—Singpur, 2—Jamsavli, 3—Mahurzari, 4—Polgaon, 5—Pisdura, 6—Daiwal, 7—Surla/Anandvan, 8—Khandala-Ashta, 9—Sindhi,10—Kurli, 11—Jhargad,12—Shibla, 13—Shankar Lodhi, 14—Naskal, 15—Rangapur, 16—Mohgaon kalan (MKWS), 17—Mohgaon kalan (MKFF), 18—Surli, 19—Ranipur, 20—Gaumukh, 21—Padwar, 22—Bagwanya, 23—Bharudpura, 24—Lakshmipur, 25—Anjar, 26—Ninama, 27—Hiradehi, 28—Kanabagoli, 29—Lalitpur, 30—Lameta Formation. (Abbreviations MKWS: Mohagaon kalan Well Section; MFFF: Mohgaon kalan Fossil Forest)

The intertrappean sediments are rich in mega- and microflora. The megaflora record from the Deccan intertrappean sediments dates back to 1829, when Coulthard recorded fossil wood from Sagar, Madhya Pradesh. Since the last three decades, an enormous amount of data on mega-plant fossils have been recorded from the Deccan intertrappean sediments (summarized in Kapgate 2005; Bonde 2008; Smith et al. 2015; Wheeler et al. 2017). Significantly, most of the megafloral record is from the eastern part of the Sahyadri Group and the Amarkantak Group of the Mandla Lobe. In comparison to the megafloral record, the palynofloral study of the Deccan intertrappean sediments is recent and started by Chitaley (1950, 1951). Since the last two decades, a significant amount of work on the palynoflora of the Deccan intertrappean sediments has been carried out, resulting in the generation of a good amount of palynological data from the Sahyadri, Amarkantak, and Malwa groups as well as from the unclassified sequence of the eastern and western parts of the Deccan volcanic sequence.

The megaflora in the Deccan trap province is represented by fossil wood, leaves, fruits, flowers, seed, and roots of angiosperms (both dicot and monocot), pteridophytes, and a few gymnosperms (summarised by Kapgate 2005; Bonde 2008; Smith et al. 2015). Palynofloral records from the Deccan intertrappean beds are qualitatively and quantitatively diverse. So far, about 26 palynomorph bearing intertrappean beds are known, of which 23 are Maastrichtian and three are Palaeocene (Fig. 1; Table 1).

3 Palynoflora and Age of the Intertrappean Sediments

The Maastrichtian palynomorph bearing intertrappean sediments of the Amarkantak Group are: Ranipur (Mathur and Sharma 1990), Padwar (Prakash et al. 1990), Gaumukh (Mishra et al. 2021); Mohgaon Kalan Well Section (MKWS), Mohgaon Kalan Fossil Forest (MKFF), (Kar and Srinivasan 1998; Kumaran et al. 1997; Thakre et al. 2017), Singpur (Samant et al. 2008), and Jamsavli (Samant et al. 2022). Palynofloral records from these intertrappean beds are given in Table 1. From the Sahyadri Group: Daiwal, Anandvan, Khandala-Ashta, Surla, Sindhi (Samant and Mohabey 2014), Mahurzari (Samant et al. 2020a), Shimbala and Shankar Lodhi (Samant et al. 2019), Kurli (Samant and Mohabey 2016), Naskal (Singh et al. 2006), and Patan (unpublished data) intertrappean beds. From the Mawla Group: Bagwanya and Bharudpura intertrappean beds (Mohabey et al. 2018; Samant et al. 2020b; Manchester et al. 2023), and from the Satpura Group: Hiradehi and Kanabagoli intertrappean beds (Sonkusare et al. 2021). Palynofloral records from the unclassified sequence of Kutch and Saurashtra are from Anjar (Dogra et al. 2004) and Lakshamipur intertrappean beds (Samant and Bajpai 2005).

All these intertrappean sediments are characterized by the presence and dominance of age marker palynomorphs such as, *Azolla cretacea, Gabonisporis vigourouxii, Aquilapollenites bengalensis, Ariadnaesporites, Farabeipollis* spp. *Jiangsupollis* spp., *Scollardia conferta, Triporoletes reticulatus,* and the Normapolles Group. The intertrappean beds have a combination of this assemblage in varying concentrations, sometimes one or two taxa are absent or scarcely represented, and sometimes one of these taxa dominates the assemblage (Fig. 2).

Detailed palynological analysis also indicates that, the intertrappean sediments of Maastrichtian, Chron 30N (Bagwanya and Bharudpura of Malwa; Mohabey et al. 2018; Samant et al. 2020b and Mohgaon Kalan Well Section (MKWS) of Amarkantak Group; Thakre et al. 2017) and Chron 29R have marker Maastrichtian palynomorphs, but the diversity of palynoflora in more in intertrappean beds of Chron 30N than in Chron 29R.

In comparison to numerous Maastrichtian palynomorph bearing intertrappean sections, there are only three known Palaeocene palynomorph bearing intertrappean localities, one is Lalitpur in Uttar Pradesh (Singh and Kar 2002), the second is Ninama in Saurashtra (Samant et al. 2014), and the third is Surli intertrappean in the southern part of Mandla Lobe (Thakre et al. 2016). Some of the important Palaeocene taxa are *Dandotiaspora* spp., *Echistephanocolpites meghalayaensis, Longapertites vaneendenburgii, Mulleripollis bolpurensis, Neocouperipollis, Palmaepollenites* spp. and *Striacolporites striatus* (Table 1). A few palynomorphs are also reported from the Palaeocene Mumbai intertrappean beds (Sukheswala 1956; Bande et al. 1988; Sethna 2003; Cripps et al. 2005), but overall, preservation and palynofloral recovery from the Mumbai intertrappean beds is poor and have been identified up to family level except for *Aquilapollenites* (Cripps et al. 2005). The Palaeocene intertrappean beds in the Deccan volcanic sequence are marked by their position in volcanostratigraphy, and absence or scarce representation of Maastrichtian palynoassemblage, and the appearance of palynomorphs, which are known from the post Deccan Paleogene sediments of western India.

Recently, diverse palynoassemblage have been recorded from the chronostratigraphically constrained Naskal intertrappean beds of Telangana (Wilson Mantilla et al. 2022). The palynoassemblage shows transitional flora in which many marker Maastrichtian palynomorphs are absent, and the palynoflora shows some similarity with the post Deccan Palaeocene palynofloral assemblage (*Striacolporites striatus, Echistephanocolpites meghalayaensis*, and *Palmaepollenites neyvelii)* of western India (Saxena 1991; Saxena and Trivedi 2006; Table 1).

4 Response of Flora to Changing Deposition Environments and Climate

The fluvio-lacustrine Lameta sediments were deposited in semiarid to arid climatic conditions with strong seasonality (Hansen et al. 2005; Sarkar et al. 1991; Mohabey and Samant 2005). Lithologically, the Lameta sediments are characterised by well-developed gypsum crystals, and calcrete profiles, suggesting highly alkaline conditions in the lake (Mohabey and Samant, 2005). The abundance of biodegraded organic matter, *Glomus* like mycorrhizal fungi, and the lake sediments suggest a fluctuation of lake level due to desiccation and evaporation. The presence of algal gyrogonites

Fig. 2 Important palynomorphs from the Deccan intertrappean sediments. 1. *Azolla cretacea* Stanley (1965), 2. *Gabonisporis vigourouxii* Boltenhagen (1967), 3. *Crybelosporites intertrappea* Samant et al. (2020a, b), 4. *Triporoletes reticulatus* Playford (1971), 5. *Aquilapollenites bengalensis* (Baksi and Deb) extended & emend. Samant et al. (2013), 6. *Jiangsupollis striatus* Song (1980), 7. *Farabeipollis minutus* Samant et al. (2013), 8. *Scollardia conferta* Venkatachala and Sharma (1974), 9. *Triporotetradites* sp., 10. *Clavatricolpites* sp., 11. *Sparganiaceaepollenites reticulatus*, Samant et al. (2022), 12. *Proxapertites cursus* Van Hoeken-Klinkenberg (1966)

of Chara (*Platycharaperlata*, *Microchara*) with mass mortality of clupeid fishes indicates a deteriorating condition of the lake, probably due to increased alkalinity (Mohabey et al. 1993).

The gymnospermous plant community of the Lameta, such as, conifers (*Podocarpus* and *Araucaria),* indicates open vegetation and a warm, temperate climate. The pollen grains of the gymnospermous herbaceous taxon *Classopollis* (Cheirolepidiaceae), suggest dry climatic conditions at the time of deposition (Srivastava 1976; Samant et al. 2014). However, the record of fossil woods of the families Sapindaceae, Lecythidaceae and Capparidaceae suggests a prevailing warm tropical climate.

The freshwater conditions during Lameta sedimentation are indicated by the presence of numerous freshwater algal remains of *Oedogonium, Lecaniella* and *Botryococcus*, fishes- *Clupeids, Phareodus, Pycnodont, Lepidotes, Dasyatis, Eoserranus, Nandidae* and *Lepisosteus* (Mohabey et al. 1993; Mohabey and Udhoji 1996), and freshwater ostracods (*Cypridopsis, Paracypretta*) and gastropods (*Physa, Lymnia, Paludina*). The Charophytes, algal remains of *Oedogonium,* and *Botryococcus* indicate the prevalence of alkaline shallow water conditions in the paleolake (Nichols 1973; Fritsch 1961).

Tracking of palynoflora from Lameta to intertrappean sediments during the Deccan transition indicates that the palynoflora of the Lameta sediments has a gymnosperm and angiosperm rich pollen assemblage. The onset of volcanic activity brought about changes in the physical environment as well as in soil type and topography. These changes are reflected in the lithology of the intertrappean sediments. The Lameta sediments have a well-developed calcrete profile, silici-clastic and carbonate dominating sediments fluvio-lacustrine, whereas the intertrappean sediments have shales, silicified shales, cherts, and limestones. A change in climatic conditions from semiarid to arid during Lameta time to humid to subhumid with the onset of volcanic activity during depostion of intertrappean beds is also indicated by a change in vegetation. The intertrappean sediments have well preserved magaflora and microflora in the Upper Cretaceous (Maastrichtian). The presence of plant families Sonneratiaceae (*Sonneratia*), Elaeocarpaceae (*Elaeocarpoxylon antiqum*), Simaroubaceae (*Ailanthoxylon ghiarense*), Lecythidaceae (*Barringtonioxylon deccanence, B. eopterocarpon*), Datiscaceae (*Tetrameleoxylon prenudiflora*), Tiliaceae (*Grewioxylon mahurzariense*), Palmae (*Palmoxylon* of *Phoenix*), Musaceae (*Musa cardiosperma* and *Heliconiaites mohgaoensis*) and Cannaceae (*Cannites intertrappea*) suggest the prevalence of a warm humid climate in central India and evergreen wet conditions and low seasonality (Wheeler et al. 2017). Frost free conditions are indicated by the presence of fossil palms (both megaflora and microflora) in the intertrappean sediments (Smith et al. 2015). The presence of a number of epiphyllous fungi, such as *Phragmothyrites* and *Callimothallus* (Samant and Mohabey 2014), in the palynomorph bearing intertrappean suggests good rainfall and humid climatic conditions (Selkirk 1975; Johnson and Sutton 2000; Limaye et al. 2007; Kalgutkar and Jansonius 2000). The record of charcoal in the intertrappean sediments of Maastrichtian Chron 30N (Mohabey et al. 2018) also suggest good vegetation. However, sometimes warm conditions and volcanism induced mock aridity (Harris and Van Couvering 1995) have also been suggested during the deposition of intertrappean beds (Khadkikar et al. 1999; Keller et al. 2009; Gertsch et al. 2011).

Palynoflora as well as megaflora associated with the intertrappean beds are also good indicators of the depositional environments. The presence of microfloral remains of the algae *Ovoidites* (*Spirogyra*) and *Lecaniella* (*Debarya*), aquatic and semiaquatic ferns (*Azolla*: Salviniaceae; *Gabonisporis*: Marsileaceae) and angiosperms (*Sparganiaceaepollenites*: Sparganiaceae/Typhaceae) in almost all the intertrappean beds suggests deposition of most of the intertrappean sediments in freshwater lacustrine conditions.

The presence of pollen grains of mangrove palm *Nypa* (*Spinizonocolpites*) in many intertrappean of central India suggests an estuarine influence in central India. Mishra et al. (2021), using the Coexistence Approach (CA) analysis of the flora from the Goumukh intertrappean beds in the Mandla lobe, suggested a tropical climate with strong seasonality in central India.

The chronostratigraphically constrained Naskal intertrappean, which was deposited close to the K-Pg boundary, shows the presence of Maastrichtian taxa (*Crybelosporites intertrappea*), Maastrichtian–Palaeocene taxa (*Gabonisporis vigourouxii*, *Mulleripollis bolpurensis*), and Palaeocene taxa (*Striacolporites striatus*, *Echistephanocolpites meghalayaensis*, *Palmaepollenites neyvelii*, *Aesculipollis*). Overall, palynoflora shows a district decline in Maastrichtian palynoflora and the appearance of a new floral assemblage, probable due to changed climatic conditions close to the K-Pg boundary (Wilson Mantilla et al. 2022).

5 Deccan Volcanic Associated Sediments and Evolution of Angiosperm Families

After the breakup of the Indian subcontinent from Gondwanaland around 120.4 My (Ali and Aitchison 2008), the Indian subcontinent moved from south of the equator to its current location. During its long journey, it served as a hotspot for the origin of many angiosperm families. Prasad et al. (2005) reported the oldest grass phytoliths from the sauropod coprolites of the Lameta Formation of the N-D Basin. The phytoliths of the rice tribe Oryzeae (Poaceae) (Prasad et al. 2011), pollen grains of Poaceae (Samant and Mohabey 2003), and Asteraceae (Samant and Mohabey 2014) have also been recorded from the Lameta Formation. Apart from the Lameta sediments, the intertrappean sediments have also recorded many oldest angiosperm families such as Lytheraceae; (Graham 2013); Euphorbiaceae (Kapgate 2017; Reback et al. 2022; Lakhanpal and Dayal 1962), Vitaceae (Manchester et al. 2013), Arecaceae (*Cocos*: Srivastava and Srivastava 2014), Coryphoid palm (Srivastava et al. 2014); Arecoidae palm (Manchester et al. 2016); Borasseae palm (Matsunga et al. 2019); Ceroxyloideae palm (Khan et al. 2020), Dipterocarpaceae (Khan et al. 2020); Torricelliaceae (Manchester et al. 2019), Zingiberales

(Smith et al. 2021); Alismatales (Matsunga et al. 2018), and Malvaceae (Malvoideae, Manchester et al. 2023) (Table 2). Pollen grains of Malvaceae (*Abutilon*: Manchester et al. 2023) has also been recorded from the Deccan intertrappean sediments (Table 2).

6 Conclusions

So far, studies on the Upper Cretaceous Lameta Formation and Upper Cretaceous-early Paleocene intertrappean sediments, helped in drawing the following conclusions:

(i) The Upper Cretaceous sediments of the Lameta Formation (infratrappean) were deposited in fluvio-lacustrine depositional environments under semiarid to arid climatic conditions with a strong seasonality. Both microflora and megaflora indicate the presence of a gymnosperm-angiosperm rich floral assemblage. Megaflora and microflorl from the Lameta sediments indicate the prevalence of open freshwater depositional conditions and a dry arid to semiarid climate.

(ii) The onset of Deccan volcanism, changed both the physical and depositional environments. This brought changes in the ecosystem, and consequently, over the newly developed lakes/ponds, new vegetation developed that had pteridophyte-angiosperms dominating vegetation, and the gymnosperms (Araucariaceae, Podocarpaceae, Cycadaceae) plant community, was severely affected due to volcanism. This newly developed flora in the intertrappean beds is distinctly different from the Lameta flora. Megaflora and microflora associated with intertrappean beds suggest deposition in warm, humid climatic conditions.

(iii) Maastrichtian palynoflora seems more stable and is dominated by taxa such as *Azolla cretacea, Gabonisporis vigourouxii, Aquilapollenites bengalensis, Ariadnaesporites, Farabeipollis* spp. *Jiangsupollis* spp., *Scollardia conferta, Triporoletes reticulatus,* and Normapolles Group pollen. These taxa are present in varying concentrations in the intertrappean sediments. The palynoflora of Maastrichtian Chron 30N is more diverse than the palynoflora of Chron 29R. In comparison to Maastrichtian intertrappean beds, Palaeocene palynomorph bearing intertrappean beds are scarce. The flora of these intertrappean beds is also different and shows less diversity.

(iv) Most of the intertrappean beds are dominated by freshwater palynotaxa, such as algal spores *Ovoidites* (*Spirogyra*) and *Lecaniella* (*Debarya*), aquatic and semiaquatic ferns (*Azolla*: Salviniaceae; *Gabonisporis*: Marsileaceae) and aquatic angiosperms, *Sparganiaceaepollenites* (Sparganiaceae/Typhaceae). In addition, the presence of palynomorphs of *Spinizonocolpites* (*Nypa*) in the intertrappean beds of central India suggests estuarine influence.

Table 2 Oldest known megaflora families/subfamilies/tribe from the Deccan volcanic associated sediments

Sr. no	Fossil taxa	Fossil type	Extant family	Occurrences	References
1	*Matleyites indium, Jainium pisdurensis, Chitaleya deccana, Pipernoa pearsalla Eliasundo lameti, Vonhueneites papillosum, Thomassonites sinuatum, Stebbinsana intertrappea* Prasad et al. (2011)	Phytoliths	Poaceae	Lameta sediments of Pisdura (Nand-Dongargaon Basin)	Prasad et al. (2005)
2	*Changii indicum, Tateokai deccana* Prasad et al. (2011)	Phytoliths	Poaceae, Oryzeae	Lameta sediments of Dhamni, Pisdura, Piraya and Polgaon (Nand-Dongargaon Basin)	Prasad et al. (2011)
3	*Graminidites* spp.	Pollen	Poaceae	Lameta sediments of Polgaon (Nand-Dongargaon Basin)	Samant and Mohabey (2003); Samant and Mohabey (2014)
4	*Compositoipollenites* sp. A	Pollen	Asteraceae	Lameta sediments of Polgaon (Nand-Dongargaon Basin)	Samant and Mohabey (2014)
5	*Indovitis chitaleyae* Manchester et al. (2013)	Fruit	Vitaceae	Mohgaon kalan, Mahurzari, Ambagholi, Shimbala	Manchester et al. (2013)
6	*Palmocarpon drypeteoides* (Mehrotra et al.) emend Manchester et al. (2016)	Fruit	Arecaceae Attaleinae of Arecoideae	Ghughua	Manchester et al. (2016)
7	*Palmoxylon ceroxyloides* Khan et al. (2020)	Wood	Arecaceae Ceroxyloideae	Dindori	Khan et al. 2020)
8	*Cocos binoriensis* Srivastava and Srivastava (2020)	Fruit	Arecaceae, *Cocos*	Binori	Srivastava and Srivastava (2020)
9	*Sabalites dindoriensis* Srivastava et al. (2014)	Leaf	Arecaceae Coryphoid palm	Umaria	Srivastava et al. (2014)
10	*Hyphaeneocarpon indicum* (Bande et al.) emended Smith, Manchester et al. (2019)	Fruit	Arecaceae Borasseae	Umaria, Keriya, Mohgaon kalan, Palasunder, Dhangaon	Matsunga et al. (2019)

(continued)

Table 2 (continued)

Sr. no	Fossil taxa	Fossil type	Extant family	Occurrences	References
11	*Momordicarpon cardiospermum* (Jain) Manchester et al. (2021)	Fruit and seeds	Zingiberales	Mohgaon kalan, Paladaun, Keriya, Mahurzari, Simbala Pudiyal Mauda, Marai Patan, Ghughua	Smith et al. (2021)
12	*Euphorbiotheca deccanensis* Reback et al. 2021 *Surangeocarpon mohgaoense* (Kapgate et al.) emend Reback et al. (2021)	Fruit	Euphorbeaceae Malpighiales	Mohgaon kalan	Reback et al. (2021)
13	*Phyllanthocarpon singpurensis* Kapgate et al. (2017)	Fruit	Phyllanthaceae Malpighiales	Singpur	Kapgate (2017)
14	*Mallotoxylon keriense* Lakanpal and Dayal (1962)	Wood	Phyllanthaceae Malpighiales	Keriya, Mandla Lobe	Lakhanpal and Dayal (1962); Reback et al. (2021)
15	*Enigmocarpon parijal* Sahni (1934)	Fruit	Lythraceae	Mohgaon kalan	Graham (2013)
16	*Dipterocarpus dindoriensis* Khan et al. (2020)	Leaf	Dipterocarpaceae	Umaria, Mandla Lobe	Khan et al. (2020)
17	*Pentacarpon deccanense* (Kapgate et al.) emended Manchester et al. (2019)	Fruit	Torricelliaceae	Bhutera, Mohgaon kalan, Keriya, Paladon, Marai Patan, Shimbala, Singpur	Manchester et al. (2019)
18	*Daberocarpon gerhardii* (Chitaley & Sheikh) emend Manchester et al. (2022) *Harrisocarpon sahnii* (Chitaley & Nambudiri) emended Manchester et al. (2022)	Fruit	Malvaceae Malvoideae	Mohgaon kalan, Paladaun, Marai Patan Mohgaon kalan, Keriya, Udadaun, Shibla, Marai Patan	Manchester et al. (2023)
19	*Malvacipolloides intertrappea*	Pollen	Malvaceae *Abutilon* of Malvoideae	Bharudpura,	Manchester et al. (2023)
20	*Malvacipolloides deccanensis*	Pollen	Malvaceae of Malvoideae	Naskal	Manchester et al. (2023)

(v) Overall, data shows that the Deccan volcanic associated sediments were the hotspot for the origin of many angiosperm families as well as subfamilies such as Poaceae, Dipterocarpaceae, Torricelliaceae, Vitaceae, and Zingiberales, Oryzeae of Poaceae, Malpighiales of Phyllanthaceae, Malvoideae of Malvaceae, and many subfamilies or tribes of Arecaceae.

Overall, the Deccan volcanic associated palynoflora plays a significant role in understanding floral biodiversity, age, and changes in paleoecology and paleoclimate during the Upper Cretaceous-early Paleocene Deccan volcanic eruptions. The flora also helps in tracking biotic changes across the Upper Cretaceous-early Palaeocene and understanding the evolution of angiosperm families.

Acknowledgements We are thankful to the Science and Engineering Research Board, New Delhi, for financial assistance (No. CRG/2020/001339) and the Head, Department of Geology, RTM Nagpur University for working facility. We are also thankful to Mr. Anup Dhobale, research student for helping with the figures of this chapter.

References

Ali RJ, Aitchison JC (2008) Gondwana to Asia: Plate tectonics, paleogeography and the biological connectivity of the Indian sub-continent from the Middle Jurassic through latest Eocene (166–35 Ma). Earth Sci Rev 88:145–166. https://doi.org/10.1016/j.earscirev.2008.01.007

Ambwani K, Sahni A, Kar RK et al (2003) Oldest known non marine diatoms (*Aulacoseira*) from the uppermost Cretaceous Deccan intertrappean beds and Lameta formation of India. Review De Micropaleo 46:67–71

Bande MB, Chanda A, Venkatachala BS et al (1988) Deccan traps floristics and their stratigraphic implications. In: Proceedings of the symposium on palaeocene of India: limits and subdivision. Indian Association of palynostratigraphers, Lucknow, pp 83–123

Beane JE, Turner CA, Hooper PR et al (1986) Stratigraphy, composition and form of the Deccan basalt, Western Ghats India. Bull Volcanol 48(1):61–83

Birader NV, Bonde SD (1979) On a fossil palm peduncle from Dongargaon District, Chandrapur Maharashtra, India. Geophytology 9:132–138

Bonde SD (2008) Indian fossil monocotyledons: current status, recent developments and future direction. Paleobotanist 57(1–3):141–164

Bonde SD, Birader NV (1981) On two palm woods from the Deccan intertrappean beds of Dongargaon District, Chandrapur, Maharashtra (India). J Univ Poona 54:247–257

Chenet AL, Fluteau F, Courtillot V et al (2007) Determination of rapid Deccan eruptions across the KTB using paleomagnetic secular variation: 1. Results from 1200 m thick section in the Mahabaleshwar escarpment. J Geophy Res 113:B04101. https://doi.org/10.1029/2006JB004635

Chenet AL, Fluteau F, Courillot V et al (2008) Determination of rapid Deccan eruption across the KTB using paleomagnetic secular variation: 1 results from 1200 m thick section in the Mahabaleshwar escarpment. J Geop Res 113

Chitaley SD (1950) Microflora of the Deccan intertrappean cherts, paleobotany in India VII. J Indian Bot Soc 29:30

Chitaley SD (1951) Fossil megaflora from the Mohgaon Kalan beds of the Madhya Pradesh, India. Proc Nat Inst Sci India 17:373–383

Cox KG, Hawkesworth CJ (1984) Relative contribution of crest and mantle to flood basalt magmatism Mahabaleshwar Area, Deccan Traps. Phil Trans Royal Soc London 310:627–641

Courtillot V, Besse J, Vandamme D et al (1986) Deccan flood basalt at the Cretaceous/Tertiary boundary. Earth Planet Sci Lett 80:361–374

Cripps JA, Widdowson M, Spicer RA, Jolley D (2005) Coastal ecosystem response to late stage Deccan Trap volcanism: the post K-T boundary (Danian) palynofacies of Mumbai (Bombay), West India. Palaeogeogr Palaeoclimatol Palaeoecol 216(3–4):303–332

Deshmukh SS (1984) Petrography and petrochemistry of the Deccan basalts in Maharashtra. Geol Surv India Sp Pub 12:56–76

Deshmukh SS, Sano T, Nair KK (1996) Geology and chemical stratigraphy of the Deccan basalts of Chikaldara and Behramghat sections from the eastern part of the Deccan Trap province, India. Gond Geol Mag 2:1–22

Dogra NN, Singh RY, Kulshreshta SK (1994) Palynostratigraphy of infratrappean Jabalpur and Lameta Formations (lower and upper cretaceous) in Madhya Pradesh, India. Cretac Res 15:205–215

Dogra NN, Singh YR, Kumar S (2010) The lithostratigraphic status of Jabalpur Formation and their palynological constraints on the age and environment. Gond Geol Magz 25(1):185–194

Dogra NN, Singh YR, Singh RY (2004) Palynological assemblage from the Anjar intertrappean beds, Kutch, District Gujarat. Curr Sci 86:1596–1597

Dutta D, Ambwani K (2007) A food for Upper Cretaceous dinosaurs of Pisdura, India. Curr Sci 92:897–899

Fritsch FE (1961) The structure and reproduction of the algae, vol 1. Cambridge University Press, London p 791

Geological survey of India (2001a) District resource map of Yavatmal District, Maharashtra. Scale 1:125,000

Geological survey of India (2001b) District resource map of Nanded District, Maharashtra. Scale 1:125,000

Geological survey of India (2001c) District resource map of Chandrapur District, Maharashtra. Scale 1:125,000

Geological survey of India (2001d) District resource map of Wardha District, Maharashtra. Scale 1:125,000

Geological survey of India (2003) District resource map of Chhindwara, District, Madhya Pradesh. Scale 1:125,000

Ghosh P, Bhattacharya SK, Sahni A et al (2003) Dinosaur coprolites from the late Cretaceous (Maastrichtian) Lameta formation of India: isotopic and other markers suggesting a C3 plant diet. Cretac Res 24:743–750

Gertsch B, Keller G, Adatte T et al (2011) Environmental effects of Deccan volcanism across the Cretaceous-Tertiary transition in Meghalaya, India. Earth Plan Sci Lett 310:272–285

Graham SA (2013) Fossil records in the Lythraceae. Bot Rev 79:48–145

Hansen HJ, Mohabey DM, Lojen S et al (2005) Orbital cycles and stable carbon isotopes of sediments associated with Deccan volcanic suite, India: implications for the stratigraphic correlation and Cretaceous/tertiary boundary. Gond Geol Mag Sp 8:5–28

Harris J, Van Couvering J (1995) Mock aridity and the paleoecology of volcanically influenced ecosystems. Geology 23:593–596

Huene FV, Matley CA (1933) Cretaceous Saurischia and Ornithischia of the Central Provinces of India. Palaeontol Indica 21:1–74

Jay AE, Widdowson M (2008) Stratigraphy, structure and volcanology of the SE Deccan continental flood basalt province: implications for eruptive extent and volumes. J Geol Soc 165:177–188

Johnson EM, Sutton TB (2000) Response of two fungi in the apple sooty blotch complex to temperature and relative humidity. Phytopathology 90:362–367

Kapgate DK (2005) Megafloral analysis of Intertrappean sediments with focus on diversity and abundance of flora of Mohgaon Kalan, Mandla and adjoining areas of Madhya Pradesh. Gond Geol Mag 20(1):31–45

Kapgate VD (2017) Fossil fruit of *Hibiscus esculentus* L. of family Malvaceae from Deccan Intertrappean cherts of India. Palaeobotanist 66:211–216

Kar RK, Mohabey DM, Shrivastava R (2004) First occurrence of angiospermous (dicot) fossil woods from the Lameta Formation (Maastrichtian), Maharashtra, India. Geophytology 33:21–27

Kar RK, Srinivasan S (1998) Late Cretaceous palynofossils from the Deccan intertrappean beds of Mohgaon Kalan, Chhindwara District, Madhya Pradesh. Geophytology 27(1):17–22

Keller G, Adatte T, Bajpai S et al (2009) K-T transition in Deccan Traps of central India marks major marine seaway across India. Earth Plan Sci Lett 282(1–4):10–23

Khadkikar AS, Sant DA, Gogte V et al (1999) The influence of Deccan volcanism on climates: insights from lacustrine intertrappean deposits, Anjar, western India. Palaeogeo Palaeocli Palaeoeco 147:141–149. https://doi.org/10.1016/S0031-0182(98)00156-4

Khan MA, Hazra M, Mahato S et al (2020) A Cretaceous Gondwana origin of the wax palm subfamily (Ceroxyloideae: Arecaceae) and its paleogeographic context. Rev Paleobot Paly 283:1–11. https://doi.org/10.1016/j.revpalbo.2020.104318

Khosla A, Verma O (2015) Paleobiota from the Deccan volcanosedimentary sequences of India: paleoenvironments, age and paleobiogeographic implications. His Bio 27(7):898–914

Kumaran KPN, Bonde SD, Kanitkar MD (1997) An *Aquilapollenites* associated palynoflora from Mohgaon Kalan and its stratigraphic implications for age and stratigraphic correlation of Deccan intertrappean beds. Curr Sci 72(8):590–592

Lakhanpal RN, Dayal R (1962) *Mallotoxylon keriense* gen. et sp. nov, a fossil dicotyledonous wood from the Deccan Intertrappean Series, India. Palaeobotanist 11:149–153

Limaye RB, Kumaran KPN, Nair KM, Padmalal D (2007) Non pollen palynomorphs as potential palaeoenvironmental indicators in the late Quaternary sediments of the west coast of India. Curr Sci 92:1370–1382

Manchester RS, Kapgate DK, Wen J (2013) Oldest fruits of the grape family (Vitaceae) from the late Cretaceous Deccan cherts of India. Am J Bot 100(9):1849–1859

Manchester SR, Bonde SD, Nipunage DS et al (2016) Trilocular palm fruits from the Deccan intertrappean beds of India. Int J Plant Sci 177(7):633–641. https://doi.org/10.1086/687290

Manchester SR, Kapgate DK, Ramteke D et al (2019) Morphology and anatomy of the angiosperm fruit *Baccatocarpon* Insertae Sedis from the Maastrichtian Deccan Intertrappean beds of India. Acta Paleobotanica 59(2):241–250

Manchester SR, Kapgate DK, Samant B et al (2023) Fruits and pollens of Malvaoideae (Malvaceae) in the Maastrichtian-Danian Deccan intertrappean beds of Central India. Int J Plant Sci 184

Mankar RS, Srivastava AK (2015) Salbardi-Blekher inland basin. A new site of Lameta sedimentation at the border of districts Amravati, Maharashtra and Betul, Madhya Pradesh, Central India. Curr Sci 109 (7):1337–1344.

Mathur AK, Mathur UB (1985) Boraginaceae (angiosperm) seeds and their bearing on the age of Lameta beds of Gujarat. Curr Sci 54:1070–1071

Mathur NS, Sharma SD (1990) Palynofossils and age of the Ranipur intertrappean beds, Gaur River, Jabalpur MP. In: Sahni A, Jolley A (eds) Cretaceous event stratigraphy and correlation of Indian non-marine strata: Chandigarh. International Geological Correlation Project 216–245, pp 58–59

Matsunaga KS, Smith SY, Manchester RS et al (2018) Reinvestigating an enigmatic Late Cretaceous monocot: morphology taxonomy and biogeography of *Viracarpon*. Peer J 6:e4580. https://doi.org/10.7717/peerj.4580

Matsunaga KK, Manchester SR, Srivastava R et al (2019) Fossil palm fruit from India indicate a Cretaceous origin of Arecaceae tribe Borasseae. Bot J Linnean Soc 20:1–21

Mishra S, Singh SP, Arif M et al (2021) Late Maastrichtian vegetation and palaeoclimate: palynological inferences from the Deccan Volcanic Province of India. Cretac Res 133(1–2):105126

Mohabey DM, Udhoji SG (1993) Palaeoenvironment interpretation of Lameta formation (late Cretaceous) of Nand area, Nagpur District Maharashtra. Gond Geol Mag 4:13–22

Mohabey DM (1996) Depositional environment of Lameta Formation (Late Cretaceous) of Nand-Dongargaon Basin, Maharashtra: the fossil and lithofacies evidence. Memo Geol Soc India 37:363–386

Mohabey DM (2001) Dinosaurs eggs and dung (faecal mass) from late Cretaceous of Central India: dietary implications, in recent advances in the field of earth sciences and their implications in national development. Geol Suv India Sp Pub 64:605–615

Mohabey DM (2005) Late Cretaceous (Maastrichtian) nests, eggs, and dung mass (coprolites) of sauropods (titanosaurs) from India. In: Tidwell V, Carpenter K (eds) Thunder lizards: the sauropodomorph dinosaurs: Bloomington. Indiana University Press, Indiana, pp 466–489

Mohabey DM, Samant BS (2003) Floral remains from Late Cretaceous faecal mass of sauropods from central India: implication to their diet and habitat. Gond Geol Mag Sp 6:225–238

Mohabey DM, Samant B (2005) Lacustrine facies association of a Maastrichtian Lake (Lameta Formation) from Deccan volcanic terrain, central India: implications to depositional history, sediment cyclicity and climates. Gond Geol Mag 8:37–52

Mohabey DM, Udhoji SG (1996) Fauna and flora from Late Cretaceous (Maastrichtian) nonmarine Lameta sediments associated with Deccan volcanic episodes, Maharashtra: its relevance to the K/T boundary problem, palaeoenvironment and palaeoclimate. Gond Geol Mag 2:349–364

Mohabey DM, Udhoji SG, Verma KK (1993) Palaeontological and sedimentological observations on non-marine Lameta Formation (Upper Cretaceous) of Maharashtra, India: Their palaeoecological and palaeoenvironmental significance. Palaeogeo Palaeoclim Paleoeco 105:83–94

Mohabey DM, Samant B, Kumar D (2018) Record of charcoal from early Maastrichtian intertrappean lake sediments of Bagh valley of Madhya Pradesh: Paleofire proxy. Curr Sci 114(7):1540–1544

Nair KKK, Bhusari B (2001) Stratigraphy of Deccan Traps: a review, in recent advances in the field of earth sciences and their implications in national development. Geol Suv India Spec Pub 64:477–491

Nair KKK, Chatterjee AK, Sano T (1996) Stratigraphy and geochemistry of the Deccan basalt along Toranmal section, western Satpura region. Gond Geol Mag Sp 2:23–48

Nichols HW (1973) Growth media-freshwater. In: Stein JR (eds) Handbook of phycological methods. Culture methods and growth measurements. Cambridge University Press, pp 7–24

Prakash T, Singh RY, Sahni A (1990) Palynofloral assemblage from the Padwar Deccan intertrappean (Jabalpur) MP. In: Sahni A, Jolly A (eds) Cretaceous event stratigraphy and correlation of Indian Non-marine strata, Chandigarh International Geological Correlation Project 216–245, pp 68–69

Prasad V, Stromberg CAE, Alimonammadian H et al (2005) Dinosaurian coprolites and the early evolution of grasses and grazers. Science 310:1177–1180

Prasad V, Stromberg CAE, Leach AD et al (2011) Late Cretaceous origin of the rice tribe provides evidence for early diversifi cation in Poaceae. Nat Comm 2:480

Renne PR, Sprain CJ, Richards MA et al (2015) State shift in Deccan volcanism at the Cretaceous-Paleogene boundary, possibly induced by impact. Science 350(6256):76–78

Reback G, Kapgate DK, Wurdack K et al (2022) Fruits of Euphorbiaceae from the late Cretaceous Deccan intertrappean beds of India. Inter J Plant Sci 183(2):128–138

Samant B, Bajpai S (2005) Palynoflora from Lakshmipur intertrappean deposits of Kutch, Gujarat, age implications. J Pal Soc India 50(2):177–182

Samant B, Mohabey DM (2005) Response of flora to Deccan volcanism: a case study from Nand-Dongargaon basin of Maharashtra: implications to environment and climate. Gond Geol Mag 8:151–164

Samant B, Mohabey DM (2009) Palynoflora from Deccan Volcano sedimentary sequence (Cretaceous-Palaeogene transition) of central India: implications for spatio-temporal correlation. J Biosci 34(5):811–823

Samant B, Mohabey DM (2014) Deccan volcanic eruption and their impact on flora: Palynological evidences. Geol Soc Am Spec Pap 505:171–191

Samant B, Mohabey DM (2016) Tracking temporal palynofloral changes close to Cretaceous-Paleogene Boundary in Deccan volcanic associated sediments of eastern part of central Deccan volcanic province. Glo Geo 19(4):205–215

Samant B, Mohabey DM, Kapgate DK (2008) Palynofloral record from Singpur Intertrappean, Chhindwara District, Madhya Pradesh: Implication for late Cretaceous stratigraphic correlation and resolution. J Geol Soc Ind 71:851–858

Samant B, Mohabey DM, Srivastava P et al (2014) Palynology and clay mineralogy of the Deccan volcanic associated sediments of Saurashtra, Gujarat: age and paleoenvironments. J Earth Syst Sci 123(1):219–232

Samant B, Yadav D, Mohabey DM et al (2019) Palynoflora from intertrappean localities in south-eastern part of Deccan volcanic province: taxonomic composition, age and paleogeographic implications. Palaeoworld 29:161–175

Samant B, Kapgate DK, Kumar D et al (2020a) Maastrichtian palynoflora from Deccan volcanic associated sediments of Mahurzari, Nagpur district, Maharashtra: age and paleoenvironment with comments on megaflora. J Geol Soc Ind 95:475–482

Samant B, Kumar A, Mohabey DM et al (2020b) *Centropyxis aculeata* (testate lobose amoebae) and associated diatoms from the intertrappean lacustrine sediments (Maastrichtian) of central India: implications in understanding paleolake ecology. Palaeo Electro. https://doi.org/10.26879/1082

Samant B, Puranik S, Kapgate DK et al (2022) Palynoflora from the Upper Cretaceous freshwater paleolake in Central India: Paleoecological implications. Cre Res 139:1–17

Sarkar A, Bhattacharya SK, Mohabey DM (1991) Stable isotope analysis of dinosaur eggshells: Paleoenvironmental implications. Geology 19:1068–1071

Saxena RK (1991) A catalogue of fossil plants from India: part 5A, B, Cenozoic (Tertiary): A Spores and Pollen B Fungi. Birbal Sahni Institute of Paleobotany, Lucknow, p147

Saxena RK, Trivedi GK (2006) A catalogue of tertiary spores and Pollen from India. Birbal Sahni Institute of Paleobotany, Lucknow, p 221

Selkirk DR (1975) Tertiary fossil fungi from Kiandra, New South Wales. Proc Linn Soc NSW 100:70–94

Schoene B, Samperton KM, Eddy MP et al (2015) U-Pb geochronology of the Deccan traps and relation to the end-Cretaceous mass extinction. Sci Exp 118:223–253

Schulte P, Alegret L, Arenillas I et al (2010) The Chicxulub asteroid impact and mass extinction at the Cretaceous-Paleogene boundary. Science 327(5970):1214–1218

Sethna SF (2003) The occurrence of acid and intermediate rocks in the Deccan volcanic province with associated high positive gravity anomalies and their probable significance. J Geol Soc India 51:813–815

Sharma N, Kar RK, Aggarwal C et al (2004) Fungi in dinosaurian (*Isisaurus*) coprolites from the Lameta Formation (Maastrichtian) and its reflection on food habit and environment. Micropaleontology 51:7–82

Singh RS, Kar R (2002) Paleocene palynofossils from the Lalitpur intertrappean beds, Uttar Pradesh. J Geol Soc India 60(2):213–216

Singh RS, Kar R, Prasad GVR (2006) Deccan intertrappean beds of Naskal, Rangareddy District, Andhra Pradesh. Curr Sci 90:1281–1285

Smith SY, Manchester SR, Samant B et al (2015) Integrating paleobotanical, paleosol, and stratigraphic data to Study critical transitions: a case study from the Late Cretaceous–Paleocene of India. In: Polly PD, Head J, Fox DL (eds) Earth-life transitions: paleobiology in the context of earth system evolution. The Paleontological Society Papers, 21, pp 137–166

Smith SY, Kapgate DK, Robinson S et al (2021) Fossil fruits and seeds of *Zingiberales* from the late Cretaceous-early Cenozoic Deccan Intertrappean beds of India. Inter J Plant Sci 182(2):91–108

Sonkusare H, Samant B, Mohabey DM (2017) Microflora from Sauropod coprolites and associated sediments of Late Cretaceous (Maastrichtian) Lameta formation of Nand-Dongargaon Basin, Maharashtra. J Geol Soc India 89:391–197

Sonkusare H, Samant B, Mohabey DM (2021) Palynoassemblage from intertrappean sediments of Satpura Group, Betul district, Madhya Pradesh: implications in understanding age and palaeoclimate. J Pal Soc India 66(1):35–54

Srivastava SK (1976) The fossil pollen grains *Classopollis*. Lethaia 9:437–457

Srivastava R, Srivastava G (2014) Fossil fruit Cocos L. (Arecaceae) from Maastrichtian-Danian sediments of central India and its phytogeographical significance. Acta Paleobot 54(1):67–75

Srivastava R, Srivastava G, Dilcher DL (2014) Coryphoid palm leaf fossils from the Maastrichtian-Danian of central India with remarks on phytogeography of the Coryphoideae (Arecaceae). PLoS ONE 9(11):e111738. https://doi.org/10.1371/journal.pone.0111738

Sukheswala RN (1956) Notes on the fi eld occurrences and petrography of the rocks of the Bombay Island, Bombay. Trans Min Geol Metall Instit India 50:101–126

Thakre D, Samant B, Mohabey DM (2016) Palynology and magnetostratigraphy of Deccan volcanic associated sedimentary sequence of Amarkantak Group in Central India: age and paleoenvironment. In: Dzyuba OS, Pestchevitskaya EF, Shurygin BN (eds) Short papers for the fourth international symposium of international geoscience programme IGCP project 608. Russian Academy of Science, Novosibirsk, Russia pp 101–102

Thakre D, Samant B, Mohabey DM et al (2017) New insight into age of intertrappean beds of Mohgaon kalan, Chhindwara District, Madhya Pradesh using palynology, megaflora, magnetostratigraphy and clay mineralogy. Curr Sci 112(11):2193–2197

Wheeler EA, Srivastava R, Manchester SR et al (2017) Surprisingly modern latest Cretaceous–earliest Paleocene woods of India. IAWA J 38(4):456–542

Willson Mantilla GPW, Renne PR, Samant B et al (2022) New mammals from the Naskal intertrappean site and the age of India's earliest eutherians. Palaeogeo Palaeoclim Palaeoeco 591:110–857

Paleogene Indian Plate Dynamics and Palaeoclimate: A Review from Palynological Perspective

Poonam Verma and Yogesh Pal Singh

Abstract At the dawn of the Paleogene that was the beginning of the Cenozoic era, the continents drifted farther apart, heading toward their modern positions. After India separated from Gondwana, it took 160million years to travel 9000 km northward from the Southern Hemisphere to Northern Hemisphere and to its modern geographical position coupled with Eurasia. During that voyage as an isolated Island continent, India experienced various climatic conditions due to latitudinal changes as well as extreme climate variations. Documentation of India's climate during its journey across the paleoequatorial region allows us to assess biotic as well as climatic changes across the low latitudes, at a time when global climate was exceptionally warm. It is also postulated that the similar situation might be experience in the future. The palynological studies of the Paleogene deposits of Indian subcontinent can provide substantial information of past vegetation and climate, and can serve as a natural laboratory to understand the relationship between biota and climate.

Keywords Paleogene · Palynology · Extreme warming events · Palaeoclimate · Palaeovegetation

1 Indian Plate Dynamics During Paleogene

The initial start of the Cenozoic Era was the Paleogene Period, which includes the Paleocene, Eocene, and Oligocene epochs, covering the time period between 66 to 23.03 Ma. It was one of the most dynamic time periods of earth's history that created intense changes in the geosphere and biosphere. Particularly, the terminal Cretaceous-early Paleogene transition in India is marked by Deccan Volcanism which is one of the largest volcanic eruptions in Earth's history and has received global attention for

P. Verma (✉) · Y. P. Singh
Birbal Sahni Institute of Palaeosciences (BSIP), 53 University Road, Lucknow 226007, India
e-mail: verma.poonam07@gmail.com

Y. P. Singh
Centre of Advanced Studies in Geology, University of Lucknow, Lucknow 226007, India

B. Samant and D. Thakre (eds.), *Applications of Palynology in Stratigraphy and Climate Studies*, Society of Earth Scientists Series, https://doi.org/10.1007/978-3-031-51877-5_6

its possible role in K-T boundary mass extinction (Keller et al. 2011; Chatterjee et al. 2013). It was an episodic event considered to have occurred in phases. Particularly, the third and last short-lived phase of the eruption (64.5 Ma) coincides with the early Paleogene (Keller et al. 2011).

The Indian subcontinent, after separation from the Gondwana land and before its collision with the Eurasian plate, took around 160 million years to cover a distance of around 9000 km during its northward journey (Chatterjee et al. 2013). In the early Paleogene, its separation from Seychelles (~65 Ma), the Indian plate was drifting as an isolated island subcontinent (Fig. 1). Interestingly, the rapid acceleration from an initial 5 cm/year to 20 cm/year drift of the Indian plate during the early Paleogene interval has been suggested through Plate reconstructions based on palaeomagnetic data. It is also considered that the drift speed slowed to 5 cm/year around 52 million years ago during the initial stage of collision with Asia and continued with the slow speed throughout its convergence (Copley et al. 2010; Chatterjee et al. 2013) that makes the Indian subcontinent unique among the other continents of Gondwana, however, the cause of rapid acceleration in drift is contentious (Chatterjee et al. 2013 and reference therein). Recently, the Réunion mantle plume has been suggested as the main source of India's rapid drift (Pusok and Stegman 2020). The northward voyage of the Indian plate as an isolated island is regarded as the "Noah's Ark" (McKenna 1973; Chatterjee and Scotese 2010) as it carried many endemic biotas with it. As a natural laboratory, Paleogene biota may hold the key to unraveling the evolutionary history of modern-day distribution and diversity of both flora and fauna.

2 Palaeoclimate During Paleogene

During early Paleogene, the earth experienced the global greenhouse condition when the climate was warmer and more humid than today. Largely, the driving forces of global climate change were oscillating insolation caused by changes in orbital configurations and alterations in atmospheric composition (Chatterjee et al. 2013). The large quantity of carbon release into the atmosphere and ocean that was accompanied by a global temperature increase of 4–8 °C (Sluijs et al. 2006, 2009; Zachos et al. 2006, 2007). It was a warm period superimposed by several transient events of intense warming. These hyperthermal events have been revealed from early Paleogene marine and continental sections across the world (Kennett and Stott 1991; Bralower et al. 1995; Thomas and Zachos 2000; Zachos et al. 2001, 2004a, b, 2008, 2010; Cramer et al. 2003; Bowen et al. 2004; Nicolo et al. 2007; Clementz et al. 2011; Westerhold and Röhl 2009; Samanta et al. 2013a, b; Chen et al. 2014). Among the various hyperthermal events, the Paleocene Eocene Thermal Maxima (PETM) occurred at ∼56 Ma coincident with Paleocene-Eocene boundary and it was followed by other hyperthermal events during the early Eocene namely, ETM2/H1, H2, L1, L2. During PETM sea surface temperature increased by ~ 5 °C in the tropics and as much as 9 °C in the high latitudes that cause major biotic turnover. The evolution and expansion of tropical vegetation, and emergence and dispersal of modern mammals

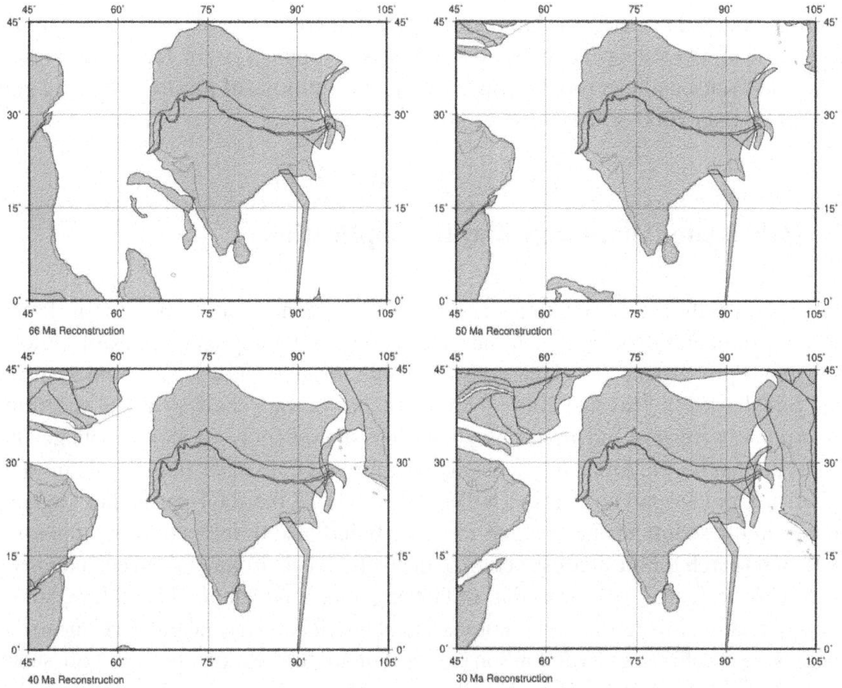

Fig. 1 Palaeogeographic positions of Indian subcontinent at different geological ages during Paleogene (Mapsource: https://www.odsn.de/odsn/services/paleomap/paleomap.html; accessed on 17.10.2022

took place during this time. After the exceptionally warm first half of the Paleogene Period, the global cooling started in the middle Eocene that approached its maximum in the earliest Oligocene. The cooling was mainly caused by rapid decline in CO_2 concentrations in the atmosphere (DeConto and Pollard 2003), and was characterized by rapid growth of the Antarctic ice sheet that caused an eustatic sea level fall of ~50 m (Zachos et al. 2001; Ivany et al. 2000).

In context to the Indian plate voyage from the Southern to the Northern Hemisphere, it experienced a range of climatic conditions. In fact, its palaeographical positions and effect of the global extreme warming events together have significantly shaped the palaeoclimate and palaeobiology of Indian subcontinent. The eventual collision of the Indian plate with the Eurasian plate, formation of the Himalayas and closure of the Tethys Sea during early Paleogene resulted in significant changes in paleooceanographic circulation patterns that in turn initiated the Indian Monsoon. From India, recent proxy records indicate the presence of a monsoonal climate during the early Paleogene (Srivastavaet al. 2012; Shukla et al. 2014; Spicer et al. 2017; Bhatia et al. 2021). In the western India, open-cast lignite mines of Cambay and Kutch Basin in the states of Gujarat preserve the Eocene hyperthermal events (Clementz et al. 2011; Samanta et al. 2013a, b; Agrawal et al. 2017). In north-east

India, Lakadong Limestone Member in East Khasi Hills, Meghalaya is also considered to be of late Paleocene in age, deposit of High Stand System Tract of 3rd order sea level cycles and can provide environmental conditions of pre-PETM (Garg and Ateequzzaman 2000; Prasad et al. 2006).

3 Paleogene Palynology and Its Application

Palynology is the science dealing with the study of organic walled microfossils found in the sedimentary rocks. The basic morphology and study techniques of various palynomorphs groups are extensively available in publications (e.g. Venkatachala and Salujha 1970; Traverse 2007). Now days, palynology is a very broad spectrum incorporate terrestrial and marine palynomorphs such as spores-pollen, dinoflagellate cysts, acritarchs, chitinozoans, scolecodonts, microscopic algal-fungal remains and other non-pollen palynomorphs (NPP). The study of the total particulate organic matter in the sediments i.e., palynofacies study has also developed as an important field of research for the reconstruction of depositional environment and evaluation of hydrocarbon source rock potential. In fact, significant developments with regard to the application aspects of palynomorphs and palynofacies have taken place due to the increasing emphasis on hydrocarbon exploration in different on-shore and off-shore sedimentary basins. Palynology has also proven its application in the commercially oil producing sedimentary basins of India—Cambay, Mumbai Offshore, Krishna-Godavari, Cauvery and Assam (Mehrotra et al. 2008). The palynological studies from Paleogene deposits have multifaceted application mainly in the following aspects-

- High resolution biostratigraphic framework and finer biozonations,
- Dating and correlation of terrestrial and marine sediments,
- Palaeoecological, palaeoclimate and palaeobiogeographical reconstructions,
- Sequence biostratigraphy, and
- Source Rock Evaluation for identifying hydrocarbon generation potential

In addition, palynofloras from this time period is very rich and diversified reflecting palaeovegetation flourishing in favorable climatic conditions that facilitated the proliferation of angiosperms particularly the tropical rain forests (Morely 2000). These sediments provide an opportunity to study the floral as well faunal changes at a crucial time of Paleogene such as the Paleocene Eocene Thermal Maximum (PETM) and Early Eocene Climatic Optimum (EECO). Indian Paleogene palynofloras also provide significant information about evolution of palaeotropical floral diversity and their biogeographical implications.

4 Indian Paleogene Deposits from Palynological Perspective

When the Indian subcontinent was located within the equatorial zone during the early Paleogene time, extensive sedimentary deposits were developed in the northern, north-eastern, western and near coastal peninsular part of India (Fig. 2).

The Paleogene rocks distributed in wide geographical areas of Lesser as well as Tethys Himalayas and sedimentary successions are referred as Subathu, Dagshai, Muree and Kasauli formations. Mainly, Subathu Formation represents the early Paleogene succession in Lesser Himalaya from Jammu and Kashmir to Himachal Pradesh mainly consists of coal, shale and lignites in small pockets possibly representing the marine transgression along the northern margin of Indian plate, as a consequence of shrinking of the Tethys Ocean prior to India-Eurasia collision. Several palynological studies from the various exposures of Subathu Formation provided the late Paleocene

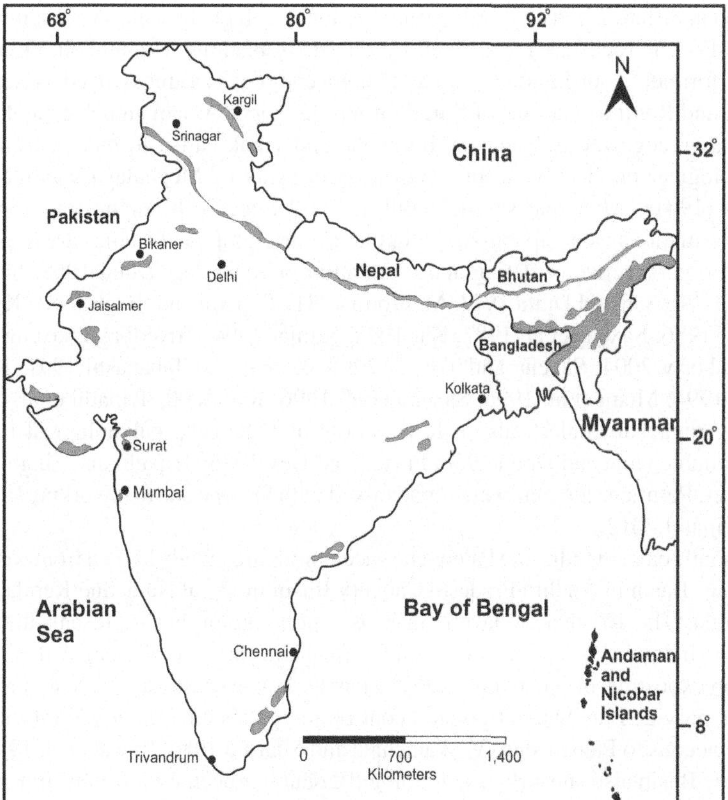

Fig. 2 Map of India showing the Paleogene deposits of India (Modified after Saxena and Tripathi 2012)

to late Eocene age and suggested shallow marine to brackish-water depositional environment, and are marked by moist deciduous and wet semi-evergreen forest types growing mostly under tropical climate (Khanna and Singh 1979; Khanna et al. 1979; Mathur and Venkatachala 1979; Misra and Kapoor 1994; Sarkar 1997; Singh and Sarkar 1987, 1988, 1990; Sarkar and Prasad 2000; Singh et al. 2003, Panwar et al. 2021). A record is also available from a section of the Dras Volcanics in south-east Ladakh, where angiospermic pollen assigned Paleocene age to exposed tuffaceous shales (Mathur and Jain 1980).

In Assam, Meghalaya and other north-eastern states of India, the Paleogene sediments are found deposited over basement rocks of Precambrian age and also over Sylhet Traps under different marginal marine depositional settings. These are well developed in Garo hills, Khasi and Jiantia hills along the southern margins of Shillong Plateau in Meghalaya, Mikir and North Cachar hills of Assam and also in the Upper Assam valleys. The deposits consist of alternating units of carbonate and terrigenous clastics deposited in tropical climate. The facies variations caused the different classification in different parts of north-eastern region. In Meghalaya and Assam, the Paleocene–Eocene rocks are represented by Therria, Lakadong, Umlatodoh, Prang and Kopili formations of Jaintia Group and Oligocene rocks are recognized as Laisong, Jenam and Renji formations of Barail Group. In Upper Assam and Nagaland, these rocks are recognized as Naogaon, Baragolai and Tikak Parbat formations. The first palynological publication from Cenozoic sediments of Meghalaya was made by Ghosh (1941). After that several publications dealing with palynostratigraphy of these formations came up emphasizing that biostratigraphical studies are important from the perspective of dating and correlation of rocks (eg. Dutta 1968; Sah and Singh 1974; Sah and Dutta 1974; Mehrotra 1981; Tripathi and Singh 1984; Kar and Kumar 1986; Saxena et al. 1987; Kar 1992; Kumar 1994; Trivedi and Saxena 2000; Chakraborty 2004; Saxena and Trivedi 2009; Kumar and Takahashi 1991; Kumar 1993, 1994; Misra et al. 1996; Saxena et al. 1996; Rao 2000; Tripathi et al. 2000). From Arunachal Pradesh also a few records of Paleogene palynological records are available (Jain and Dutta 1978; Prasad and Dey 1986; Tripathi and Singh 1992) however, Permian elements were found mixed with Eocene due to reworking (Saxena and Tripathi 2012).

In south-eastern India, the Paleogene successions are mainly known from Krishna-Godavari Basin in Andhra Pradesh, Cauvery Basin in Tamil Nadu and Kerala Basin in Kerala. The Krishna-Godavari Basin is a pericratonic basin present in eastern coast of India and its extensions are also found in the Bay of Bengal Basin. The outcrop exposures of Paleogene are very patchy however; subsurface palynological studies provided the biostratigraphic control and palaeoenvironmental information of Cretaceous to Pleistocene (Venkatachala and Sharma 1984; Rawat et al. 1991). In Cauvery Basin also subsurface palynological data has been used for biostratigraphy and palaeoenvironmental reconstructions. The Cenozoic palynoflora of Cauvery Basin represents Paleocene to Miocene age. The recovered palynofloral indicated that the area experienced a warm and humid climate conditions with heavy rain fall and deposition took place in marginal marine mangrove swamp depositional settings

(Venkatachala and Rawat 1972, 1973). In the same basin, Neyveli lignites and associated successions belonging to Cuddalore Sandstone Formation have been extensively studied from palynological perspective (Navale 1962; Ramanujam 1963, 1996a, b, 1967, 1982; Deb 1972; Deb et al. 1973; Venkatachala 1973; Navale and Misra 1979; Ramanujam et al. 1984, 1985, 1988; Ramanujam and Reddy 1984; Siddhanta 1986; Saxena 1992). Saxena (1992) founded synchronicity by correlating the palynological biozones of Neyveli lignites with those of late Paleocene-middle Eocene deposits of western Indian- Kutch and Rajasthan basins, southern Indian- Cauvery Basin, eastern Indian-Bengal Basin and Garo, Khasi and Jaintia Hills deposits of Meghalaya. A tropical warm and humid climate with heavy rainfall conditions have been reconstructed. The coastal depositional environments ranging from back mangrove to mangrove was inferred. In addition, a pulse of transgressive phase was also inferred before deposition of lignites (Saxena 1992).

In the south-western India, the Paleogene deposits are mainly found in offshore regions at the continental shelf. India's one of the major oil producing basins, the Bombay Offshore Basin also continental shelf basin of Paleocene to recent age. The lithostratigraphic units in these blocks are -Panna Formation of Paleocene-early Eocene that is the oldest Cenozoic litho-unit succeeded by Alibagh, Mahim, Tapti and Chinchini formations of Eocene to early Oligocene age. Palynological biozonation of Panna Formation has been used as a tool for dating and correlation. The palynological studies also suggested marginal marine to fluvial swampy depositional settings for this formation (Rawat et al. 1996a, b; Mehrotra et al. 2001).

In western India, the Paleogene deposits are well developed in Rajasthan, Kutch and Cambay basins. The sedimentary sequences of Rajasthan are mainly intracratonic forming part of Western Rajasthan Shelf. Palynological works in Barmer Basin of Rajasthan have been mainly done by Bose 1952; Jain et al. 1973; Naskar and Baksi 1978; Tripathi 1994, 1995, 1997; Tripathi et al. 2003; Tripathi et al. 2009; Tripathi and Srivastava 2010. In Bikaner-Nagaur Basin mainly Singh and Dogra 1988; Kar 1995, 1996; Ambwani and Singh 1996; Kar and Sharma 2001; Tripathi et al. 2008 Shivanna and Singh 2016; Singh et al. 2017 have done the palynological studies and Palana beds have been explored by Rao and Vimal 1950, 1952; Sah and Kar 1974 from palynological perspective. Rock samples from open-cast lignite mines like Giral and Akli lignite mine in Barmer have been palynologically investigated and the palynoassemblages have been found over dominated by angiosperm's pollen, particularly the morphologically diverse palm pollen (Tripathi et al. 2003, 2009; Tripathi and Srivastava 2010; Singh and Tripathi 2010). Further, palynological studies are required to quantify the nature and diversity of palaeovegetation.

Paleogene sequences of Cambay Basin, Gujarat have been deposited unconformably over the weathered basement of Deccan Trap. Within Paleogene sequences, Olpad Formation is the oldest Formation overlain by Cambay Shales of Paleocene-early Eocene. These sequences are found well-exposed in the vast open-cast mines in the western and southern Gujarat region. Extensive faunal studies have been carried out from Vastan lignite mine (Bhandari 1998; Rana et al. 2004; Ali mohammadian et al. 2005; Bajpai et al. 2005a, b; Bhandari et al. 2005; Sahni et al. 2006; Nolf et al. 2006; Rust et al. 2010). Spore-pollen studies, age diagnostic dinoflagellate cyst, sequence stratigraphic studies and palynofacies studies (Mandal and Guleria 2006; Garg et al. 2008; Tripathi and Srivastava 2012; Prasad et al. 2013; Rao et al. 2013) have been extensively worked out. In addition, Rajpardi and Bhavnagar lignite mines of Cambay Basin have also been explored for palynology and inferred to be of early Paleogene (Kar and Bhattacharya 1992; Kumar 1996; Samant and Phadtare 1997; Samant 1994, 2000). Most of the palynotaxa from Cambay Basin represent tropical to subtropical families (Fig. 3). Palynofacies investigations show that the upper part of the mine was formed under entirely anoxic circumstances, but the lower and middle parts were deposited under less anoxic settings (Prasad et al. 2013).

Kutch Basin is a pericratonic rift basin in the western continental margin of India encloses one of the best developed Mesozoic-Cenozoic sequences. A major part of Cenozoic rocks is found in the offshore region extending up to the present continental shelf. These rocks unconformably overlie the Deccan basalts and Mesozoic rocks with a maximum thickness of around 300 m and encompass enormous lignite deposits. According to the lithostratigraphic classification of Cenozoic rocks in Kutch (Biswas 1992), Paleocene sedimentary rocks are represented by Matanomadh Formation (=Madh Series) that is overlain by Early to Middle Eocene Naredi, Harudi and Fulra Limestone formations (=Berwali Series). The Oligocene succession is classified as Maniyara Fort Formation (=Bermoti Series). Detailed palynology and biostratigraphy on the Paleogene sediments of Kutch have been carried out by Mathur (1963, 1966), Venkatachala and Kar (1969), Sah and Kar (1969, 1970), Saxena (1978, 1979, 1980, 1981), Kar (1978, 1979, 1985), Mandal (2000). The terrestrial palynomorphs exhibited highly diverse and rich assemblages indicating a warm and exceptionally humid climate with a dense tropical rain forest in the area (Fig. 3). In western Kutch Basin, lignite-bearing successions are found in subsurface open-cast mines unconformably overly almost peneplained terrain of the Deccan Traps in the Panandhro-Lakhpat. These are considered to be equivalent to the lower and middle members of the Naredi Formation (Biswas and Raju 1971; Biswas 1992). Saraswati et al. (2014) and Sharma and Saraswati (2015) construed a middle Eocene age ranging from Lutetian to Early Bartonian for the succession exposed at Panandhro based on foraminifers and dinoflagellate cysts. However, based on palynological studies that including marine dinoflagellate cysts suggested an early Eocene (Ypresian) age for the lignitic succession and an early Lutetian age for the post-lignitic succession at Panandhro (Agrawal et al. 2017) and Akri lignite mine (Verma et al. 2020). The palynological study of Maniyara fort Formation (Oligocene) was attempted by Poddar (1963) and later by Kar (1978, 1979). Recently, calcareous nannofossil and dinoflagellate cyst assemblages dated the Lumpy clay Member of Maniyara fort Formation as Middle Rupelian (~31 Ma) and suggested a shallow marine neritic depositional setting, occasionally swept by open oceanic water (Verma and Singh 2019).

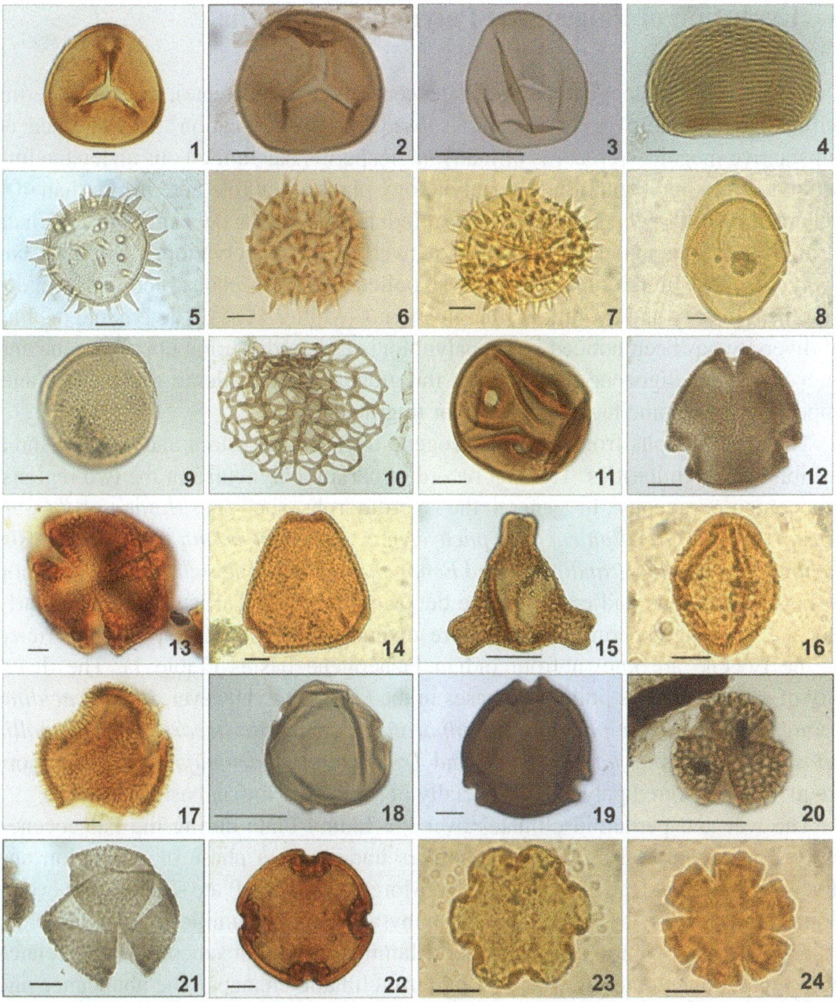

Fig. 3 Significant sporomorphs from Paleogene deposits of India (scale bar = 10 μm; modified after Rao et al. 2013; Agrawal et al. 2017; Verma et al. 2020). **1.** *Dandotiaspora dilate* Sah et al. (1971), **2.** *Dandotiaspora telonata* Sah et al. (1971), **3.** *Lygodiumsporites lakiensis* Sah and Kar (1969),**4.** *Schizaeoisporites palanensis* (Sah and Kar) Kar (1985),**5.** *Spinizonocolpites prominatus* (McIntyre) Stover and Evans (1973), **6.** *Acanthotricolpites bulbospinosus* Kar (1985), **7.** *Spinomonosulcites brevispinosus* (Biswas) Kumar (1996), **8.** *Proxapertites operculatus* (van der Hammen) van der Hammen (1956), **9.** *Proxapertites microreticulatus* Jain et al. (1973), **10.** *Retipollenites confusus* González Guzmán (1967), **11.** *Tricolporopollis decorus* Dutta and Sah (1970), **12.** *Pellicieroipollis langenheimii* Sah and Kar (1970), **13.** *Kielmeyerapollenites eocenicus* Sah and Kar (1970), **14.** *Proteacidites protrudus* Sah and Kar (1970), **15.** *Grevilloideaepites eocenicus* Biswas emend. Singh and Misra (1991), **16.** *Yeguapollis indicus* Mandal and Guleria (2006), **17.** *Tripilaorites triangulus* (Sah and Kar) Kar (1985), **18.** *Lakiapollis ovatus* Venkatachala and Kar (1969), **19.** *Dermatobrevicolporites dermatus* (Sah and Kar) Kar (1985), **20.** *Margocolporites tsukadae* Ramanujam (1966), **21.** *Dipterocarpuspollenites retipilatus* (Kar and Jain) Kar (1992), **22.** *Meliapollis ramanujamii* Sah and Kar (1970), **23.** *Pseudonathofagidites kutchensis* Venkatachala and Kar (1969), **24.** *Ctenolophonidites costatus* (van Hoeken-Klinkenberg) van Hoeken-Klinkenberg (1966)

5 Evolution of Palaeogene Palynoflora in India

During the early Paleogene period (Paleocene and early Eocene), the angiosperms witnessed maximum expansion and subsequent diversification. The absence of substantive megafossil records of many angiosperms has limited our understanding of their origin and evolution. The palynology plays a key role here. More than 400 genera and 1000 species are known from Indian Palaeogene deposits (Venkatachala et al. 1996). The abundance of pollen as well as the preservation capabilities has overcome these limitations, making fossil pollen useful for reconstruction of biogeographical history and evolution of angiosperms (Muller 1981). The gradual decline of diversity has been noticed in the palynological records during upper Eocene and later in lower Oligocene in pace with the global changes. Some groups of plants appeared while some faced extinction or migrated.

The sporomorphs from early Palaeogene from north-eastern and western India exhibit broad similarities. The first record of certain palynotaxa in the two regions, are contemporaneous. In general the sporomorphs such as, *Matanomadhiasulcites maximus*, *Grevilloideaepites pachyexinus*, *Recemonocolpites thanjinathensis*, *Echimonoporopollis grandiporus* and *Echimonoporopollis neyveliensis* are common to both Palaeocene and early Eocene deposits (Table 1). Particularly during early Eocene, the angiosperms became more diverse. More than 33 species restricted to the Eocene are known from Indian Palaeogene basins (Table 1). The diversity of angiospermous pollen decreases in the Oligocene, however, *Bombacacidites triangulus*, *Compositoipollenites tricolporatus*, *Coneopollis decorus*, *Lacrimapollis pilosus*, *Meyeripollis naharkotensis* and *Trisyncolpites ramanujamii* became more frequently recovered palynotaxa specially in north-east Indian basins (Table 1).

The wet and perhumid climates over the Indian Plate during the Palaeocene–early Eocene, several angiosperm families underwent a phase of adaptation and diversification that lead to high diversity. Moreover, Indian Plate served as an evolutionary hotspot for the diversification many angiosperm families during the early Paleogene. The evolution of angiosperm families in Indian subcontinent is related with the plate tectonics as well as long term climatic changes. The abundance and diversity of tropical megathermal families such as Arecaceae, Bombacaceae, Dipterocarpaceae, Anacardiaceaea, Alangiaceae Annonaceae, Thymeleaceae increased during early Paleogene warming phases. The studies also suggested that the tropical rainforest was able to persist under elevated temperatures and high levels of atmospheric carbon dioxide, in contrast to speculations that tropical ecosystems were severely compromised by heat stress. Palms, being one of the most dominant components of early Paleogene floral assemblages, are also characteristic and ecologically important components of tropical rain forests (TRF) worldwide. The palms represent a model plant group for investigation of TRF evolution. During the Palaeocene–early Eocene, apart from increased diversity of palms, *Alangium* (Cornaceae), *Durio* (Bombacaceae), *Caesalpinoidae* (Fabaceae) and *Ctenolophon* (Ctenolophonacaea) also diversified. The mangroves became abundant, swamp forests developed and the entire subcontinent was covered by multi-storied rain forests (Morely 2000).The

Table 1 Stratigraphic ranges of important palynotaxa from the Paleogene deposits of India (modified after Venkatachala et al. 1988, 1996; Kar 1992)

TAXA	PALEOCENE	EOCENE	OLIGOCENE
Lycopodiumsporites speciosus	—	—	
Dandotiaspora dilata	—	—	
Matanomadhiasulcites maximus	—	—	
Grevilloideaepites pachyexinus	—	—	
Racemonocolpites thanjinathensis	—	—	
Echimonoporopollis grandiporus	—	—	
Echimonoporopollis neyveliensis	—	—	
Umbelliferoipollenites spp.	—	—	
Spinomonosulcites achinatus	—	—	
Ctenolophonidites stellatus	—	—	
Iugopollis tetraporites	—	—	
Kielmeyerapollenites eocenicus	—	—	
Lakiapollis ovatus	—	—	
Retitribrevicolporites matanomadhensis	—	—	
Longapertites vaneedenbergii	—	—	
Meliapollis neyvelii	—	—	
Pellicieroipollis langenbeimii	—	—	
Polycolpites flavatus	—	—	
Polycolpites pedaliaceoides	—	—	
Proxapertites assamicus	—	—	
Proxapertites cursus	—	—	
Proxapertites operculatus	—	—	
Proxapertites microreticulatus	—	—	
Spinizonocolpites echinatus	—	—	
Psilodiporites hammenii	—	—	
Psilodiporites erdtmanii	—	—	
Psilodiporites bengalensis	—	—	
Retistephanocolpites williamsii	—	—	
Striacolporites cephalus	—	—	
Striacolporites striatus	—	—	
Matanomadhiasulcites kutchensis	—	—	
Proteacidites protrudes	—	—	
Grevilloideaepites bellus	—	—	
Arecipites bellus	—	—	
Acanthotricolpites bulbospinosus	—	—	
Acanthotricolpites kutchensis	—	—	

(continued)

Table 1 (continued)

TAXA	PALEOCENE	EOCENE	OLIGOCENE
Ghoshiacolpites globatus	—	—	—
Striacolporites ovatus	—	—	—
Verrucolporites verrucus	—	—	—
Anacolosidites luteoides	—	—	—
Neocouperipollis rarispinosus	—	—	—
Ctenolophonidites costatus	—	—	—
Dicolpopollis kalewensis	—	—	—
Marginipollis kutchensis	—	—	—
Marginipollis concinnus	—	—	—
Margocolporites tsukadae	—	—	—
Meliapollis ramanujamii	—	—	—
Meliapollis quadrangularis	—	—	—
Monocolpopollenites kutchensis	—	—	—
Monocolpopollenites nadhamunii	—	—	—
Monocolpopollenites ovatus	—	—	—
Monocolpopollenites eocenicus	—	—	—
Polybrevicolporites cephalus	—	—	—
Polycolpites granulatus	—	—	—
Pseudonothifagidites kutchensis	—	—	—
Tricolpites reticulatus	—	—	—
Clavaperiporites clavatus		—	
Cheilanthoidspora enigmata		—	
Ctenolophonidites neyveliensis		—	
Septacolpites radiatus		—	
Palaeomalvaceaepollis rudis		—	
Meliapollis tamilii		—	
Meliapollis pachydermis		—	
Meliapollis simplex		—	
Ornatetradites droseroides		—	
Pseudonothifagidites cerebrus		—	
Retipollenites confusus		—	
Cryptopolyporites cryptus		—	
Tripilaorites triangulus		—	
Tricolporopilites robustus		—	
Tricolporocolumellites pilatus		—	
Cruciferoipollenites elonagtus		—	
Icacinoipollenites spinulatus		—	
Liguliforaedites pilatus		—	
Parumbelliferoipollenites dulcis		—	

(continued)

Table 1 (continued)

TAXA	PALEOCENE	EOCENE	OLIGOCENE
Plumbaginacipites neyvelii		—	
Arengapollenites achinatus		—	
Liliacidites baculatus		—	
Alangiopollis arcotensis		—	
Alangiopollis gemmatus		—	
Angulocolporites microreticulatus		—	
Calophyllumpollenites rotundus		—	
Densiverrupollenites eocenicus		—	
Dermatobrevicolporites dermatus		—	
Foveotricolporites reticuloidus		—	
Plicatiaperturites retipilatus		—	
Thymelaepollis crotonoidis		—	
Varispinitriporites ratariensis		—	
Hippocrateaceaedites constrictus		—	
Dakshinipollenites tripakshi		—	
Polyporina multiporosa		—	
Coramandalipollis pauciornatus		—	—
Graminidites media		—	—
Marginipollis grandis		—	—
Meyeripollis laudabilis		—	—
Polygalacidites clarus		—	—
Quilonipollenites sahnii		—	—
Trilatiporites erdtmanii		—	—
Bombacacidites triangulatus			—
Compositoipollenites tricolporatus			—
Coneopollis decorus			—
Coneopollis reticulatus			—
Lacrimapollis pilosus			—
Meyeripollis naharkotensis			—
Trisyncolpites ramanujamii			—
Crassoretitriletes vanraadshoovenii			—
Pinuspollenites crestus			—
Abiespollenites cognatus			—
Piceapollenites excellensus			—

swamp forest composition have been fairly characterized by swamp dwellers plants such as *Barringtonia* (*Marginipollis*), *Durio* (*Lakiapollis*), *Gonystylus* (*Cryptopolyporites*), *Nypa* (*Spinizonocolpites*) as well as associated evergreen forest representatives such as Ctenolophonaceae (*Ctenolophonidites* and *Retistephanocolpites*), Olacaceae (*Anacolosidiles*), Caesalpiniaceae (*Margocolporites* and *Striacolporites*). During warm early Paleogene, high rainfall and low seasonality in the low latitude equatorial region were the main factor in wide distribution of the TRF in the Indian subcontinent (Prasad et al. 2009). These highly diversified early Palaeogene equatorial rain forest vegetation once widespread in the Indian subcontinent, are now restricted in a small area as a refugia in the southernmost part of the Western Ghats area (Prasad et al. 2009).

Subsequent northward voyage and a sharp drop in global temperature around 34 Ma (Eocene–Oligocene boundary) triggered the transformation of the extreme tropical climate of Paleogene to subtropical climate of Oligo-Miocene on Indian subcontinent. The megathermal plants representatives like, *Spinizonocolpites*, *Proxapertites*, *Lakiapollis*, *Dermatobrevicolporites* and *Ctenolophonidites* along with other forms either declined or migrated. In India, the characteristic feature of Oligocene palynoflora is the appearance of new pollen types related with families such as Fabaceae, Bombacaceae, Chenopodiaceae/Amaranthaceae and Rubiaceae. A few of them indicate the aridity. Although warm and humid conditions continued to prevail almost throughout Late Eocene and Oligocene time but during this period existing flora became less diversified, certain palynomorphs characteristic of early Paleogene became less represented and differentiation in regional floras progressed. The collision of Indian Plate with Asian Plate also resulted in both way migrations of plants that transformed the flora of both India and SE Asia. The changes in both geodynamics and climate during late Paleogene resulted retraction of Paleogene tropical megathermal rainforest and initiated the evolution and expansion of subtropical deciduous vegetation types in different parts of Indian subcontinent.

6 Conclusions

During northward journey of Indian subcontinent till its collision with Eurasia, it resided within the equatorial zone during Paleocene-Eocene time. Its equatorial positioning coupled with global extreme warm climate resulted in the changes of various tropical biota that are well preserved in marine and terrestrial rocks exposed in the northeastern region (Garo-Khasi-Jaintia Hills in Meghalaya), in the western margin (Barmer, Jaisalmer and Bikaner in Rajasthan; Kutch and Cambay in Gujarat) and also along the eastern margin of India (Krishna-Godavari Basin in Andhra Pradesh and Cauvery Basin in Tamil Nadu). Extremely wet and humid climate prevailed during early Paleogene that provided sustenance of the dense tropical evergreen forest. The situation also resulted into formation of coastal lowland peats swamps and those with time transformed into lignite and coal found in western and north-eastern parts. Open-cast lignite mines of Rajasthan, Gujarat and Kutch provide opportunity to study the biotic changes during the Early Paleogene. The studies in different disciplines in Indian early Paleogene have been and continue to be in search of extensive palaeofloral and palaeofaunal contents to trace the changes in flora, fauna and also in ecosystems. Although, numerous studies have been made on the early Paleogene sequences but further attention is need for tracking the climatic changes of India and quantification of climate parameters. As Indian plate traveled across the equator during early Paleogene, it offers an opportunity to comprehend low-latitude climate during an episode of extreme global warming that is similar to the future climate condition which we may heading now in near future.

Acknowledgements Authors would like to thank the Director of the Birbal Sahni Institute of Palaeosciences (BSIP) for providing constant encouragement, infrastructural facilities, and the Research Development Co-ordination Cell BSIP, Lucknow for the necessary permission for publication (BSIP/RDCCPublication No. 01/2023-24). PV contributes the paper to the in-house project no. 3 at the BSIP. YPS is grateful to the Science and Engineering Research Board (SERB), Government of India (Grant No. EMR/2016/005983) for financial assistance (JRF/SRF fellowship).

References

Agarwal S, Verma P, Rao MR et al (2017) Lignite deposits of the Kutch Basin, western India: carbon isotopic and palynological signatures of the early Eocene hyperthermal event ETM2. J Asian Earth Sci 146:296–303

Ali mohammadian H, Sahni A, Patnaik R et al (2005) First record of an exceptionally diverse and well preserved amber-embedded biota from lower Eocene (c.52 Ma) lignites, Vastan, Gujarat. Curr Sci 89:1328–1330

Ambwani K, Singh RS (1996) *Clavadiporopollenites raneriensis* gen. et sp. nov. from the tertiary sediments of Bikaner District, Rajasthan, India. Palaeobotanist 43:139–142

Bajpai S, Kapur VV, Das DP et al (2005a) Early Eocene land mammals from Vastan Lignite Mine, District Surat (Gujarat), western India. J Palaeontol Soc India 50:101–113

Bajpai S, Kapur VV, Thewissen JGM et al (2005b) First fossil marsupials from India: Early Eocene Indodelphis n.gen. and Jaegeria n. gen. from Vastan lignite mine, District Surat, Gujarat. J Palaeontol Soc India 50:147–151

Bhandari A (1998) Non-marine ostracoda from the subsurface of the Cambay Shale (Eocene) of Gujarat, India and their paleoecological significance. Rev Español Micropaleont 30:1–10

Bhandari A, Singh H, Rana RS (2005) A note on the occurrence of Ostracoda from the Vastan Lignite Mine, Gujarat. J Palaeontol Soc India 50:141–146

Bhatia H, Khan MA, Srivastava G et al (2021) Late Cretaceous-Paleogene Indian monsoon climate vis-à-vis movement of the Indian plate, and the birth of the South Asian Monsoon. Gondwana Res 93:89–100

Biswas SK (1992) Tertiary stratigraphy of Kutch. J Palaeontol Soc India 37:1–29

Biswas SK, Raju DSN (1971) Note on the rock stratigraphic classification of the Tertiary sediments of Kutch. Quart J Geol Min Met Soc India 43:177–180

Bose MN (1952) Plant remains from Barmer District, Rajasthan. J Sci Ind Res 118:185–190

Bowen GJ, Beerling DJ, Koch PL et al (2004) A humid climate state during the Paleocene/Eocene thermal maximum. Nature 432:495–499

Bralower TJ, Zachos JC, Thomas E et al (1995) Late Paleocene to Eocene paleoceanography of the equatorial Pacific Ocean: stable isotopes recorded at ocean drilling program Site865, Allison Guyot. Paleoceanography 10:841–865

Chakraborty M (2004) Palynology of the Lakadong Sandstone (late Paleocene) exposed around Bhalukurung, North Cachar Hills, Assam. Palaeobotanist 53:113–121

Chatterjee S, Goswami A, Scotese C (2013) The longest voyage: tectonic, magmatic and palaeoclimatic evolution of the Indian plate during its northward flight from Gondwana to Asia. Gondwana Res 23:238–267

Chatterjee S, Scotese CR (2010) The wandering Indian plate and its changing biogeography during the late cretaceous-early tertiary period. In: Bandopadhyay S (ed) New aspects of mesozoic biogeography. Springer-Verlag, Berlin Heidelberg, Germany, pp 105–126
Chen Z, Ding Z, Tang Z et al (2014) Early Eocene carbon isotope excursions: evidence from the terrestrial coal seam in the Fushun Basin, Northeast China. Geophys Res Lett 41:3559–3564
Clementz M, Bajpai S, Ravikant V et al (2011) Early Eocene warming events and the timing of terrestrial faunal exchange between India and Asia. Geology 39:15–18
Copley A, Avouac JP, Royer JY (2010) Indian-Asia collision and the Cenozoic slowdown of the Indian plate: implication for the forces driving plate motions. J Geophys Res 115(B03410):1–14
Cramer BS, Wright JD, Kent DV et al (2003) Orbital climate forcing of $\delta^{13}C$ excursions in the late Paleocene-early Eocene (chrons C24n–C25n). Paleoceanography 18:1097
Deb (1972) Some pollen grains from the Neyveli lignite. In: Ghosh AK et al (eds) Proceedings of the seminar on paleopalynology and Indian stratigraphy Calcutta 1971. Botany Department, Calcutta University, Kolkata, 220228
Deb U, Baksi SK, Ghosh AK (1973) On the age of Neyveli lignite-a palynological approach. Quart J Geol Min Met Soc India 45:23–38
DeConto RM, Pollard D (2003) A coupled climate ice-sheet modeling approach to the early Cenozoic history of the Antartic ice-sheet. Palaeogeogr Palaeoclimatol Palaeoecol 198:39–52
Garg R, Ateequzzaman K (2000) Dinoflagellate cysts from the Lakadong Sandstone, Cherrapunji area: Biostratigraphical and Palaeoenvironmental significance and relevance to sea level changes in the Upper Paleocene of the Khasi Hills, South Shillong Plateau, India. Palaeobotanist 49:461–484
Garg R, Ateequzzaman K, Prasad V et al (2008) Age-diagnostic dinoflagellate cysts from the lignite bearing sediments of the Vastan lignite mine, Surat district, Gujarat, Western India. J Palaeontol Soc India 53:99–105
Ivany LC, Patterson WP, Lohmann KC (2000) Cooler winters as a possible cause of mass extinctions at the Eocene/Oligocene boundary. Nature 407:887–890
Jain KP, Dutta SK (1978) Lower tertiary dinoflagelates, spores and pollen grains from Siang District, Arunachal Pradesh. J Palaeontol Soc India 21–22:106–111
Jain KP, Kar RK, Sah SCD (1973) A palynological assemblage from Barmer, Rajasthan. Geophytology 3:150–165
Kar RK (1978) Palynostratigraphy of the Naredi (Lower Eocene) and the Harudi (Middle Eocene) formations in the district of Kutch, India. Palaeobotanist 25:161–178
Kar RK (1979) Palynological fossils from the Oligocene sediments and their biostratigraphy in the district of Kutch, western India. Palaeobotanist 26:16–49
Kar RK (1985) The fossil floras of Kachchh-IV tertiary palynostratigraphy. Palaeobotanist 34:1–280
Kar RK (1992) Stratigraphical implications of Tertiary palynological succession in north-eastern and western India. Palaeobotanist 40:336–344
Kar RK (1995) Some new spore-pollen genera from early Eocene sediments of Rajasthan. J Palynol 31:161–170
Kar RK (1996) On the Indian origin of *Ocimum* (Lamiaceae): a palynological approach. Palaeobotanist 43:43–50
Kar RK, Bhattacharya M (1992) Palynology of Rajpardi lignite, Cambay Basin and Gujra Dam and Akri lignite, Kutch Basin. Palaeobotanist 39:250–263
Kar RK, Kumar M (1986) Paleocene palynostratigraphy of Meghalaya. Pollen Spores 28:177–217
Kar RK, Sharma P (2001) Palynostratigraphy of late Paleocene and early Eocene sediments of Rajasthan, India. Palaeontographica Abt B 256:123–157
Keller G, Bhowmick PK, Upadhyay H et al (2011) Deccan volcanism linked to the Cretaceous-Tertiary boundary mass extinction: New evidence from ONGC wells in the Krishna-Godavari Basin. J Geol Soc India 78:399–428
Kennett JP, Stott LD (1991) Abrupt deep-sea warming, palaeoceanographic changes and benthic extinctions at the end of the Paleocene. Nature 353:25–229

Khanna AK, Singh HP (1979) Palynological evidences in determination of age and environment of deposition of the Subathu Formation, Simla Hills. Himalayan Geol 9:292–303
Khanna AK, Singh HP, Sah SCD (1979) Palynostratigraphic correlation of the Subathu Formation, Simla Hills. Himalayan Geol 9:255–272
Kumar M (1996) Palynostratigraphy and palaeoecology of Early Eocene palynoflora of Rajpardi lignite, Bharuch District, Gujarat. Palaeobotanist 43:110–121
Kumar A (1993) Biozonation and depositional environment of the Tertiary outcrop section long the Silchar–Haflong Road, Cachar District, southern Assam. In: Biswas SK et al (eds) Proceedings of the second seminar on petroliferous basins of India 1. Indian Petroleum Publishers, Dehradun, pp 459–476
Kumar A (1994) Palynology of the Tertiary sediments exposed along the Silchar-Haflong Road Section, southern Assam. Palaeontogr Indica 2:1–241
Kumar A, Takahashi K (1991) Palynology of the Tertiary sediments of southern Assam, India. Bull Fac Liberal Arts Nagasaki Univ (nat Sci) 31:515–659
Mandal J (2000) Occurrence of *Pilatrisyncolpites* from the early Eocene of Kutch Basin and its implication. Geosci J 21:69–72
Mandal J, Guleria JS (2006) Palynology of Vastan lignite (Surat District), Gujarat: its age, palaeoecology and depositional environment. Palaeobotanist 55:51–66
Mathur K (1963) Occurrence of Pediastrum in Subathu formation (Eocene) of Himachal Pradesh, India. Sci Cult 29:250
Mathur YK (1966) On the microflora in the Supra-Trappeans of western Kutch, India. Quart J Geol Min Metal Soc India 38:33–51
Mathur YK, Jain AK (1980) Palynology and age of the Dras volcanics near Shergol, Ladakh, Jammu & Kashmir, India. Geosci Lett 1:55–74
Mathur YK, Venkatachala BS (1979) Palynological studies of the Cenozoic sediments of the Himalayan foothills. Misc Publ Geol Surv India 41:103–110
McKenna MCC (1973) Sweepstakes, filters, corridors, Noah's Ark, and beached Viking funeral ships in paleobiogeography. In: Tarling DH, Runcorn SK (eds) Implications of continental drift to earth sciences. Academic Press, London, pp 295–398
Mehrotra NC (1981) Palynological correlation of Mikir formation with lower Paleogene sediments of Shillong Plateau. Geophytology 11:133–142
Mehrotra NC, Swamy SN, Rawat RS (2001) Reworked Carboniferous palynofossils from Panna Formation, Bombay Offshore Basin: clue to a hidden target for hydrocarbon exploration. J Geol Soc India 57:239–248
Misra BK, Singh A, Ramanujam CGK (1996) *Trilatiporate* pollen from Indian Paleogene and Neogene sequences: evolution, migration and continental drift. Rev Palaeobot Palynol 91:331–352
Misra CM, Kapoor PN (1994) Paleocene to Middle Miocene palynoflora, age and palaeoenvironment of the Jwalamukhi- B, Himalayan Foothills. In: Biswas SK et al (eds) Proceedings of the second seminar on petroliferous basins of India 3. Himalayan Foothills and Gondwana Basins, Geoscientific studies and Hydrocarbon Exploration Techniques. Indian Petroleum Publishers, Dehradun, pp 147–160
Morely RJ (2000) Origin and evolution of tropical rain forests. John Wiley & Sons Ltd., p 362
Muller J (1981) Fossil pollen records of extant angiosperm. Bot Rev 47:1–47
Naskar P, Baksi SK (1978) Palynological investigation of Akli lignite, Rajasthan, India. Palaeobotanist 25:314–329
Navale GKB (1962) Pollen and spores from Neyveli lignite, South India. Palaeobotanist 10:87–90
Navale GKB, Misra BK (1979) Some new pollen grains from Neyveli lignite, Tamil Nadu, India. Geophytology 8:226–239
Nicolo MJ, Dickens GR, Hollis CJ et al (2007) Multiple early-Eocene hyperthermals: their sedimentary expression on the New Zealand continental margin and in the deep sea. Geology 35:699–702

Nolf D, Rana RS, Singh H (2006) Fish otoliths from the Ypresian (early Eocene) of Vastan, Gujarat, India. Bull Inst R Sci Nat Belg Sci Terre 76:105–118
Panwar R, Thakur OP, Dogra NN (2021) Palynological and palynofacies assemblage from the Subathu Formation (Eocene) of Northwestern Himalaya, Nilkanth, Uttarakhand, India. Curr Sci 121:667–675
Poddar MC (1963) Geology and oil possibilities of the Tertiary rocks of western India. Min Res Div ser 18(1):226-236
Prasad B, Dey AK (1986) The occurrence of Eocene sediments in Arunachal Pradesh: palynological evidence. Bull ONGC 23:67–74
Prasad V, Farooqui A, Tripathi SKM, Garg R, Thakur B (2009) Evidence of Late Palaeocene–Early Eocene equatorial rain forest refugia in southern Western Ghats, India. J Biosci 34(5):777–797
Prasad V, Garg R, Ateequzzaman K et al (2006) *Apectodinium* acme and palynofacies characteristics in the latest Paleocene-Earliest Eocene of North Eastern India: biotic response to the Paleocene-Eocene Thermal Maxima (PETM) in low latitude. J Palaeontol Soc India 51:75–91
Prasad V, Singh IB, Bajpai S et al (2013) Palynofacies and sedimentology-based high-resolution sequence stratigraphy of the lignite-bearing muddy coastal deposits (early Eocene) in the Vastan Lignite Mine, Gulf of Cambay, India. Facies 59:737–761
Pusok AE, Stegman DR (2020) The convergence history of India-Eurasia records multiple subduction dynamics process. Sci Adv 6:eaaz8681
Ramanujam CGK (1966) Palynology of the Miocene lignite from South Arcot District, Madras, India. Pollen Spores 8:149–203
Ramanujam CGK (1967) Pteridophytic spores from the Miocene lignite of South Arcot district, Madras. Palynol Bull 2–3:29–40
Ramanujam CGK (1982) Tertiary palynology and palynostratigraphy of southern India. Cenozoic Stratigr Palynol India, Palaeontol Soc India Spl Publ 1:57–64
Ramanujam CGK (1996a) Tertiary floristic complexes of southern India-a critical appraisal. Geophytology 25:1–14
Ramanujam CGK (1996b) When did *Ctenolophonidites* disappear from India? Geophytology 26:133–134
Ramanujam CGK, Reddy PR (1984) Palynoflora of the Neyveli lignite-floristic and palaeoenvironmental analysis. J Palynol 20:58–74
Ramanujam CGK, Reddy PR, Sarma PS (1985) Additions to the palynoflora of Neyveli lignite, Tamil Nadu. J Palaeontol Soc India 30:49–53
Ramanujam CGK, Reddy PR, Sarma PS (1988) *Marginipollis* from the clay and lignite of South Arcot District, Tamil Nadu. Geol Surv India Spec Publ 11:271–276
Ramanujam CGK, Sarma PS, Reddy PR (1984) Quantification of the palynoassemblages of the first and second mine areas of Neyveli lignite. In: Badve RM et al (eds) Proceedings of the 10th Indian colloquium on the indian micropalaeontology and stratigraphy, Pune, 1982. Maharashtra Association for the Cultivation of Science Pune, pp 269–275
Rana RS, Kumar K, Singh H (2004) Vertebrate fauna from the subsurface Cambay Shale (Lower Eocene), Vastan Lignite Mine, Gujarat, India. Curr Sci 87:1726–1733
Rao AR, Vimal KP (1950) Plant microfossils from Palana Lignite (Eocene), Bikaner. Curr Sci 19:82–84
Rao AR, Vimal KP (1952) Preliminary observations on the planet microfossils contents of some lignites from Warkalli in Travancore. Curr Sci 21:302–305
Rao MR (2000) Palynological investigation of the Kherapara Formation (Oligocene) exposed along Tura-Dalu Road near Kherapara, West Garo Hills District, Meghalaya, India. Palaeobotanist 49:293–309
Rao MR, Sahni A, Rana RS, Verma P (2013) Palynostratigraphy and depositional environment of Vastan Lignite Mine (Early Eocene), Gujarat, western India. J Earth Syst Sci 122:289–307
Rawat MS, Sharma KD, Juyal NP (1991) A standard palynozonation for Krishna-Godavari Basin. In Pandey J, Banerjie V (eds) Proceedings of the conference on integrated exploration—research achievements and perspectives. Oil and Natural Corporation, Dehradun, pp 281–297

Rawat RS, Misra CM, Sh Arma J et al (1996a) Palynostratigraphy and palaeoenvironment of Basal Clastic (Panna Formation) sequence of Mahim area, Bombay Offshore. In: Pandey J et al (eds) Contributions to the 15th Indian colloquium on micropalaeontology and stratigraphy, Dehradun, 1996. Oil and Natural Gas Corporation, Dehradun, pp 657–662

Rawat RS, Swamy SN, Juyal NP et al (1996b) Palynostratigraphic and paleoenvironmental interpretation of Panna formation in the east of Bombay High. Bull ONGC 33:123–135

Rust J, Singh H, Rana RS et al (2010) Biogeographic and evolutionary implications of a diverse biota in amber from early Eocene of India. Proc Natl Acad Sci USA 107:18360–18365

Sah SCD, Dutta SK (1974) Palynostratigraphy of the sedimentary formations of Assam-3. Biostratigraphic zonation of the cherra formation of South Shillong Plateau. Palaeobotanist 21:42–46

Sah SCD, Kar RK (1969) Pteridophytic spores from the Laki Series of Kutch, Gujarat, India. In: Santapau H et al (eds) Sen memorial volume. Botanical Society of Bengal, Calcutta, pp 109–122

Sah SCD, Kar RK (1970) Palynology of the Laki sediments in Kutch-3. Pollen from the boreholes around Jhulrai, Baranda and Panandhro. Palaeobotanist 18:127–142

Sah SCD, Kar RK (1974) Palynology of the Tertiary sediments of Palana, Rajasthan. Palaeobotanist 21:163–188

Sah SCD, Kar RK, Singh RY (1971) Stratigraphic range of Dandotiaspora, gen.nov., in the Lower Eocene of India. Geophytology 1:54-63

Sah SCD, Singh RY (1974) Palynological biostratigraphy of the Tura Formation in the type area. In: Symposium stratigraphical palynology, Birbal Sahni Institute of Palaeobotany, Lucknow, special publication No 3, pp 76–98

Sahni A, Saraswati PK, Rana RS et al (2006) Temporal constraints and depositional palaeoenvironments of the Vastan lignite sequence, Gujarat: analogy for the Cambay Shale hydrocarbon source rock. Indian J Petrol Geol 15:1–20

Samant B, Phadtare NR (1997) Stratigraphic palynoflora of the Early Eocene Rajpardi lignite, Gujarat and the lower age limit of the Tarkeshwar Formation of South Cambay Basin. India. Palaeontographica Abt B 245:1–108

Samant B (1994) Age of the Bhaunagar lignite deposit of Gujarat. J Palynol 30:49

Samant B (2000) Palynostratigraphy and age of the Bhavnagar lignite, Gujarat, India. Palaeobotanist 49:101–118

Samanta A, Bera MK, Ghosh R et al (2013a) Do the large carbon isotopic excursions in terrestrial organic matter across Paleocene-Eocene boundary in India indicates intensification of tropical precipitation? Palaeogeogr Palaeoclimatol Palaeoecol 387:91–103

Samanta A, Sarkar A, Bera MK et al (2013b) Late Paleocene-early Eocene carbon isotope stratigraphy from a near terrestrial tropical section and antiquity of Indian mammals. J Earth Syst Sci 122:163–171

Saraswati PK, Khanolkar S, Raju DSN et al (2014) Foraminiferal biostratigraphy of lignite mines of Kutch, India, age of lignite and fossil vertebrates. J Palaeogeogr 3:90–98

Sarkar S (1997) Palynostratigraphy and palaeoenvironment of the Subathu Formation (Eocene) of Lesser Himalaya, Himachal Pradesh, India. Indian J Petrol Geol 6:99–115

Sarkar S, Prasad V (2000) Palaeoenvironmental significance of dinoflagellate cysts from the Subathu Formation (Late Ypresian-Middle Lutetian) of Koshalia Nala Section, Shimla Hills, India. Himalayan Geol 21:167–176

Saxena RK (1978) Palynology of the Matanomadh formation in type area, north-western Kutch, India (Part 1) systematic description of Pteridophytic spores. Palaeobotanist 25:448–456

Saxena RK (1979) Palynology of the Matanomadh Formation in type area, North-Western Kutch, India (Part 2): systematic description of gymnospermous and angiospermous pollen grains. Palaeobotanist 26:130–143

Saxena RK (1980) Palynology of the Matanomadh formation in type area, North-Western Kutch, India (Part 3)-Discussion. Palaeobotanist 26:279–296

Saxena RK (1981) Stratigraphy of the area around Matanomadh in north-western Kutch with special reference to the Matanomadh formation. Palaeobotanist 27:300–313

Saxena (1992) Neyveli lignites and associated sediments—their palynology, palaeoecology, correlation and age. The Palaeobotanist 40:345-353

Saxena RK, Singh HP, Rao MR (1987) Palynology of the Tatrot-Pinjor sequence exposed between Masol and Kiratpur, Ambala District, Haryana. Geophytology 17(2):270-284

Saxena RK, Tripathi SKM (2012) Seven decades of Indian Tertiary spore-pollen floras: a compendium. AASP Contribution Series number 45, AASP Foundation, Dallas, Texas, U.S.A

Saxena RK, Tripathi SKM, Prasad V (1996) Palynofloral investigation of the Tura formation (Paleocene) in Nongwal Bibra area, East Garo Hills, Meghalaya. Geophytology 26:19–31

Saxena RK, Trivedi GK (2009) Palynological investigation of the Kopili formation (Late Eocene) in North Cachar Hills, Assam, India. Acta Palaeobotanica 49:1–61

Sharma J, Saraswati PK (2015) Lignites of Kutch, western India: dinoflagellate biostratigraphy and palaeoclimate 58:107–119

Shivanna M, Singh H (2016) Depositional environment and hydrocarbon potential of marginal marine sediments of Eocene from western India: a palynofacies perspective. Mar Pet Geol 73:311–321

Shukla A, Mehrotra RC, Spicer RA et al (2014) Cool equatorial terrestrial temperatures and the South Asian Monsoon in the early Eocene: evidence from the Gurha Mine, Rajasthan, India. Palaeogeogr Palaeoclimatol Palaeoecol 412:187–198

Siddhanta BK (1986) The age of Neyveli lignite with reference to stratigraphy and palynology. Indian Miner 40:61–82

Singh A, Shivanna M, Mathews RP et al (2017) Paleoenvironment of Eocene lignite bearing succession from Bikaner-Nagaur Basin, Western India: organic petrography, palynology, palynofacies and geochemistry. Int J Coal Geol 181:87–102

Singh H, Prasad M, Kumar K et al (2015) Early Eocene macroflora and associated palynofossils from the Cambay Shale Formation, Western India: phytogeographic and palaeoclimatic implications. Palaeoworld 24:293–323

Singh H, Tripathi SKM (2010) Fungal remains from the early Palaeogene subsurface sediments of Barkha, Barmer, Western Rajasthan, India. Geophytology 39:9–15

Singh H, Shivanna M (2015) Palynofacies characterization for evaluation of hydrocarbon source rock potential of Lower Paleogene (Thanetian-Ypresian) sub-surface sediments of Barmer Basin, Western Rajasthan, India. Mar Pet Geol 59:422–450

Singh HP, Sarkar S (1987) Significant palynozones of Subathu Formation (Himachal Pradesh) and their bearing on stratigraphy. The Palaeobotanist 35(3):249-257

Singh HP, Sarkar S (1988) Palynostratigraphy of Subathu Formation (Eocene) in the Banerhi-Bagrhan area, Sirmaur District, Himachal Pradesh, India. Palaeobotanist 35:204–210

Singh HP, Sarkar S (1990) Vegetational dynamics of tertiary Himalaya. Palaeobotanist 38:333–344

Singh YR, Dogra NN, Singh RY (2003) Age and palaeoenvironment constraints of Subathu Formation of Dharampur and Koti areas of Solan District, Himachal Pradesh—a palynological approach. Gondwana Geol Mag Spec 6:195–205

Singh RY, Dogra NN (1988) Palynological zonation of Paleocene of India with special reference to western Rajasthan. In: Maheshwari HK (eds) Paleocene of India. Proceedings of the symposium on paleocene of India: limits and subdivisions, 1986. Indian Association of Palynostratigraphers, Lucknow, pp 51–64

Sluijs A, Schouten S, Pagani M et al (2006) Subtropical Arctic ocean temperatures during the Palaeocene/Eocene thermal maximum. Nature 441:610–613

Sluijs A, Schouten S, Donders TH, Schoon PL, Röhl U, Reichart GJ, Sangiogi F, Kim JH, Damste JSS, Brinkhuis H (2009) Warm and wet conditions in the Arctic region during Eocene Thermal Maximum 2. Nat Geosci(2):777–780

Spicer RA, Yang J, Herman AB et al (2017) Paleogene monsoon across India and South China: drivers of biotic change. Gondwana Res 49:350–363

Srivastava G, Spicer RA, Spicer TEV et al (2012) Megaflora and palaeoclimate of a Late Oligocene tropical delta, Makum Coalfield, Assam: evidence for the early development of the South Asian Monsoon. Palaeogeogr Palaeoclimatol Palaeoecol 342:130–142

Thomas E, Zachos JC (2000) Was the late Paleocene thermal Maximum a unique event? GFF 122:169–170

Traverse A (2007) Paleopalynology. Unwin Hyman, Boston, MA, p 600

Tripathi SKM (1994) New angiosperm pollen from subsurface early Paleogene sediments of Barmer District, Rajasthan, India. Palaeobotanist 42:61–65

Tripathi SKM (1995) Palynology of subsurface Paleocene-Eocene sediments near Kapurdi, Barmer District, Rajasthan, India. Palaeobotanist 43:45–53

Tripathi SKM (1997) Palynological changes across subsurface Paleocene-Eocene sediments at Barmer, Rajasthan, India. Palaeobotanist 46:168–171

Tripathi SKM, Kumar M, Srivastava D (2009) Palynology of lower paleogene (Thanetian-Ypresian) coastal deposits from the Barmer Basin (Akli formation, Western Rajasthan, India): palaeoenvironmental and palaeoclimatic implications. Acta Geol 7:147–160

Tripathi SKM, Mathur SC, Nama SL et al (2008) Palynological studies from early eocene sequence exposed near Matasukh, Nagaur District, western Rajasthan, India. In: Trivedi PC (ed) Palaeobotany to modern botany. Pointer Publishers, Jaipur, India, pp 49–56

Tripathi SKM, Saxena RK, Prasad V (2000) Palynological investigation of the Tura Formation (Early Eocene) exposed along Tura-Dalu Road, West Garo Hills, Meghalaya, India. The Palaeobotanist 49(2):239-251

Tripathi SKM, Singh HP (1984) Palynostratigraphical zonation and correlation of the Jowai-Sonapur Road Section (Paleocene-Eocene), Meghalaya, India. In: Tiwari RS et al (eds.) Proceedings of the 5th Indian geophytological conference, Lucknow, 1983. Special Publication, Palaeobotanical Society, Lucknow, pp 316–328

Tripathi SKM, Singh T (1992) Record of early tertiary palynotaxa from Siang District, Arunachal Pradesh, India. Palaeobotanist 32:149–154

Tripathi SKM, Singh UK, Sisodia MS (2003) Palynological investigation and environmental interpretation on Akli formation (late Paleocene) from Barmer Basin, Rajasthan, India. Palaeobotanist 52:87–95

Tripathi SKM, Srivastava D (2010) Palynological investigations, facies analysis and palaeoenvironmental interpretations from Late Paleocene to Early Eocene lignites and associated sediments of Barmer, western India. Palaeobotanist 59:1–32

Tripathi SKM, Srivastava D (2012) Palynology and palynofacies of the early Paleogene lignite bearing succession of Vastan, Cambay Basin, Western India. Acta Paleobotanica 52:157–175

Trivedi GK, Saxena RK (2000) Palynofloral investigation of the Kopili Formation (Late Eocene) exposed near Umrongso in North Cachar Hills District, Assam, India. Palaeobotanist 49:269–280

Venkatachala BS, Caratini C, Tissot C et al (1988) Paleocene-Eocene marker pollen from India and tropical Africa. Palaeobotanist 37:1–25

Venkatachala BS, Saxena RK, Singh HP et al (1996) Indian Tertiary angiosperm pollen: a critical assessement. Palaeobotanist 42:106–138

Venkatachala BS (1973) Palynological evidence on the age of Cuddalore Sandstone. Geophytology 3:145–149

Venkatachala BS, Kar RK (1969) Palynology of the tertiary sediments of Kutch-1: Spores and pollen from borehole no. 14. Palaeobotanist 17:157–178

Venkatachala BS, Rawat MS (1972) Palynology of the tertiary sediments in the Cauvery Basin-1. Paleocene-eocene palynoflora from the subsurface. In: Ghosh AK et al (eds.) Proceedings of the seminar on paleopalynology and Indian stratigraphy, Calcutta, 1971. Botany Department, Calcutta University, Kolkata, pp 292–335

Venkatachala BS, Rawat MS (1973) Palynology of the tertiary sediments in the Cauvery basin-2 Oligocene-Miocene palynoflora from the subsurface. Palaeobotanist 20:238–263

Venkatachala BS, Salujha SK (1970) Palynological fossils and their exploration values. Selected Lectures on Petroleum Exploration. KDMIPE, ONGC, Dehradun, p 1

Venkatachala BS, Sharma KD (1984) Palynological zonation in subsurface sediments in Narsapur Well No. l, Godavari-Krishna Basin, India. In: Badve RM et al (eds.) Proceedings of the 10th

Indian colloquium on the Indian micropalaeontology and stratigraphy, Pune, 1982. Maharashtra Association for the Cultivation of Science, Pune, pp 445–465
Verma P, Garg R, Rao MR (2020) Palynoflora diversity and palaeoenvironments of early Eocene Akri lignite succession, Kutch Basin, western India. Palaeobiodivers Palaeoenviron 100:605–627
Verma P, Singh A (2019) Early Oligocene (Rupelian) dinoflagellate cysts and calcareous nannofosiils from Lumpy Clay Member of Maniyara Fort Formation, Kutch, Gujarat, India. Curr Sci 116:1571–1578
Westerhold T, Röhl U (2009) High resolution cyclostratigraphy of the early Eocene-new insights into the origin of the Cenozoic cooling trend. Clim past 5:309–327
Zachos JC, Bohaty SM, John CM et al (2007) The Paleocene-Eocene carbon isotope excursion: constraints from individual shells planktonic foraminfera records. Philos Trans R Soc A 365:1829–1842
Zachos JC, Dickens GR, Zeebe RE (2008) An early Cenozoic prespective on greenhouse warming and carbon–cycle dynamics. Nature 451:279–283
Zachos JC, Kroon D, Blum P (2004a) Scientific shipboard party, leg 208, site 1262 initial report. In: Proceedings of the ocean drilling program. Ocean Drill Program, College Station, Texas, Initial Report 208, pp 1–112
Zachos JC, Kroon D, Blum P (2004b) Scientific shipboard party, leg 208, site 1263 initial report. In: Proceedings of the ocean drilling program. Ocean Drill Program, College Station, Texas, Initial Report 208, pp 845–954
Zachos JC, McCarren H, Murphy B et al (2010) Tempo and scale of late Paleocene and early Eocene carbon isotope cycles: implications for the origin of hyperthermals. Earth and Plant Sci Lett 299:242–249
Zachos JC, Pagani M, Sloan L, Thomas E, Billups K (2001) Trends, rhythms, and aberrations in global climate 65 Ma to present. Science 292(5517):686-693
Zachos JC, Schouten S, Bohaty SM et al (2006) Extreme warming of mid-latitude coastal ocean during the Paleocene-Eocene thermal maxima: inferences from TEX_{86} and isotope data. Geology 34:737–740

Palynological Perspective on Understanding Climate Change in India Over the Pre-industrial Common Era (CE; Past Ca. 2000 Years): a Comprehensive Review and a Critical Evaluation

Mohammad Firoze Quamar

Abstract Palynology (and pollen analysis) has proven to be a reliable tool in reconstructing the past vegetation (dynamics) and contemporary climatic changes. The chapter reviews the role of pollen in reconstructing the palaeoclimate in India over the pre-industrial Common Era (CE; past ca. 2000 years [2 ka]; past 2 millennia; Late Holocene), comprising the global climatic events, such as the Roman Warm Period (RWP), Dark Ages Cold Period (DACP), Medieval Climate Anomaly (MCA), Little Ice Age (LIA) and the Current Warm Period (CWP). In addition, the hydro-climatic variability, which influences the socio-economic welfare of one-third of the human population through the agricultural output and Gross Domestic Product (GDP) of this densely-populated country, are also analyzed. This chapter also discusses the solar forcing and the other forcing factors behind the hydroclimatic variability during the past 2 millennia in India.

Keywords Palynology · Climate change · Common Era (CE) · Hydro-climate variability · Indian Summer Monsoon (ISM) · Solar forcing · Inter Tropical Convergence Zone (ITCZ) · India

1 Introduction

Vegetation (the sum total of all the plants/plant population growing in an area or covering a region) plays a pivotal role as a reservoir for Earth's climate. Consequently, studies that utilize vegetation to glean insights into past climates stand as potent and reliable methods for comprehending climate change and hydroclimatic changes/variability/fluctuations. In the context of India, monsoonal dynamics

M. F. Quamar (✉)
Birbal Sahni Institute of Palaeosciences, Lucknow 226007, Uttar Pradesh, India
e-mail: mohdfiroze_quamar@bsip.res.in

B. Samant and D. Thakre (eds.), *Applications of Palynology in Stratigraphy and Climate Studies*, Society of Earth Scientists Series, https://doi.org/10.1007/978-3-031-51877-5_7

are predominantly characterized by precipitation, making vegetation-based palaeoclimate investigations a significant avenue for exploring hydroclimatic variability. Climate, coupled with factors, such as soil composition, topography, and elevation, exerts substantial influence over the global distribution and makeup of vegetation. Moreover, vegetation forms a fundamental component of the biosphere, adapting in response to climatic and soil conditions. It plays a crucial role in essential ecosystem functions, including nutrient and food production, climate change mitigation, and the generation and conservation of soil and freshwater resources (Faucon et al. 2017). Terrestrial plants serve as a crucial carbon sink, playing a role in mitigating the warming of atmospheric and ocean temperatures by absorbing carbon through photosynthesis. Additionally, plants contribute to cooling through the evapotranspiration process, which elevates water vapor levels. Alterations in land cover also impact albedo, exerting either positive (warming) or negative (cooling) radiative forcing effects. Thus, vegetation forms an integral component of the biogeochemical and physical interactions between the land surface and the atmosphere (Foley et al. 2003). The dynamics of vegetation are pivotal indicators in the regulation of terrestrial carbon balance and climate shifts (Zhang et al. 2020). Consequently, changes in vegetation patterns in response to climate fluctuations are preserved within sediment-based pollen records. Pollen grains and spores, generated by plants and subsequently dispersed, released, transported, and mixed by wind and/or water, compose a pollen assemblage that reflects the vegetation characteristics and contemporaneous climate or sediment environment at a specific time and area (Erdtman 1952; Birks and Birks 1980; Faegri and Iversen 1989). Pollen-derived vegetation records from a given region unveil fluctuations in monsoonal rainfall, such as the Indian Summer Monsoon (ISM) or Southwest Monsoon (SWM), and/or northeast Monsoon (NEM) (Kar and Quamar 2020a, b; Mohanty et al., 2024; Quamar 2019, 2022a, b, c; Quamar and Kar 2020; Quamar and Bera 2020, Quamar et al., 2023, 2024, and references therein). As an agricultural nation, India heavily relies on monsoonal rainfall to sustain its agricultural output, economy, and overall societal well-being (Webster et al. 1998; Gadgil 2003; Gadgil and Gadgil 2006). The country experiences a diverse range of climates and hosts an array of vegetation types (Quamar and Kar 2020). The great diversity of vegetational types in India is possible due to the variations in annual rainfall, altitude, temperature and substratum (edaphic factor; soil). Indian sub-continent has been divided into nine (9) vegetational belts namely (1) Western Himalaya; (2) Eastern Himalaya; (3) West Indian Deserts (Indus Plain); (4) Gangetic Plain; (5) Assam; (6) central India; (7) Malabar; (8) The Deccan; and (9) Andamans. Additionally, India is also divided into 4 climatic zones which correspond with the great rainfall divisions; (1) Wet zone (average 80 inches; 1 inch = 25.4 mm; 1 mm = 0.1 cm; 1 cm = 10 mm); (2) Intermediate zone (40–80 inches); (3) Dry zone (20–40 inches) and (4) Arid zone (<20 inches) (Sharma 1995).

In India, the distribution of natural vegetation is determined by three key factors: the annual average rainfall, the duration of the dry season, and the mean temperature during the coldest month (Gunnell 1997). In mountainous regions, vegetation patterns also follow temperature changes along the altitudinal gradient. However, below 900 m, the amount and seasonal distribution of annual rainfall emerge as the most crucial factors shaping vegetation distribution (Bonnefille et al. 1999). As a result, the impact of the monsoon on vegetation becomes evident, along with the reciprocal response of vegetation to the monsoon. Therefore, pollen-based records of vegetation in a given area will reflect variations in monsoonal precipitation, encompassing its frequency and intensity. Such investigations can be carried out through the analysis of profiles or cores from wetland regions such as lakes, swamps, sedimentary (depositional) basins, ice cores, and similar sources, involving palynological studies conducted on samples extracted from the relevant profiles or cores.

The present communication reviews the studies carried out on vegetation-based climate change and the ISM rainfall variability in India over the pre-industrial Common Era [CE (earlier AD-Anno Domini; Medieval Latin "anno Domini nostri Jesu Christi" means "in the year of the Lord") (https://en.wikipedia.org/wiki/CommonEra; searched on 14.03.2022); past 2000 years (2 ka), past 2 millennia; Late Holocene] with much emphasis on global contextualization. During the Late Holocene (the Meghalayan age: 4.2 kyr BP to Present) as identified), a discernible weakening trend in the summer monsoon was observed (Walker et al. (2019). This trend was punctuated by a sudden reduction in rainfall around 4.2 ka (thousand years before present) (Berkelhammer et al. 2012; Dixit et al. 2014a, b, 2018). The significant reduction in monsoonal precipitation at 4.2 ka had profound societal consequences. It prompted the migration of the population of the Indus-Harappan Valley Civilization to new regions towards the east, seeking viable locations for domestic and agricultural purposes. This shift was driven by the disappearance of the previously reliable seasonal flooding of the river system around the periphery of the Thar desert (Giosan et al. 2012). Concurrently, the agricultural productivity of the Harappan settlements dwindled due to the weakened summer monsoon (Berkelhammer et al. 2012; Dutt et al. 2018; Gupta et al. 2020). Furthermore, the weakening of the Indian Summer Monsoon (ISM) and the associated 4.2 kyr drought event were potentially connected to a heightened phase of El Niño and a shift in the Indian Ocean Dipole (IOD) towards a strongly negative state (Morrill et al. 2003; MacDonald 2011). However, the predominant drivers influencing the ISM variability throughout the Holocene Epoch included solar insolation, the migration of the Inter Tropical Convergence Zone (ITCZ), the North Atlantic Oscillation (NAO), and the El Niño and Southern Oscillation (ENSO) (Fleitmann et al. 2003; Gupta et al. 2003; Dutt et al. 2015).

For ease of presentation, the chapter has discussed climate change studies over the past 2 millennia with global contextualization, such as the Roman Warm Period (RWP), Dark Ages Cold Period (DACP), Medieval Climate Anomaly (MCA), Little Ice Age (LIA) and Current Warm Period (CWP). The inclusion of the RWP (recorded globally between ca. 2500–1600 calyr BP: ~550 BCE/BC to CE/AD ~ 350, Wang et al. 2012) in the present review lies with the fact that the signatures of this warm event have been recorded around the studied areas (Bemetara and Mahasamund districts of Chhattisgarh State, central India), although roughly encompassing the jurisdiction of this book chapter (palaeoclimate over the Common Era [the past 2000 years; the past 2 millennia; the 2 ka-thousand years ago; the 2kyr B.P.-B.P.: before present-1950 is considered as present]). The first millennium of the CE has also been subdivided into the Dark Ages Cold Period (DACP) (Helama et al. 2017; Ljungqvist 2010), or Late Antique Little Ice Age (LALIA) (Büntgen et al. 2011),occurred within ca.CE400–800 and lastly the Roman Warm Period (RWP) (Rothlisberger 1986; Ljungqvist 2010), which covers the first few centuries of the CE.

2 Palaeoclimatic Studies Over the Past 2 Millennia: Regional and Global Contextualization

An overview and recent advances on the pollen-based Quaternary palaeoclimatic studies in India was made by Kar and Quamar (2020a). Moreover, Kar and Quamar (2020b) also presented a palynological perspective on the Late Pleistocene–Holocene vegetation and climate change from the Western and Eastern Himalaya, India. Dixit and Tandon (2016) reviewed and assessed the hydroclimatic variability on the Indian sub-continent in the past millennium. Misra et al. (2019) and Banerji et al. (2020) analyzed many Holocene records from various climatic archives across India. Quamar et al. (2024) provided an overview of the hydroclimatic changes since the Last Glacial Maximum (LGM) from the central Indian Core Monsoon Zone (CMZ), based on palynological records. They have also correlated their pollen based findings with that of the continental and marine records in global contextualization. In the present review work, however, palynological perspective on understanding the climate change, as well as the ISM rainfall variability (hydroclimatic variability) over the past 2 millennia in regional and global contextualization, falling within the time limit, such as the Roman Warm Period (RWP), Dark Age Cold Period (DACP), Medieval Climate Anomaly (MCA), Little Ice Age (LIA) and Current Warm Period (CWP)in the Indian sub-continent are discussed below (Fig. 1). Table 1 shows the details of the pollen analytical studies included in this review article. As a significant shift in the vegetation occurred over the last 2k yr in the Indian subcontinent and meticulously documented the above-said global climatic events; therefore, the present aspect (review work) is being discussed under the heads of global climatic events in regional and global contextualization, as below.

2.1 Roman Warm Period (RWP)

The Roman Warm Period (RWP), also known as the Roman Climate Optimum (RCO), denotes an era characterized by elevated temperatures that were observed on a global scale from 2500 to 1600 cal yr BP (~550 BCE/BC to ~350 CE/AD) (Wang et al. 2012). However, the extent of this climatic phase during the RWP didn't exhibit uniformity across the globe; instead, it displayed notable disparities among

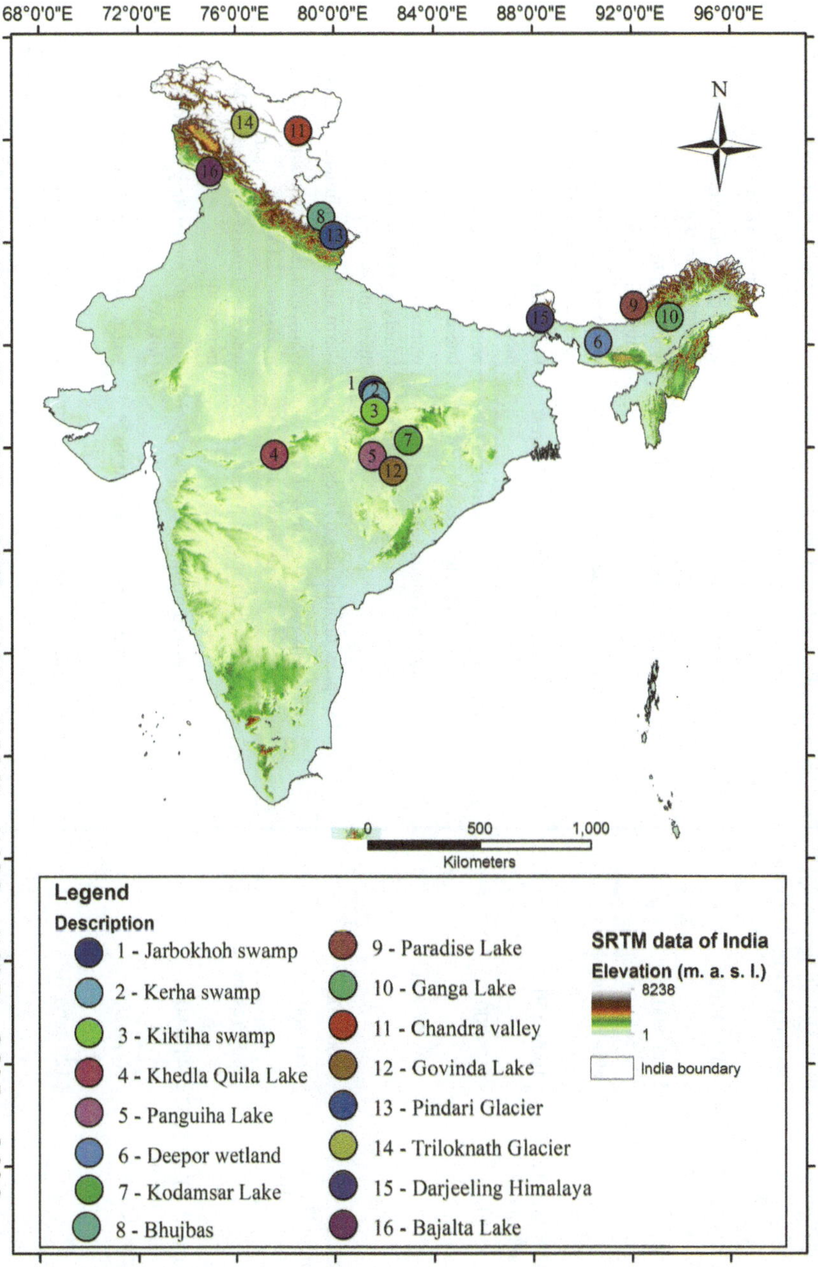

Fig. 1 Shuttle radar topographic mission (SRTM) digital elevation map (DEM) of India showing the locations of the palaeoclimatic studies. (For interpretation of the references to colour in this figure legend, the reader is advised to see the web version of this article). The details of the studies are given in Table 1 *Source* This figure is made using ArcGIS 10.3

Table 1 Detailed vegetation-based climate change studies and hydro-climatic variability (incorporated in the review article) in India over the past ca. 2000 years

Area/locality	Period (yr BP; unless otherwise stated)	Vegetation dynamics	Inferred climate change and hydro-climate variability
Jarbokhoh swamp (24° 12′ 28.8″ N: 81° 34′ 1.2″ E), Sidhi District; Chauhan (2004)	1500–1160	dense tropical deciduous forest	warm and moist with increased summer monsoon
	160–800	open tropical deciduous sal forest	warm and less moist with weak summer monsoon
	800-Present	dense tropical deciduous sal forest	warm and moremoist with strong summer monsoon (CWP)
Kerha swamp (24° N: 81° 41′ E), Sidhi District; Chauhan (2005)	1600–850	Open tropical deciduous sal forest	warm and moist with moderate monsoon
	850–400	more open and less diversified tropical deciduous forest	warm and dry climate, attributable to late and week summer monsoon
	400-Present	tropical deciduous sal forest	warm and moist (more) climate with timely arrival of strong summer monsoon (CWP)
Kiktiha swamp (23° 25′ 30” N: 81° 38′ 38.4” E), Shahdol District; Chauhan and Quamar (2010)	1600–700	tropical deciduous sal forest	warm and moist climate (MCA)
	700–300	mixed tropical deciduous forest (sparse and less diversified)	warm and less moist with weak SW monsoon (LIA)
	300-Present	tropical deciduous sal forest	warm and moist with active SW monsoon
Khedla Quila Lake (21° 43′ 25.585” N: 77° 35′ 15.23” E; ~ 400 m a.s.l.; KQL), Betul District; Quamar and Chauhan (2014)	1416–506	open mixed tropical deciduous forest	warm and moderately humid climate (MCA)
	506–120	dense mixed tropical deciduous forest	warm and humid climate (LIA)
	120 calyr BP-Present	open mixed tropical deciduous forest	warm and less humid climate

(continued)

Table 1 (continued)

Area/locality	Period (yr BP; unless otherwise stated)	Vegetation dynamics	Inferred climate change and hydro-climate variability
Panguiha Lake (21° 39.849′ N: 81° 33.249′ E; 273 m a.s.l.; PL), Bemetra District; Quamar et al. (2021)	ca. 3000–2600 calyr BP	tree savannah	cool and dry with reduced monsoon rainfall
	ca. 2600–2200 calyr BP	open mixed tropical deciduous forest	warm and moderately humid climate with moderate monsoon rainfall
	ca. 2200–2000 calyr BP	mixed tropical deciduous forest	warm and humid climate with increased monsoon rainfall
	ca. 2000–1800 calyr BP	dense mixed tropical deciduous forest	warm and relatively more humid climate with further increase in monsoon rainfall (RWP)
Kodamsar Lake (22° 16.681′ N: 82° 59.389′ E; ~ 300 m a.s.l.; KL), Korba District; Quamar and Bera (2021)	8378–1198 calyr BP	open mixed tropical deciduous forest	warm and moderately humid climate with moderate monsoon precipitation
	1198–478 calyr BP	mixed tropical deciduous forest	warm and humid climate with increased monsoon precipitation (MCA)
	478 calyr BP-Present	dense mixed tropical deciduous forest	warm and relatively more humid climate with further increase in monsoon precipitation (CWP)
Govinda Lake (21° 05.225′ N: 82° 22.415′ E; 389 m a.s.l.; GL), Mahasamund District; Quamar (2022a)	3600–2500 calyr BP	open forest vegetation	dry climate with reduced monsoon
	2500–1650 calyr BP	mixed tropical deciduous forest	warm and humid climate with increased monsoon (RWP)
	1650–950 calyr BP	open mixed tropical deciduous forest	warm and relatively less humid; reduced monsoon (DACP)
	950 calyr BP- Present	mixed tropical deciduous forest	warm and relatively more humid climate with increased monsoon (MCA) of strong summer monsoon

(continued)

Table 1 (continued)

Area/locality	Period (yr BP; unless otherwise stated)	Vegetation dynamics	Inferred climate change and hydro-climate variability
Deeporwetland (26°03′ N–26°09′N and long. 90°36’–90°41’E), Kamrup District, Assam; Tripathi et al. (2020)	13,040 and 11,700 cal BP	tropical tree savannah vegetation	cool and dry climate
	10,170 and 8390 cal BP	inception of tropical mixed deciduous forest	relatively less cool and dry climate
	8390 and 4640 cal BP	enrichment of tropical mixed deciduous forest	warm and moderate humid climate
	4640 and 3450 cal BP	relative decline in Sal associates	relatively less humid climate
	3450 and 2250 cal BP	decline in forest tree taxa	less warm and humid climate
	2250 cal. BP onwards	approaching towards the settlement of tropical mixed deciduous forest	warm and moderately humid climate
	1500 and 710 cal BP	tropical mixed deciduous forest	warm and increased humid climate (MCA)
	710 cal BP to Present	deterioration of tropical mixed deciduous forest	warm and relatively dry climate (LIA)
Bhujbas (30–31°N: 79.5°E), Gangotri Glacier, Uttarkashi District Uttarakhand; Kar et al. (2002)	ca. 2000	open *juniperus-betula* forests	comparatively cooler and moist
	ca. 1700	increase in local arboreals (*juniperus*, *betula*, *salix*) and extra local elements (mainly *pinus*)	further amelioration of climate with increase of both precipitation and temperature
	ca. 850	a shift in the vegetational pattern with sharp increase in *ephedra* and other	drier climatic conditions (LIA)
	during recent times	steppe elements notably *artemisia* and members of asteraceae	warm and moist

(continued)

Table 1 (continued)

Area/locality	Period (yr BP; unless otherwise stated)	Vegetation dynamics	Inferred climate change and hydro-climate variability
Paradise Lake (27° 30.324′ N: 92° 06.269′ E), Sela Pass, Arunachal Pradesh, Eastern Himalaya; Bhattacharyya et al. (2007)	ca. 1800 yrs BP (around AD 245)	conifer/broad-leaved forests	warm and moist
	ca. 1100 years BP (CE 985)	sub-alpine forests	more warmer (MWP/MCA)
	ca. 550 yrs BP (ca. CE 1400)	decrease in the values of *Tsuga*, *Juniperus* and *Quercus*	cooler and less moist (LIA)
Ganga Lake (27° 04′ 28″-27° 04′ 36″-N: 93° 34′ 04″–93° 34′ 22″ E), Itanagar, Arunachal Pradesh; Achyuthan et al. (2013)		5 warmer phases, intercepted by cold and moist climatic conditions, occurring throughout the LIA. The warmer events are of shorter duration during the LIA, ultimately indicating that LIA was highly variable and was not globally synchronous cold period	
Chandra valley (33° 79′ 40.5″ N: 78° 34′ 27.4″E), Lahaul, NW Himalaya; Rawat et al. (2015)	ca. 12,880–11,640	–	weakening of ISM corresponding to theYD cold event
	ca. 11,640–8810	–	increased strength of ISM
	ca. 10,398–9778	–	cold and dry closely linked with bond event
	ca. 8810–8117	–	cold-dry corresponding to global 8.2 ka cold event
	ca. 4808- 4327	–	cold-dry; 4.2 ka cold-arid
	ca. 1303–1609	–	period corresponding to little ice age (LIA) event
	ca. 6732–3337	–	warm and wet Holocene climate optimum (HCO) period
	ca. 1158–647	expansion of thermophillous broad leaved taxa and effective growth of meadow vegetation	warm and moist climate corresponds with the medieval warm period (MWP/MCA)

(continued)

Table 1 (continued)

Area/locality	Period (yr BP; unless otherwise stated)	Vegetation dynamics	Inferred climate change and hydro-climate variability
Pindari Glacier area (30° 15′ 36″ N: 79° 59′ 52″ E); Bali et al. (2015)	ca. 7 ka	glacial sediments	cold and dry
	7.0–4.9 ka	conifers and broad-leaved taxa	amelioration of climate
	4.9–1.7 ka	acceleration of alpine herbaceous taxa; decline in conifers and broad-leaved taxa	cold and dry
	1.7–0.9 ka	increase in conifers and broad-leaved taxa	warm and moist (MCA)
	0.9–0.2 ka	decline in conifers and broad-leaved taxa	cold-dry (LIA)
	0.2-Present	conifers, broad-leaved and herbaceous taxa	warm and moist
Triloknath Glacier (32° 39′ 25.51″–32° 36′ 19.57″ N: 76° 39′ 32.71″- 76° 36′ 09.14″E), Lahaul Valley, NW Himalaya; Bali et al. (2015)	6300–5379	broad-leaved birch (*Betula*) forests	warm and moderately humid
	5379–3167	decline of broad-leaved birch (*Betula*) forests	warm and less humid
	3167–2228	coniferous forest	severe climate (cold and dry)
	2228–962	expansion of broad-leaved birch (*Betula*) forests and improvement of coniferous forest	warm and humid
	962–300	further increase in broad-leaved taxa	warm and more humid (MCA)
	ca. 300	Sharp decline in broad-leaved taxa	cold and dry (LIA)

(continued)

Table 1 (continued)

Area/locality	Period (yr BP; unless otherwise stated)	Vegetation dynamics	Inferred climate change and hydro-climate variability
Bajalta Lake (32° 45.621' N: 74° 57.026' E), Jammu; Jammu and Kashmir, Quamar (2022b)	ca. 3205–2485	mixed broad-leaved/conifer forests	warm and humid with increased monsoon precipitation
	ca. 2485–1585	mixed conifer/broad-leaved forests	cool and dry with reduced monsoon precipitation
	ca. 1585–865	further increase in conifers	deterioration of climate, reduced monsoon precipitation (DACP)
	since ca.865 calyr BP to Present (CE1085 onwards)	increase in broad-leaved taxa and simultaneous reduction of conifers	warm and humid with increased monsoon precipitation (partly corresponding to the Medieval Warm Period (MWP/MCA) between CE 740 and 1150)
Darjeeling Himalaya (27°01′ 45.24″ N: 88° 19′18.71″ E); Ghosh et al. (2018)	during 364 BCE to 131 CE between 131 CE and 624	dense broad-leaved evergreen forest	Humid (MCA) comparatively drier condition
	during 1118 CE between 1367 CE and 1802	a thinning of the forest cover	wet phase (increase in monsoonal strength)

different geographic regions. As a consequence, its occurrence is considered asynchronous. For instance, in the North Sea region, this warm period spanned from 2400 to 1600 cal yr BP (Hass 1996). On the Qinghai-Tibet Plateau, the timeframe was approximately 2500 to 2100 cal yr BP (Ji et al. 2005). Southwest Greenland experienced the RWP from around 2700 to 1600 calibrated years before the present (Seidenkrantz et al. 2007). Similarly, in southern Spain, the warm period persisted from 2600 to 1600 cal yr BP (Martin-Puertas et al. 2009).

Quamar et al. (2021) proposed the existence of a warm and humid climate approximately spanning from 2600 to 1800 cal yr BP (750 BCE/BC to 150 CE/AD) around a lacustrine region located in the Bemetara District of Chhattisgarh State, central India. Furthermore, their research sheds light on the gradual strengthening of the monsoon (the Indian Summer Monsoon rainfall: ISMR) occurring between roughly 2600 and 1800 cal yr BP, which contrasts the prevailing trend of weakening during the Late Holocene. Specifically, the investigation indicates the presence of a warm and moderately humid climate with moderate monsoon rainfall in the study area around 2600 to 2200 cal yr BP. This climate gradually transformed into a warmer and comparatively more humid environment, accompanied by a heightened intensity in monsoon precipitation. This transformation supported the growth of dense mixed tropical deciduous forest from approximately 2000 to 1800 cal yr BP. The study further suggests that the climate evolved into a warm and humid state with increased monsoon precipitation, fostering the development of mixed tropical deciduous forest within the time frame of 2200 to 2000 cal yr BP in the vicinity of the study area. Additionally, Quamar (2022a) has pointed out a period of warm and humid climate accompanied by increased monsoon precipitation approximately spanning from 2500 to 1650 cal yr BP (~550 BCE to 300 CE). This climatic condition fostered the growth of mixed tropical deciduous forest in the landscape surrounding the lacustrine region situated in the Mahasamund District of Chhattisgarh. This warming trend and the increased Indian Summer Monsoon (ISM) rainfall, which is correlatable with the Roman Warm Period (RWP), were also concurrent with the ascendance of the Roman Empire in Europe and the Maurya Empire in the Indian subcontinent (McCormick et al. 2012; Singh 2008). In the Indian subcontinent, the increased strength of the ISM rainfall led to an augmentation in agricultural production, consequently fostering socio-economic prosperity during two time intervals: firstly, between approximately 2600 and 1800 cal yr BP (750 BCE/BC to 150 CE/AD) in the Bemetara District, and secondly, from around 2500 to 1650 cal yr BP (~550 BCE to 300 CE) in the Mahasamund District (Gupta et al. 2006; McCormick et al. 2012; Singh 2008). In the Indian subcontinent, the agricultural production boosted as a result of increase in the strength of the ISM rainfall, which brought socio-economic prosperity during the time interval [between ca. 2600–1800 cal yr BP (750 BCE/BC to CE/AD 150 in the Bemetara District) and between ca. 2500 and 1650 cal yr BP (~550 BCE to CE 300 in the Mahasamund District] (Gupta et al. 2006; McCromick et al. 2012; Singh 2008).

This period is famously referred to as the 'Golden Age of India,' marking a pinnacle in the region's economy owing to extensive agricultural growth and surplus trade (McCrindle 1877; Singh et al. 2020). Prominent hubs of trade, industry, and commerce emerged in the Indian subcontinent, and trade connections expanded to Europe, Mesopotamia, Egypt, and even western regions extending to the African coast. This led to improved societal well-being within this densely populated global region (Singh et al. 2020). The RWP has been characterized as an epoch when the circum-North Atlantic vicinity experienced warmth, potentially attributed to variations in solar activity or intrinsic North Atlantic dynamics (Martin-Puertas et al. 2009; Büntgen et al. 2011). Furthermore, Solanki et al. (2004) have proposed that the heightened rainfall during the RWP coincided with increased solar insolation and a concurrent northward shift of the Inter-Tropical Convergence Zone (ITCZ).

The observed phenomena, such as heightened wind intensity in the Arabian Sea (Gupta et al. 2003), increased negativity in oxygen isotopes within speleothems from Uttarakhand (Kotlia et al. 2015; Kathayat et al. 2017), elevated stream discharge in Ladakh's Tso Moriri Lake (Dutt et al. 2018), and a notable proliferation of leguminous plants in the Banni grassland (Pillai et al. 2018) are also indicative of the concurrent amplified rainfall conditions during the RWP. Singh et al. (2020) employed multiple proxy, such as grain size, $\delta^{13}C$, $\delta^{15}N$, as well as TOC and TN analyses, to document the occurrence of the RWP from approximately 550 BCE to 450 CE in the Himachal Pradesh (northwestern Himalaya), India. Similarly, Gupta et al. (2003) noted intensified wind activity in the Arabian Sea, while Kotlia et al. (2015) and Kathayat et al. (2017) observed more negative oxygen isotopes in speleothems from Uttarakhand. Similarly, Dutt et al. (2018) highlighted high stream discharge in Ladakh's Tso Moriri Lake, and Pillai et al. (2018) reported an abundance of leguminous plants in the Banni grassland. All these findings further substantiate the association of the RWP with increased rainfall conditions. Naidu et al. (2020) correlated the RWP with wetter Indian Summer Monsoon (ISM) rainfall during roughly 1950 to 1550 cal yr BP. This connection was drawn from the higher $\delta^{18}O_w$ values obtained from the Bay of Bengal, India. Banerji et al. (2021) also put forth the idea of intensified Indian Summer Monsoon (ISM) activity during the RWP, specifically spanning from 2000 to 1300 cal yr BP. They drew this conclusion from an array of geochemical investigations conducted on samples extracted from the mudflats of Diu Island in Southern Saurashtra, Gujarat, Western India. However, a different perspective was presented by Huang et al. (2023) regarding the climatic conditions during the RWP (CE 50–400) in southern China. Their findings, derived from milti-proxy study, such as grain size, elemental ratios (Zr/Rb and Rb/Sr), total organic carbon (TOC), total nitrogen (TN), and C/N ratios, along with precise $^{137}Cs/^{210}Pb$ dating and accelerator mass spectrometry (AMS) ^{14}C ages of a sediment core from Lake Baxian, suggested a comparatively arid climate prevailing during this period.

2.2 *Dark Ages Cold Period (DACP)*

The DACP, spanning from CE 400 to 765 (1185 to 1550 cal yr BP), denotes a period characterized by cold climate and reduced Indian Summer Monsoon rainfall (ISMR) (Helama et al. 2017). They also propose the existence of a broader cooling phase, known as the 'Late Antique Little Ice Age' (LALIA), which coincides with the DACP and is believed to be influenced by volcanic aerosols and solar irradiance variations. These factors contributed to the climatic downturn from CE 536 onward. According to McCrindle (1877) and Singh et al. (2020), this interval of weakened ISM rainfall aligns with significant dynastic upheavals and the decline of the Gupta dynasty in India. Furthermore, in Europe, a transition from Late Antiquity to the Early Middle Ages accompanied by large-scale human migrations was observed (Wanner et al. 2011). The period is also associated with the occurrence of the North Atlantic ice-rafting event around 1400 years ago (Bond et al. 2001; Helama et al. 2017). Solar proxy data further indicate reduced solar activity during this phase, which is linked to the negative phase of the North Atlantic Oscillation (NAO) and/or the El Niño–Southern Oscillation (ENSO), contributing to colder conditions in the Northern Hemisphere (Gray et al. 2010; Helama et al. 2017; Singh et al. 2020).

Quamar (2022a) indicated a warm and relatively less humid climate marked by decreased monsoon rainfall roughly spanning from 1650 to 950 cal yr BP (CE 300–1000). This climate fostered the presence of an open mixed tropical deciduous forest in the vicinity of a lacustrine area situated in the Mahasamund District, Chhattisgarh. The observed reduction in humidity and diminished monsoon rainfall correlates with the DACP. Furthermore, Quamar (2022b) proposed a cool and dry climate, signifying a move towards aridity, accompanied by reduced monsoon precipitation during the period of approximately 1585 to 865 cal yr BP (CE 400–765; 1185–1550 cal yr BP). This climate supported a mixed coniferous and broad-leaved forest in the landscape surrounding Bajalta Lake in the Jammu District of Jammu and Kashmir, India. This timeframe corresponds with the DACP. However, Singh et al. (2020) indicated weak Indian Summer Monsoon (ISM) conditions spanning from CE 450 to 950. They referred to this phase of reduced ISM as the Dark Age or Migration Period Cooling. This categorization was based on analyses of grain size, stable isotope ratios ($\delta^{13}C$ and $\delta^{15}N$), and organic geochemistry (TOC and TN) data collected from Himachal Pradesh (northwestern Himalaya), India. Naidu et al. (2020) similarly proposed diminished ISM rainfall between 1550 and 1250 yr BP, establishing this link based on higher $\delta^{18}O_w$ values obtained from the Bay of Bengal, India. They associated this phenomenon with the DACP. Banerji et al. (2021), based on various geochemical studies, suggested the presence of a weakened ISM during the DACP, encompassing the period from 1300 to 800 cal yr BP from the Diu Island of Southern Saurashtra, Gujarat (Western India). Sodhi et al. (2022) proposed the DACP as a localized extreme arid hydrological event occurring around ~ CE 700 to 900, based on data collected from the Gulf of Khambhat in Western India. Conversely, Sanwal et al. (2013) suggested a gradual strengthening of the Indian Summer Monsoon (ISM) during the DACP (CE 210–810) using evidence from the Dharamshala cave in Uttarakhand (Indian Central Himalaya). Mc Dermott et al. (2001) also identified increased rainfall during the DACP through a high-resolution speleothem $\delta^{18}O$ record from SW Ireland. These inconsistencies in the DACP records underscore its lack of coherence within the Indian subcontinent, suggesting that this period is not universally synchronized as a cold phase. Reddy et al. (2022) indicated a wetter climate prevailing during the DACP from approximately CE ~ 250 to 800, based on findings from the Kadapa cave in Peninsular India. They also asserted that the Intertropical Convergence Zone (ITCZ) remained stable and positioned further northward, leading to pronounced La Niña conditions in the tropical Pacific. This phenomenon contributed to the wet ISM conditions during the DACP. Additionally, a wetter ISM phase during the DACP was observed in sediment records from the northern Arabian Sea (von Rad et al. 1999), in the Andaman region (Laskar et al. 2013), and from the Himalayas using stalagmite records (Sanwal et al. 2013). Huang et al. (2023), based on multiproxy study, encompassing grain size, elemental ratios (Zr/Rb and Rb/Sr), total organic carbon (TOC), total nitrogen (TN), and C/N ratios, alongside high-precision $^{137}Cs/^{210}Pb$ and accelerator mass spectrometry (AMS) ^{14}C ages of a sediment core extracted from Lake Baxian in southern China, also proposed a wetter DACP period occurring around CE 500–900.

2.3 Medieval Climate Anomaly (MCA)

The Medieval Climate Anomaly (MCA), also recognized as the Medieval Warm Period (MWP) and Medieval Climate Optimum (MCO), denotes a period of elevated temperatures in the North Atlantic region, spanning approximately from CE 950 to 1250 (Lamb 1965; Bradely 1985; Bradely et al. 2003; Mann et al. 2009). This climatic event was succeeded by a cooler phase known as the Little Ice Age (LIA) in the North Atlantic region. The global uniformity of the MCA is a topic of debate, as various perspectives exist regarding its spatial and temporal records. For instance, the MCA was documented in India between approximately CE 750 (800) and 1200, while in regions like Greenland, Russia, and North America, it extended from CE 950 to 1200 (1150). In the majority of Europe, the MCA was observed between CE 1150 and 1300, and in New Zealand, it occurred from CE 1200 to 1400 (Lamb 1965; Bradely 1985; Bradley et al. 2003; Mann et al. 2009).

Pollen records retrieved from the Kiktiha Swamp, located in the Shahdol District of Madhya Pradesh (Chauhan and Quamar 2010), offer insights into a warm and humid climate characterized by increased monsoon precipitation from 1600 to 700 yr BP (CE 350 to 1250). This climatic condition supported the presence of tropical deciduous sal (*Shorea robusta* Gaertn. f.) forests within the study area and corresponds with the globally recognized MCA occurring approximately from CE 750 [740] to 1200 [1150] (Lamb 1977; Anderson et al. 2002). Quamar and Chauhan (2014) proposed a warm and moderately humid climate with moderate monsoon precipitation extending from 1416 to 506 yr BP (CE 534 to 1444). This climatic setting facilitated the growth of open mixed tropical deciduous forests in the vicinity of the Khedla Quila Lake in the Betul District of southwestern Madhya Pradesh, central India. This period of time partially overlaps with the MCA. Additionally, Quamar and Bera (2021) identified a warm and humid climate characterized by increased monsoon precipitation between approximately 1200 and 500 cal yr BP (CE 750–1450). This climatic state supported the existence of mixed-tropical deciduous forests around the study area (Kodamsar Lake) in the Korba District of Chhattisgarh, central India. This inferred climate condition falls within the timeframe of the MCA. Furthermore, Quamar (2022a) indicated a relative improvement in climate due to the intensification of Indian Summer Monsoon (ISM) rainfall since around 950 cal yr BP to the present (CE 1000 onwards). This climatic enhancement supported the growth of mixed tropical deciduous forests around the area of Govinda Lake in the Mahasamund District of Chhattisgarh State, central India. This ameliorated climate partially coincides with the MCA and aligns with the northward shift of the Intertropical Convergence Zone (ITCZ) (Haug et al. 2001). Tripathi et al. (2020) documented the MCA during 1500 to 710 cal yr BP in the Deepor wetland of the Kamrup District, Assam, Northeast India. Quamar et al. (2023) also suggested the MCA between 731 and 576 cal yr BP (CE 1219–1374) from the Western Ghats, India. Sharma and Chauhan (2001a, b) identified the MCA from Kupup Lake in the Sikkim Himalaya between ~1450 and 450 cal yr BP. Kar et al. (2002) proposed the occurrence of the MCA between ~1750 and 850 cal yr BP from the Gangotri Glacier, Uttarakhand, India. Bhattacharyya et al. (2007) highlighted a relatively warmer climatic condition ~1100 cal yr BP from the Paradise Lake, Arunachal Pradesh, and correlated this event with the MCA. The MCA was identified to have taken place between 1750 and 900 cal yr BP from the Pindari Glacier, Kumaun (Bali et al. 2015). Additionally, Bali et al. (2017) suggested the occurrence of the MCA between ~962 and 300 cal yr BP from the Triloknath Glacier, Lahaul. Rawat et al. (2015) pinpointed the MCA from 1158 to 647 cal yr BP. The MCA was proposed to have spanned from 364 BCE to 131 CE from the Darjeeling Himalaya (Ghosh et al. 2018). Kotlia and Joshi (2013) outlined the MCA between ~920 and 440 cal yr BP (CE 1080–1560) from the Garhwal Himalaya. Additionally, a warm climatic phase corresponding to CE 1300 was identified in the Higher Central Himalaya (Rühland et al. 2006), and a similar period of warmth and moisture was suggested from the Naychhuddwari bog in Himachal Pradesh, spanning between CE 700 and 1250 (Chauhan 2006). Quamar (2022b) proposed a warm and humid climate accompanied by increased monsoon precipitation since 865 yr BP to the present from the Bajalta Lake area of the Jammu District, Jammu and Kashmir, India. This climate supported the growth of a broad-leaved and coniferous forest, which was partly associated with the MCA. The MCA has also been indicated in various regions, including central India (Sinha et al. 2011), the Ganga basin (Singh et al. 2015), the northwestern Himalaya (Dutt et al. 2018; Pillai et al. 2018; Singh et al. 2020), the western Arabian Sea (Gupta et al. 2005), the Bay of Bengal (Naidu et al. 2020), the Diu Island in Southern Saurashtra, Gujarat, Western India (MCA: 1300–800 cal yr BP; Banerji et al. 2021), the Kadapa cave in peninsular India (~900 and 1300; Reddy et al. 2022), the Gulf of Khambhat in Western India (CE 900–1400; Sodhi et al. 2022), and the southeastern Tibetan Plateau (Li et al. 2014a, b).

It is noteworthy to acknowledge that the spatial and temporal records of the MCA have displayed notable heterogeneity. The MCA exhibits variability in its documentation across the Indian sub-continent (Kotlia and Joshi 2013; Dixit and Tandon 2016; Kar and Quamar 2020a, b). For instance, Sanwal et al. (2013) proposed a drier climatic condition during the MCA in the Indian Central Himalaya, whereas Chauhan and Quamar (2010) and Quamar and Chauhan (2014) put forward evidence for a warm, humid climate with intensified monsoonal rainfall during the MCA in central India (specifically eastern and southwestern Madhya Pradesh). Furthermore, Kotlia et al. (2015) reported drier climatic conditions during the MCA from the Sainji Cave, Uttarakhand, Central Himalaya, India. du Plessis et al. (2021) observed a relatively dry and potentially slightly cooler climate compared to the present during the MCA (CE 900 to 1300) in the southern Cape coast of South Africa (MCA CE ~950–1250). In contrast, Chen et al. (2006) suggested a warm and arid climate during the MCA from Bosten Lake, south of Xinjiang, China, whereas Zhang et al. (2009) projected a humid climate in the north of Xinjiang. Additionally, Prasad et al. (2014) and Xu et al. (2016) highlighted dry Indian Summer Monsoon (ISM) climates during the MCA, respectively from central India and the northern Tibetan Plateau. These findings stemmed from variations in moisture patterns contributing to regional ISM dynamics. Huang et al. (2023) also indicated a dry climate during the MCA (CE 1000–1300) from southern China (Fig. 2).

2.4 Little Ice Age (LIA)

The Little Ice Age (LIA) marked a cooling phase, which conventionally extends from CE 1350 to 1850, succeeding the Medieval Climate Anomaly (MCA) (Anderson et al. 2002). Matthes (1939) coined the term "LIA" to characterize Late Holocene glacier growth in the Sierra Nevada Mountains. Furthermore, the concept of the LIA was employed to describe inferred Late Holocene glacial advances across numerous regions, notably the European Alps (Grove 1988; Matthews and Briffa 2005). The temporal occurrence of the LIA was not uniform globally, leading to varying perspectives on its spatial and chronological records. Masson-Delmotte et al. (2013) and Mann et al. (2009) indicated that the LIA roughly spanned between CE 1300–1850. Lamb (1982, 1995) proposed the duration of the LIA as CE 1300–1800. Bradley (1985) suggested the LIA occurred from around CE 1440–1850, whereas Bradley and Jones (1993) proposed a time frame of CE 1550–1850 (for LIA). Sun et al. (2020) pointed to the LIA as approximately spanning from CE 1400 to ~1900. The LIA corresponds to Bond Event 0 (CE 1500) and is recognized as one of the coldest multi-decadal to multi-century periods within the Holocene (Bradley et al. 2003). These Bond Events are instances of North Atlantic ice-rafting events during the Holocene, influenced by solar cycle dynamics (Bond et al. 2001).

Pollen records spanning from 700 to 300 yr BP (CE 1250 to 1650) reveal a warm and relatively less humid climate with weakened southwest monsoon rainfall, which supported a sparser and less diverse mixed tropical deciduous forest near the Kiktiha Swamp in the Shahdol District (Madhya Pradesh), central India (Chauhan and Quamar 2010). This climatic interpretation falls within the timeframe of the LIA. Quamar and Chauhan (2014) proposed that between 506 and 120 years BP (CE 1444–1830), a warm and humid climate prevailed around the Khedla Quila Lake in the Betul District (southwestern Madhya Pradesh), India, fostering a dense mixed tropical deciduous forest. However, their attempt to correlate this climatic event with

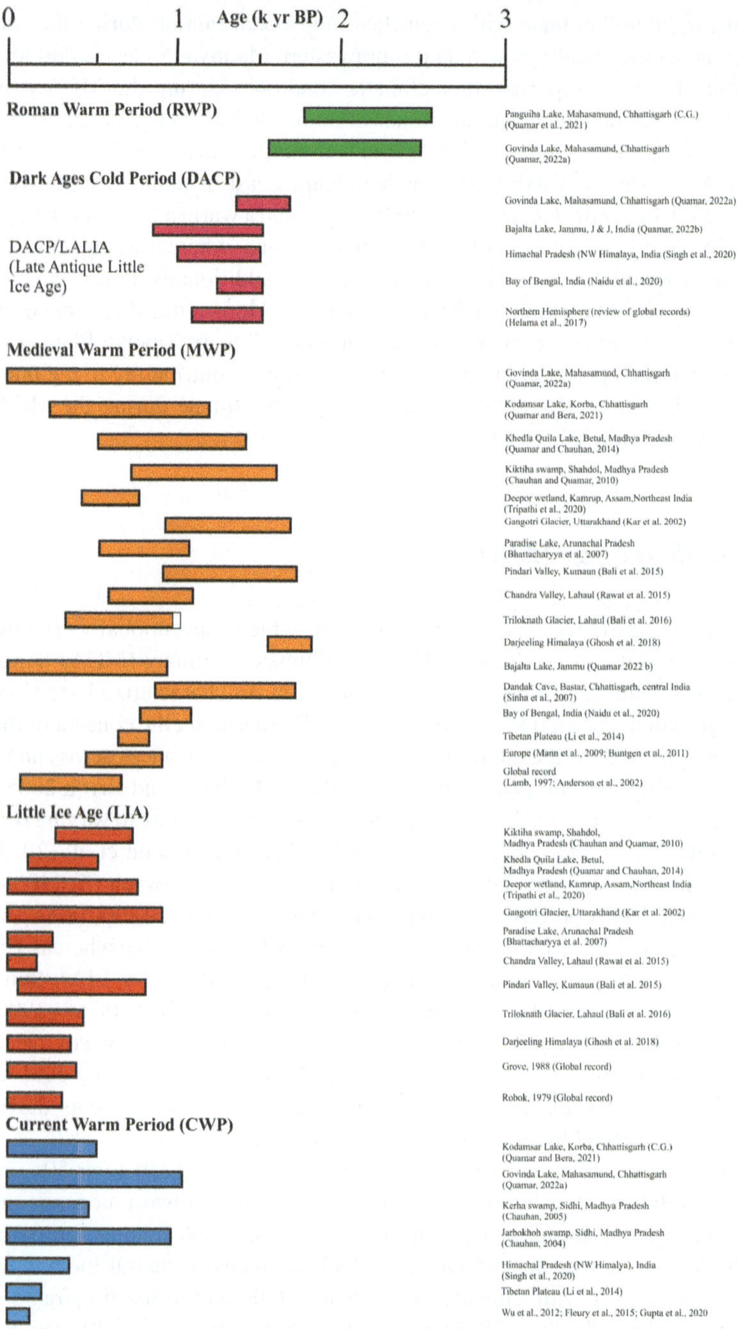

Fig. 2 Correlation diagram showing the spatio-temporal variation in the palaeoclimatic records in global contextualization (the RWP, DACP, MCA, LIA and the CWP)

the global LIA might be incorrect, potentially stemming from the coincidental timeframe. Nevertheless, this alignment should not be taken for granted. The LIA has also been documented in the Deepor wetland of Kamrup District (Assam), Northeast India, from 710 cal yr BP to the present (Tripathi et al. 2020). Sharma and Chauhan (2001a, b) recorded the LIA between ~450 and 200 yr BP at Kupup Lake in the Sikkim Himalaya. The LIA was also identified from the Gangotri Glacier (Uttarakhand) between ~850 years BP to the present (Kar et al. 2002). Bhattacharyya et al. (2007) observed a cooler and less humid climate ~550 yr BP at the Paradise Lake, Arunachal Pradesh, which they correlated with that of the LIA. Additionally, Bali et al. (2015) documented the LIA between 900 and 200 yr BP from the Pindari Glacier, Kumaun Himalaya. Rawat et al. (2015) identified the period of the LIA during 647 to 341 cal yr BP (CE 1303 to 1609). Bali et al. (2017) detected signatures of the LIA since ~300 cal yr BP (CE ~1650) to the present from the Triloknath Glacier, Lahaul. From the Darjeeling Himalaya, Ghosh et al. (2018) proposed the timeframe for the LIA as CE 1367 to 1802. Banerji et al. (2021) also recognized signs of the LIA (evidenced by weak Indian Summer Monsoon) between 800 and 200 cal yr BP, based on the various geochemical proxy records from the Diu Island of Southern Saurashtra, Gujarat, Western India. Yadav and Singh (2002) reported a cooler phase during the LIA between CE 1600 and 1950 in the Western Himalaya, while Cook et al. (2003) identified a similar cooling trend in Nepal between CE 1605 and 1770 using tree ring records. However, contrasting perspectives emerge as well. Kotlia and Joshi (2013) and Kotlia et al. (2015) and Sanwal et al. (2013) suggested a wetter LIA period in the Himalayas, attributed to the influence of strong westerlies. Xu et al. (2016) indicated a wet climate during the LIA in the northern Tibetan Plateau, linking it to spatio-temporal variations in moisture patterns associated with regional Indian Summer Monsoon dynamics. Huang et al. (2023) also put forth evidence for a wet climate during the LIA (CE 1400–1850) in southern China. Quamar et al. (2023) also suggested wet (moist) LIA between CE 1671 and 1942 from the Western Ghats, India.

However, it is imperative to acknowledge that the LIA has not exhibited spatial and temporal coherence and consistency either. The records of the LIA display variation in the Indian sub-continent (Kotlia and Joshi 2013; Quamar and Chauhan, 2014; Dixit and Tandon 2016; Kar and Quamar 2020a, b; Quamar et al., 2023 and references therein). For instance, Kotlia et al. (2015) indicated increased precipitation in the Indian Central Himalaya during the LIA, while Sinha et al. (2010) suggested mega drought in the central Indian Core Monsoon Zone (CMZ). Additionally, Quamar and Chauhan (2014) observed a humid climate with increased monsoonal rainfall during the LIA in southwestern Madhya Pradesh, central India, falling in the CMZ. Consequently, in terms of inconsistency, the findings of Quamar and Chauhan (2014) stands right and valid, and justified as well. Misra et al. (2015) and Ruhland et al. (2006) also proposed relatively higher rainfall during the LIA. Singh et al. (2021) identified increased precipitation during the LIA in Kishtwar, Jammu and Kashmir, in the Northwest Himalaya, India. Achyuthan et al. (2013) pointed to shorter, warmer events occurring during the LIA in Itanagar, Arunachal Pradesh, India. Similarly, du Plessis et al. (2021) described cool and dry climatic conditions prevailing during the LIA (~CE 1600 to 1850; LIA ~CE 1300–1850) in the southern Cape coast of South Africa. The potential reasons for the divergence in LIA records across the Indian sub-continent could be attributed to the distinct moisture sources these regions receive. The Indian Central Himalaya is subject to precipitation from both the Arabian Sea (AS) and the Bay of Bengal (BoB) branches of the Indian Summer Monsoon (ISM) (Kotlia et al. 2015). In contrast, central India, which constitutes the core of the monsoon zone (CMZ), predominantly receives its rainfall from the AS branch (Sinha et al. 2005). Furthermore, Kishtwar (Jammu and Kashmir), Northwest Himalaya, India, experience the combined influence of both the ISM and Western Disturbances (WDs) (Kar and Quamar 2020a, b; Singh et al. 2021; Quamar et al., 2024).

Dixit and Tandon (2016) proposed a connection between the cold and dry period along with a weakened Indian Summer Monsoon (ISM) during the LIA and the intensified Westerlies over Northwest India. This intensification was attributed to the strengthened Asian westerly jet stream over the Middle East during a positive phase of the North Atlantic Oscillation (NAO), along with the migration of the Asian jet to lower latitudes during the warm phase of El Niño-Southern Oscillation (ENSO). The reduced ISM rainfall significantly impacted agricultural productivity, leading to economic collapse in the Indian subcontinent (Dutt et al. 2020). This economic decline, coupled with foreign invasions, plundering, and internal conflicts over resources, resulted in significant socio-economic disruptions and the establishment of foreign rule in India. The weakest phase of the LIA spanned approximately 100 years and coincided with the Maunder Minimum (CE 1645 to 1710), a period marked by very low temperatures and minimal sunspot activity (Gupta et al. 2019; Eddy 1976; Mann and Jones 2003). Furthermore, the subdued ISM across the Indian subcontinent during the LIA may be attributed to the southward shift of the Intertropical Convergence Zone (ITCZ) owing to increased northward energy flux across the equator during a colder northern hemisphere. Various studies, including Singh et al. (2020), Dutt et al. (2018), Bischoff and Schneider (2014), Kathayat et al. (2017), Laskar et al. (2013), Rawat et al. (2015), and Das et al. (2017), support this hypothesis. In particular, El Niño-Southern Oscillation (ENSO) plays a role in delaying the northward movement of the ITCZ, thereby prolonging the onset of the summer monsoon and leading to reduced rainfall (Sinha et al. 2011).

2.5 *Current Warm Period (CWP)*

The Current Warm Period (CWP) marks a period of warming and increased Indian Summer Monsoon rainfall (ISMR) conditions, documented globally from CE 1800 to the present day (post ca. CE 1900; Wu et al. 2012; Fleury et al. 2015; Neukom et al. 2019; Gupta et al. 2020). The increase in the ISMR during the CWP is attributed to rising global temperatures, which coincide with the era of industrialization and other human-induced factors (Masson-Delmotte et al. 2013). This phenomenon is coupled with an associated rise in surface evaporation and convection in the Indian Ocean

(Anderson et al. 2002; Wang et al. 2005a, b; Sinha et al. 2011). The strength of the ISM over the last 1300 years has predominantly been influenced by solar insolation and the intensity of El Niño events (Cook et al. 2010; Sinha et al. 2011).

Quamar and Bera (2021) suggested a warm and more humid climate since ~478 cal yr BP (CE/AD 1472) onwards, which supported a dense mixed tropical deciduous forest ecosystem around the Kodamsar Lake in the Korba District (Chhattisgarh), central India. This warming phase falls within the time span of the Current Warm Period (CWP; AD [CE] 1800 to the present; Gupta et al. 2020). Singh et al. (2020), based on analyses of grain size, $\delta^{13}C$ and $\delta^{15}N$, as well as TOC and TN from Himachal Pradesh (NW Himalaya), India, also identified a period of CWP occurring ~1600–2000 AD (CE). Li et al. (2014a, b), using AMS ^{14}C dating, along with LOI, TOC, TN, grain size, and magnetic susceptibility (MS) analyses of sediments from Basomtso Lake, southeastern Tibetan Plateau, proposed that elevated levels of LOI, TOC, TN, coarse silt, sand, and magnetic susceptibility in the sediment indicated heightened sediment input possibly due to warmer climatic conditions and increased glacial meltwater input during AD (CE) 1790–2012, corresponding well with the CWP. The intensification of the monsoon during the CWP is attributed to escalating global temperatures, coinciding with the industrial revolution and other human-induced factors, in addition to the correlated escalation of surface evaporation and convection in the Indian Ocean (Anderson et al. 2002; Wang et al. 2005a, b; Sinha et al. 2011). Solar insolation and El Niño intensity stand out as primary driving forces influencing the strength of the Indian Summer Monsoon (ISM) over the past 1300 years (Cook et al. 2010; Sinha et al. 2011).

Chauhan (2004) proposed a period of warm and increased moist climate since ~800 yr BP (AD/CE 1150) onwards around the Jarbokhoh Swamp in the Sidhi District of Madhya Pradesh, central India. This climatic condition fostered a dense tropical deciduous sal (*Shorea robusta* Gaertn. f.) around the landscape of the study area. Similarly, Chauhan (2005) noted a warm and moist climate with the timely onset of a strong summer monsoon since 400 yr BP (CE 1650) onwards. This phase supported the growth of tropical deciduous sal (*Shorea robusta* Gaertn. f.) forest around the Kerha Swamp in the Sidhi District. It is worth mentioning that Chauhan (2004, 2005) didn't draw a connection between this warm and moist climate and the CWP, even though it indeed falls within the timeframe of the CWP (CE 1800 to the present).

3 Conclusions

Pollen grains serve as one of the most important and robust proxies used to reconstruct past climates. This comprehensive review article delves into pollen-based studies of climate changes over the last two millennia in the context of global climate change scenarios within the Indian subcontinent. Investigations into paleoclimates during the Quaternary Period play a pivotal role in comprehending climatic processes and are indispensable for understanding future climatic shifts. These studies hold the potential to enhance our understanding of future climate trends through the simulation of paleoclimatic models, thereby shedding light on the potential impacts on the Indian subcontinent and their societal implications (Royer 2008; Cai et al. 2010; Singhvi et al. 2010). The traces left by significant global climatic events, such as the Roman Warm Period (RWP), Dark Ages Cold Period (DACP), Medieval Climate Anomaly (MCA), Little Ice Age (LIA), and Current Warm Period (CWP) across various regions of the Indian subcontinent illustrate the influence of these phenomena on contemporary climate and vegetation patterns. However, the hydroclimatic variations and climatic trends observed over the past two millennia exhibit inconsistencies and a lack of synchronous response due to various factors, including solar influences, the latitudinal movement (north–south) of the Intertropical Convergence Zone (ITCZ), human activities, and more. These factors collectively contribute to the spatiotemporal variability observed in these climatic records.

To summarize, it can be affirmed that: (i) a strong ISM precipitation was evident during the RWP, MCA, and the CWP periods; and (ii) the ISM precipitation was notably weak during the DACP and the LIA, largely due to solar minima. However, these globally recognized climatic events exhibit spatiotemporal disparities (asynchronicity) in their records, both within the Indian context and worldwide.

The available data underscores the necessity for fine-resolution palaeoclimatic investigations using a multi-proxy approach within the Indian subcontinent. Such studies are essential for establishing a comprehensive and reliable database, which can play a crucial role in developing accurate palaeoclimatic models. These models are instrumental in predicting future climatic trends for a scientifically sound policy planning with a key aspect of societal relevance.

Acknowledgements I am thankful to Dr. (Mrs.) Vandana Prasad, Director, Birbal Sahni Institute of Palaeosciences (BSIP), Lucknow, India for providing the infrastructure facilities to complete the review work, and also for the permission to publish. Thanks are also due to Dr. Bandana Samant, Professor, P.G. Department of Geology, R.T.M. Nagpur University, Nagpur, India for the invitation to contribute this review article as a chapter in the book entitled "*Application of palynology in stratigraphy and climate studie*s" published by the Springer, Switzerland, and edited by her.

References

Achyuthan H, Farooqui A, Eastoe C et al (2013) A five-century long limnological and environmental record from northeastern India. In: Kotlia BS (ed) Holocene, Nova Science Publisher Inc., pp 129–143. ISBN: 978-1-62257-722-4

Anderson DM, Overpeck JT, Gupta AK (2002) Increase in the Asian southwest monsoon during the past four centuries. Science 297:596–599

Bali R, Ali SN, Bera SK et al (2015) Impact of anthropocene vis-à-vis Holocene climatic changes on central Indian Himalayan Glaciers. In: Lollino G et al (ed) Engineering geology for society and territory, vol 1. Springer International Publishing Switzerland. https://doi.org/10.1007/978-3-319-09300-0-89

Bali R, Chauhan MS, Mishra AK et al (2017) Vegetation and climate change in the temperate-subalpine belt of Himachal Pradesh since 6300 cal. yrs. BP, inferred from pollen evidence of Triloknath palaeolake. Quat Int 444:11–23. https://doi.org/10.1016/j.quaint.2016.07.057

Banerji US, Arulbalaji P, Padmalal D (2020) Holocene climate variability and Indian Summer Monsoon: an overview. The Holocene 30(5):744–773

Banerji US, Bhushan R, Joshi KB et al (2021) Hydroclimatic variability during the last two millennia from the mudflats of Diu Island, Western India. Geol J 56:3586–3604. https://doi.org/10.1002/gj.4116

Berkelhammer M, Sinha A, Stott L et al (2012) An abrupt shift in the Indian monsoon 4000 years ago. Am Geophys Union Geophys Monogr Ser 198:75–87

Bhattacharyya A, Sharma J, Shah SK, Chaudhary V (2007) Climatic changes during the last 1800 yrs BP from Paradise Lake, Sela Pass, Arunachal Pradesh, Northeast Himalaya. Curr Sci 93(7):983–987

Birks HJB, Birks HH (1980) Quaternary palaeoecology. University Park Press, p. 289

Bischoff T, Schneider T (2014) Energetic constraints on the position of the intertropical convergence zone. J Clim 27(13):4937–4951

Bond G, Kromer B, Beer J et al (2001) Persistent solar influence on North Atlantic climate during the Holocene. Science 294(5549):2130–2136

Bonnefille R, Anupama K, Barboni D et al (1999) Modern pollen spectra from tropical South India and Sri Lanka: Altitudinal distribution. J Biogeogr 26:1255–1280

Bradely RS (1985) Quaternary palaeoclimatology: methods of palaeoclimatic reconstruction. Boston Allen and Unwin, London

Bradley RS, Jones PD (1993) "Little Ice Age" summer temperature variations: their nature and relevance to recent global warming trends. The Holocene 3:367–376

Bradley RS, Hughes MK, Diaz HF (2003) Climate in medieval time. Science 302:404–405. https://doi.org/10.1126/science.1090372

Büntgen U, Tegel W, Nicolussi K et al (2011) 2500 years of European climate variability and human susceptibility. Science 331(6017):578–582

Cai Y, Tan L, Cheng H et al (2010) The variation of summer monsoon precipitation in central China since the last deglaciation. Earth Planet Sci Lett 291:21–31

Chauhan MS (2004) Late-Holocene vegetation and climatic changes in Eastern Madhya Pradesh. Gondwana Geol Mag 19(2):165–175

Chauhan MS (2005) Pollen record of vegetation and climatic changes in northeastern Madhya Pradesh during last 1600 years. Trop Ecol 46(2):265

Chauhan MS (2006) Late Holocene vegetation and climate change in the alpine belt of Himachal Pradesh, India. Curr Sci 91:1572–1578

Chauhan MS, Quamar MF(2010) Vegetation and climate change in southeastern Madhya Pradesh during Late Holocene, based on pollen evidence. J Geol Soc India 76(2):143–150. https://doi.org/10.1007/s12594-010-0084-y

Chen F, Huang X, Zhang J, Holmes JA, Chen J (2006) Humid little ice age in arid central Asia documented by Bosten Lake, Xinjiang, China. Sci China Ser D Earth Sci 49:1280–1290

Cook ER, Krusic PJ, Jones PD (2003) Dendroclimatic signals in long tree-ring chronologies from the Himalayas of Nepal. Int J Climatol 23:707–732

Cook ER, Anchukaitis KJ, Buckley BM et al (2010) Asian monsoon failure and mega drought during the last millennium. Science 328:486–489

Das M, Singh RK, Gupta AK et al (2017) Holocene strengthening of the Oxygen Minimum Zone in the northwestern Arabian Sea linked to changes in intermediate water circulation or Indian monsoon intensity? Palaeogeogr Palaeoclimatol palaeoecol 483:125–135

Dixit Y, Tandon SK (2016) Hydroclimatic variability on the Indian subcontinent in the past millennium: Review and assessment. Earth Sci Rev 161:1–15

Dixit Y, Hodell DA, Petrie CA (2014a) Abrupt weakening of the summer monsoon in Northwest India ~4100 yr ago. Geology 42(4):339–342

Dixit Y, Hodell DA, Sinha R et al (2014b) Abrupt weakening of the Indian summer monsoon at 8.2 kyr BP. Earth Planet Sci Lett 391:16–23
Dixit Y, Hodell DA, Giesche A et al (2018) Intensified summer monsoon and the urbanization of Indus Civilization in Northwest India. Sci Rep 8(1):4225
du Plessis N, Chase BM, Quick LJ et al (2021) Vegetation and climate change during the Medieval Climate Anomaly and the Little Ice Age on the southern Cape coast of South Africa: Pollen evidence from Bo Langvlei. The Holocene 30(12):1716–1727
Dutt S, Gupta AK, Clemens SC et al (2015) Abrupt changes in Indian summer monsoon strength during 33,800 to 5500 years BP. Geophys Res Lett 42(13):5526–5532
Dutt S, Gupta AK, Wünnemann B et al (2018) A long arid interlude in the Indian summer monsoon during ~4,350 to 3,450 cal. yr BP contemporaneous to displacement of the Indus valley civilization. Quat Int 482:83–92
Dutt S, Gupta AK, Cheng H et al (2020) Indian summer monsoon variability in or the eastern India during the last two millennia. Quatern Int. https://doi.org/10.1016/j.quaint.2020.10.021
Eddy JA (1976) The maunder minimum. Science 192:1189–1202
Erdtman G (1952) Pollen morphology and plant taxonomy of angiosperms. Waltham (MA)
Faegri K, Iverson J (1989) Textbook of pollen analysis. John Wiley & Sons, Chichester
Faucon MP, Houben D, Lambers H (2017) Plant functional traits: soil and ecosystem services. Trends Plant Sci 22:385
Fleitmann D, Burns SJ, Mudelsee M et al (2003) Holocene forcing of the Indian monsoon recorded in a stalagmite from southern Oman. Science 300:1737–1739
Fleury S, Martinez P, Crosta X et al (2015) Pervasive multidecadal variations in productivity within the Peruvian Upwelling System over the last millennium. Quat Sci Rev 125:78–90
Foley JA, Costa MH, Delire C et al (2003) Green surprise? How terrestrial ecosystems could affect earth's climate. Front Ecol Environ 1:38–44
Gadgil S (2003) The Indian monsoon and its variability. Am Rev Earth Planet Sci 31:429–467
Gadgil S, Gadgil S (2006) The Indian monsoon, GDP and agriculture. Econ Polit Weekly 41(47):4887–4895
Giosan L, Clift PD, Macklin MG et al (2012) Fluvial landscapes of the Harappan civilization. Proc Natl Acad Sci USA 109:E1688–E1694
Ghosh R, Biswas O, Paruya DK et al (2018) Hydroclimatic variability and corresponding vegetation response in the Darjeeling Himalaya, India over the past ~2400 years. CATENA 170:84–99
GrayLJ, Beer J, Geller M et al (2010) Solar influences on climate. Rev Geophys 48(4)
Grove JM (1988) The little ice age. Methuen, London
Gunnell Y (1997) Relief and climate in South Asia: the influence of the Western Ghats on the current climate pattern of Peninsular India. Int J Climatol 17:1169–1182
Gupta AK, Anderson DM, Overpeck JT (2003) Abrupt changes in the asian southwest monsoon during the Holocene and their links to the north Atlantic Ocean. Nature 421:354–357
Gupta AK, Das M, Anderson DM (2005) Solar influence on the Indian summer monsoon during the Holocene. Geophys Res 32(17)
Gupta AK, Anderson DM, Pandey DN et al (2006) Adaptation and human migration, and evidence of agriculture coincident with changes in the Indian summer monsoon during the Holocene. Curr Sci 90(8):1082–1090
Gupta AK, Dutt S, Cheng H et al (2019) Abrupt changes in Indian summer monsoon strength during the last~ 900 years and their linkages to socio-economic conditions in the Indian subcontinent. Palaeogeogr Palaeoclimatol Palaeoecology 536:109347
Gupta AK, Prakasam M, Dutt S et al (2020) Evolution and Development of the Indian Monsoon. In: Gupta N, Tandon SK (eds) Geodynamics of the Indian Plate. Springer Geology. https://doi.org/10.1007/978-3-030-15989-4
Helama S, Jones PD, Briffa KR (2017) Dark Ages Cold Period: a literature review and directions for future research. The Holocene 27(10):1600–1606
Hass HC (1996) Northern Europe climate variations during the Late Holocene: evidence from marine Skagerrak. Palaeogeogr Palaeoclimatol Palaeoecol 123(1–4):121–145

Haug GH, Hughen KA, Sigman DM et al (2001) Southward migration of the intertropical convergence zone through the Holocene. Science 293(5533):1304–1308

Huang Y, Ziao J, Wen R et al (2023) High- and low- latitude forcing on hydroclimate variability in southern China over the last two millennia. Palaeogeogr Palaeoclimatol Palaeoecol 610:111352

Ji J, Shen J, Balsam W et al (2005) Asian monsoon oscillations in the northeastern Qinghai-Tibetan Plateau since the last glacial as interpreted from visible reflectance of Qinghai Lake sediments. Earth Planet Sci Lett 233(1–2):61–70

Kar R, Quamar MF (2020a) Pollen-based Quaternary palaeoclimatic studies in India: an overview of recent advances. Palynology 43(1):76–93

Kar R, Quamar MF (2020b) Late Pleistocene-Holocene vegetation and climate change from the Western and EasternHimalaya (India): palynological perspective. Curr Sci 119(2):195–218

Kar R, Ranhotra PS, Bhattacharayya A et al (2002) Vegetation vis-à-vis climate and glacial fluctuations of the Gangotri glacier since last 2000 years. Curr Sci 82:347–351

Kathayat G, Cheng H, Sinha A et al (2017) The Indian monsoon variability and civilization changes in the Indian subcontinent. Sci Adv 3(12):1701296

Kotlia BS, Joshi LM (2013) Late Holocene climatic changes in Garhwal Himalaya. Curr Sci:911–919

Kotlia BS, Singh AK, Joshi LM et al (2015) Precipitation variability in the Indian central Himalaya during last ca. 4,000 years inferred from a speleothem record: impact of Indian summer monsoon (ISM) and westerlies. Quat Int 371:244–253

Lamb HH (1965) The early medieval warm epoch and its sequel. Palaeogeogr Palaeoclimatol Palaeoecol 1:13–37

Lamb HH (1977) Climate: present, past and future. Methuen, London

Lamb HH (1982) Climate, history and the modern world. Routledge, Boca Raton (FL)

Lamb HH (1995) Climate history and the modern world, 2nd edn. Routledge, New York

Laskar AH, Yadava MG, Ramesh R et al (2013) A 4 kyr stalagmite oxygen isotopic record of the past Indian Summer Monsoon in the Andaman Islands. Geochem Geophys 14(9):3555–3566

Li C, Qi J, Yang L et al (2014a) Regional vegetation dynamics and its response to climate change-a case study in the Tao River Basin in Northwestern China. Environ Res Lett 9:125003

Li K, Liu X, Herzschuh U et al (2014b) Rapid climate fluctuations over the past millennium: evidence from a lacustrine record of Basomtso Lake, southeastern Tibetan Plateau. Sci Rep 6:24806

Ljungqvist FC (2010) A new reconstruction of temperature variability in the extra-tropical Nothern Hemisphere during the last two millennia. Geografiska Annaler Series: A Phys Geogr 92:339–351

MacDonald RE (2011) Understanding the impact of climate change on Northern Hemisphere extra-tropical cyclones. Clim Dyn 37:1399–1425

Mann ME, Jones PD (2003) Global surface temperatures over the past two millennia. Geophys Res Lett 30:1820.https://doi.org/10.1029/2003GL017814.

Mann ME, Zhang Z, Rutherford S et al (2009) Global signatures and dynamical origins of the Little Ice Age and Medieval Climate Anomaly. Science 326:1256–1260

Martín-Puertas C, Valero-Garcés BL, Brauer A et al (2009) The Iberian-Roman Humid Period (2600–1600 calyr BP) in the Zoñar Lake varve record (Andalucía, southern Spain). Quatern Res 71:108–120

Masson-Delmotte V et al (2013) In climate change 2013: the physical science basis contribution of working group I to the fifth assessment report of the intergovernmental panel on climate change. In: Stocker TF et al (eds) Cambridge University Press, pp 383–464

Matthes FE (1939) Report of Committee on Glaciers. EOS Trans Am Geophys Union 20:518–523

Matthews JA, Brifa KR (2005) The 'little ice age': re-evaluation of an evolving concept. Geografiska Annaler Ser A Phys Geogr 87:17–36

McCormick M, Büntgen U, Cane MA et al (2012) Climate change during and after the Roman Empire: reconstructing the past from scientific and historical evidence. J Interdiscip Hist 43(2):169–220

McCrindle JW (1877) Ancient India as described by Megasthenes and Arrian. Thacker, Spink
McDermott F, Mattey DP, Hawkesworth C (2001) Centennial-scale Holocene climate variability revealed by a high-resolution speleothem ^{18}O Record from SW Ireland. Science 294(5545):1328–1331
Misra P, Tandon SK, Sinha R (2019) Holocene climate records from lake sedimentsin India: assessment of coherence across climate zones. Earth-Sci Rev 190:370–397
Misra KG, Yadav RR, Misra S (2015) SatlujRiver flow variations since AD 1660 based on tree-ring network of Himalayan cedar from western Himalaya, India. Quatern Int 371:135–143
Mohanty RB, Mishra AK, Mishra K, Yadav AK, Quamar MF, Barua IC, Kar R (2024) Early onset of aridity in the past millennium: insights from vegetation dynamics and climate change in the alpine, cold-desert region of Trans Himalaya, India. PLOS ONE 19(1):e0295785
Morrill C, Overpeck JY, Cole JE (2003) A synthesis of abrupt changes in the Asian summer monsoon since the last deglaciation. The Holocene 13(4):465–476
Naidu PD, Ganeshram R, Bollasina MA et al (2020) Coherent response of the Indian monsoonal rainfall to Atlantic multi-decadal variability over the last 2000 years. Sci Rep 10:1302. https://doi.org/10.1038/s41598-020-58265-3
Neukom R,Steiger N, José Gómez-Navarro J et al (2019) No evidence of globally coherent warm and cold periods over the pre-industrial Common Era
Pillai AA, Anoop A, Prasad V et al (2018) Multi-proxy evidence for an arid shift in the climate and vegetation of the Banni grasslands of western India during the mid-to late-Holocene. Holocene 28(7):1057–1070
Prasad S, Anoop A, Riedel N, Sarkar S, Menzel P, Basavaiah N, Krishnan R, Fuller D, Plessen B, Gaye B, Röhl U, Wilkes H, Sachse D, Sawant R, Wiesner M, Stebich M (2014) Prolonged monsoon droughts and links to Indo-Pacific warm pool: a Holocene record from Lonar Lake, Central India. Earth Planet Sci Lett 391:171–182. https://doi.org/10.1016/j.epsl.2014.01.043
Quamar MF (2019) Vegetation dynamics in response to climate change from the wetlands of Western Himalaya, India: Holocene Indian Summer Monsoon variability. The Holocene 29(2):345–362
Quamar MF (2022a) Holocene vegetation and climate change from central India: An updated and a detailed pollen-based review. In: Kumaran KPN, Padmalal D (eds) Holocene climate change and environment. London, Elsevier
Quamar MF (2022b) Monsoonal climatic reconstruction from central India during the last ca. 3600 calyr: signatures of global climatic events, based on lacustrine sediment pollen records. Palynology 46(1):930605
Quamar MF (2022c) Late Holocene vegetation and monsoonal climate change from the Western Himalaya, India. Acta Palaeobotanica 62(1):36–49
Quamar MF, Bera SK (2020) Pollen records of vegetation dynamics, climate change and ISM variability since the LGM from Chhattisgarh State, central India. Rev Palaeobot Palynol:104237.https://doi.org/10.1016/j.revpalbo.2020.104237
Quamar MF, Bera SK (2021) A 8400-year pollen record of vegetation dynamics and Indian Summer Monsoon climate from central Indian Core Monsoon Zone: Signatures of global climatic events. J Palaeontol Soc India 66(1):12–22
Quamar MF, Chauhan MS (2014) Signals of Medieval Warm Period and Little Ice Age from southwestern Madhya Pradesh (India): A pollen-inferred Late-Holocene vegetation and climate change. Quat Int 325:74–82
Quamar MF, Kar R (2020) Prolonged warming over the last ca. 11,700 years from the central Indian Core Monsoon Zone: Pollen evidence and a synoptic overview. Rev Palaeobot Palynol 104159. https://doi.org/10.1016/j.revpalbo.2020.104159
Quamar MF, Kar R, Thakur B (2021) Vegetation response to the Indian Summer Monsoon (ISM) rainfall variability during the Late Holocene from the central Indian Core Monsoon Zone. The Holocene 31(7):1197–1211
Quamar MF, Mir IA, Jaiswal J, Bharti N, Dabhi AJ, Bhushan R, Prasad N, Javed M (2023) Hydroclimatic variability and consequent vegetation response during CE 1219–1942 from the Western Ghats, India. Catena 232:107448

Quamar MF, Banerji US, Thakur B, Kar R (2024) Hydroclimatic changes in the core monsoon zone of India since the last glacial maximum: an overview of the palynological data and correlation with the marine and continentalrecords. Palaeogeogr Palaeoclimatol Palaeoecol 633:111844.

Rawat S, Gupta AK, Sangode SJ et al (2015) Late Pleistocene-Holocene vegetation and Indian summer monsoon record from the Lahaul, Northwest Himalaya, India. Quat Sci Rev 114:167–181

Reddy AP, Gandhi N, Yadava MG (2022) The Indian monsoon variability during the last two millennia and links to the tropical equatorial Pacific. Clim Dyn. https://doi.org/10.1007/s00382-022-06513-9

Rothlisberger F (1986) 10,000 Jahre Gletschergeschichteder Erde(Sauerlan der, Aarau)

Royer DL (2008) Linkages between CO^2, climate and evolution in deep time Proc. Natl Acad Sci Unit States Am 105(2):407–408

Ruhland K, Phadtare NR, Pant RK et al (2006) Accelerated melting of Himalayan snow and ice triggers pronounced changes in a valley peatland from Northern India. Geophys Res Lett 33:15

Sanwal J, Kotlia BS, Rajendran C et al (2013) Climatic variability in central Indian Himalaya during the last ~1800 years:evidence from a high resolution speleothem record. Quatern Int 304:183–192

SeidenkrantzMS SAS, Sulsbrück H et al (2007) Hydrography and climateof the last 4400 years in a SWGreenland fjord: implications for Labrador Sea palaeoceanography. The Holocene 3:387–401

Sharma SN (1995) An advanced competitive botany. Students Friends, Govind Mitra Road, Patna, 800004, Bihar, India

Sharma C, Chauhan MS (2001a) Late Holocenevegetation andclimate from Kupup Lake, Sikkim Himalaya, India. J Palaeontol Soc India 46:51–58

Singh U (2008) A history of ancient and early Medieval India: from the stone age to the 12th century (PB). Pearson Education India

Singh S, Gupta A, Dutt S et al (2020) Abrupt shifts in the Indian summer monsoon during the last three millennia. Quat Int. https://doi.org/10.1016/j.quaint.2020.08.033

Singh DS, Gupta AK, Sangode SJ et al (2015) Multiproxy record of monsoon variability from the Ganga Plain during 400–1200 AD. Quat Int 371:157–163

Singh V, Misra KG, Singh AD et al (2021) Little Ice Age revealed in tree-ringbased precipitation record from the northwest Himalaya, India. Geophys Res Lett 48: e2020GL091298.https://doi.org/10.1029/2020GL091298

Singhvi AK, Bhattacharyya A, Kale VS et al (2010) Instrumental, terrestrial and marine records of the climate of south Asia during the Holocene: present status, unresolved problems and societal aspects. In: Global environmental changes in South Asia, Springer, The Netherlands, pp 54–124

Sinha A, Cannariato KG, Stott LD et al (2005) Variability of southwest Indian summer monsoon precipitation during the Bolling-Allerod. Geology 33(10):813–816. https://doi.org/10.1130/G21498

Sinha A et al (2010) A global context for megadroughts in monsoon Asia during the past millennium. Quatern Sci Rev. https://doi.org/10.1016/j.quascirev.2010.10.005

Sinha ABM, Stott L et al (2011) The leading mode of Indian Summer Monsoon precipitation variability during the last millennium. Geophys Res Lett 38:L15703. https://doi.org/10.1029/2011GL047713

Sharma C, Chauhan MS (2001b) Late Holocene vegetation and climate from KupupLake, Sikkim Himalaya, India. J Palaeontological Soc India 46:51–58

Sodhi A, Das A, Prozomwala SP (2022) Centennial-scale linkages between the Indian Summer Monsoon and the solar irradiation from the Gulf of Khambhat (Western India). Quatern Int 631:82–92

Solanki SK, Usoskin IG, Kromer B et al (2004) Unusual activity of the Sun during recent decades compared to the previous 11,000 years. Nature 431(7012):1084–1087

Sun A, Yang A, Wu et al (2020) Climate change on the northeastern Tibetan Plateau during the past ~600 years inferred from peat pollen records. Rev Palaeobot Palynol 276:104194

Tripathi S, Thakur B, Nautiyal CM et al (2020) Floristic and climatic reconstruction in the Indo-Burma region for the last 13,000 cal. yr: a palynological interpretation from the endangered wetlands of Assam, northeast India. The Holocene 30(2):315–331

von Rad U, Schaaf M, Michels KH et al(1999) A 5000-yr record of climate change in varved sediments from the oxygen minimum zone of Pakistan, Northeastern Arabian Sea. Quat Res 51:39–53.https://doi.org/10.1006/qres.1998.2016

Walker M, Gubbard P, Head MJ et al (2019) Formal subdivision of the holocene series/epoch: a summary. J Geol Soc India 93:135–141

Wang Y, Cheng H, Edwards RL et al (2005a) The Holocene Asian monsoon: links to solar changes and North Atlantic climate. Science 308:854–857

Wang P, Clemens S, Beaufort L et al (2005b) Evolution and variability of the Asian monsoon system: state of the art and outstanding issues. Quatern Sci Rev 24(5):595–629

Wang T, Surge D, Mithen S (2012) Seasonal temperature variability of the Neoglacial (3300–2500 BP) and Roman warm period (2500–1600 BP) reconstructed from oxygen isotope ratios of limpet shells (*Patella vulgata*), Northwest Scotland. Palaeogeogr Palaeoclimatol Palaeoecol 317:104–113

Wanner H, Solomina O, Grosjean M et al (2011) Structure and origin of Holocene cold events. Quat Sci Rev 30(21–22):3109–3123

Webster PJ, Magana VO, Palmer TN (1998) Monsoons: processes, predictability, and the prospects for prediction. J Geophys Res Oceans 103(C7):14451–14510

Wu W, Tan W, Zhou L et al (2012) Sea surface temperature variability in southern Okinawa trough during last 2700 years. Geophys Res Lett 39:L14705. https://doi.org/10.11029/2012GL052749

Xu H, Jianghu L, Sheng E, Liu B, Yu K, Ye Y, Shi Z, Cheng P, Wang X, Zhou X, Yeager KM (2016) Hydroclimatic contrasts over Asian monsoon areas and linkages to tropical Pacific SSTs. Sci Rep 6:33177. https://doi.org/10.1038/srep33177

Yadav RR, Singh J (2002) Tree-ring-based spring temperature patterns over the past four centuries in Western Himalaya. Quatern Res 57:299–330

Zhang Y, Zhao CK, Shun Y, Zen JY et al (2009) Medieval Warm Period" on the northern slope of central Tianshan Mountains, Xinjiang, NW China. Geophys Res Lett 36:L11702. https://doi.org/10.1029/2009GL037375

Zhang W, Wang L, Xiang F, Qin W, Jiang W (2020) Vegetation dynamics and the relations with climate change at multiple time scales in the Yangtze river and Yellow River Basin, China. Ecological Indicatots 110:105892. https://doi.org/10.1016/j.ecolind.2019.105892

Palynology: A Tool to Decipher the Impact of Anthropogenic Activity on Palaeo-Vegetation—A Review Based on Fossil Pollen Records from India

Nivedita Mehrotra, Santosh K. Shah, and Ratan Kar

Abstract The applications of palynology to understand palaeo-environment and palaeo-vegetation changes have been well established from deposits ranging from the Palaeozoic to the Quaternary. The Holocene epoch has best witnessed the impact of anthropogenic activities lead by a quantum jump in human population. Vegetation distribution and its variations due to exogenous influences are evident in the fossil and modern pollen records globally. Palynology has proved to be a tested tool to decipher these changes in the environmental dynamics and changing floral compositions, as a consequence of anthropogenic activities. We have made an attempt to understand the pattern of preservation of these anthropogenic marker taxa in the pollen records across the different physiographic regions of India. The spatio-temporal distribution patterns of the occurrence of anthropogenic marker taxa have been discussed and their use in deciphering related human activities. We provide an overview of the palynological records which evidently state the presence of the marker taxa with a clear reporting of anthropogenic influence in the different regions. Here we review the Holocene pollen records across India to specify the presence of anthropogenic pollen in the vastly populated and agriculture dominated regions of the country.

Keywords Holocene · Palynology · Paleoenvironment · Paleoclimate · Anthropogenic activity

1 Introduction

Human activities have changed the ecological landscape globally altering the land-use and land-cover dynamics consequently disturbing the various forest types. The complex entanglement of climate, vegetation and human activities have caused alteration of the floral compositions in the various biomes. The presence of evidence of

N. Mehrotra (✉) · S. K. Shah · R. Kar
Birbal Sahni Institute of Palaeosciences, 53-University Road, Lucknow 226007, Uttar Pradesh, India
e-mail: nivedita_mehrotra23@hotmail.com; nivedita.mehrotra@bsip.res.in

B. Samant and D. Thakre (eds.), *Applications of Palynology in Stratigraphy and Climate Studies*, Society of Earth Scientists Series, https://doi.org/10.1007/978-3-031-51877-5_8

such vast human induced environmental changes has been recorded in various proxy records (Ruddiman 2000). The paleo-ecological records based on fossil pollen have introduced a spatio-temporal understanding of the anthropogenic impacts on the vegetation dynamics. Palaeoclimate reconstructions based on pollen yield from plant taxa sensitive to winter frost was one of the earliest applications of the pollen indicator approach (Iversen 1944). In central and western Europe was the region where originally the comprehensive anthropogenic pollen-indicators were developed (Behre 1981) and further applied elsewhere. The abundance of anthropogenic pollen indicators in the pollen records formulated the bases of application of palynology to reconstruct the human impact on environments.

The role of non-native plant species ('adventives') in a pollen record to identify human introduced (cultivars or weeds) plants as indicators of past human activities is well established (Behre 1981, 1988; Huntley and Webb 1988). Plantation of native species favored by humans adds limited, but significant information, through the pollen of these taxa known as 'apophytes'. Apophytes change the natural landscape and create human-induced vegetation changes favored by environmental factors (Behre 1981; Lang 1994; Deza-Araujo et al. 2020). Multi-proxy evidence, such as pollen, macrofossils, megafossils, a DNA combined with archeological records can provide detailed information of the local distribution of cultivated plants and weeds. Thus, relying on these evidence, regions specific understanding of species distribution and occurrences, particular plant taxa can be considered as adventives or apophytes (Castri et al. 1990; Lang 1994; Conedera et al. 2004; van Leeuwen et al. 2008; Krebs et al. 2019; Deza-Araujo et al. 2020). Complexities of natural and anthropogenic changes in vegetation are significant to decipher palaeoecological variations. The common methodology applied, since long, is the species indicator approach, but there have been recent development of alternate methodologies and proxies (Sugita 2007; Fyfe et al. 2010). There is a dearth of methodological approaches and standard paraphernalia for pollen-based reconstructions of land use (Deza-Araujo et al. 2020).

We attempt to understand the limited use of anthropogenic pollen indicator approach in the Holocene pollen records of India. We have assembled these pollen-based palaeo-vegetation reconstructions and have attempted to track the anthropogenic taxa or species adopted in the Indian scenario to understand the advent of human interventions in the natural vegetation. India is still known for its vast agricultural practices across the lengths and breadths of its vast areas. Land-use and land cover changes are still lesser known through the long-standing anthropogenic pollen indicator approach, alongside other proxy records such as macrofossils, megafossils, ancient DNA together with archeological evidence. We have selected studies that specifically mention the presence or impact of human-induced changes in the vegetation, from the various regions of India. We have attempted to incorporate all major findings clearly identifying anthropogenic indicator taxa and their relationship to the dynamics of anthropogenic scenario in the region. We do discuss a few studies, which give an outline of similar palynological evidence but explain the human impact in a limited manner. The overall aim was to highlight the significance of palynological records towards deciphering the role of humans alongside the environmental factors inducing vegetation changes.

2 Distribution of Pollen Records Across India: Anthropogenic Indicator Pollen; Human Impact on Vegetation; Spatio-Temporal Disparities in Fossil Pollen Records

The palynological records are well distributed across India (Kar and Quamar 2019; Mehrotra et al. 2014) presenting an overall understanding of the palaeo-vegetation and palaeoclimatic scenario from the various regions. We attempt to understand the variations and similarities in the records (Fig. 1) that clearly mention the presence of anthropogenic indicator taxa and/or presence of human impact signals in the pollen data.

2.1 Northeast India

Presence of Poaceae or large sized grass pollen (cultivated/ wild/ human infused) varieties have been reported in the fossil pollen records from northeastern India. This region has been known to be as the corridor for the movement of Poaceae (and subfamilies) across Southeast Asia and the Indian subcontinent since the Oligocene (Srivastava et al. 2019). Impacted by the human population this ecologically sensitive region has vegetation variations from high-altitude scrubs to tropical and mangrove taxa (Dikshit and Dikshit 2014). In the northern parts of Assam, large sized grass pollen grains and primitive agricultural impacts were recorded in the sediment from Ledo-Matikhad, Tinsukia, and Digboi region (Bhattacharya and Chanda 1988). The palynological assemblage from upper Assam recorded northeast Asian elements such as *Carya*, *Pterocarya*, and *Tsuga* indicating a trans-migratory link between Southeast Asia and northeast India. These taxa did not exist later probably due to natural factors of climate variations in the region (Bhattacharya and Chanda 1992).

In Assam, the Mikir Hills, Kaki Forest Division, the profile sediments yielded Cerealia (10–12%) along with other culture pollen viz., Asteraceae, Caryophyllaceae, Chenopodiaceae/Amaranthaceae, etc., around 13,000 years B.P. indicating forest clearance and subsequent agricultural practices (Bera 2003). In the pollen record from Dabaka Swamp, Nagaon District, in Lower Assam, Cerealia, *Plantago, Chenopodiaceae, Xanthium, Celosia argentea, Artemisia*, were reported during 11600–8310 cal years B.P. (Dixit and Bera 2012a). But the authors attributed the dominance of pastoral activities in the region along with reduction of monsoon strength and the deterioration of tropical mixed deciduous forests since 768 cal years B.P. to the enhanced Cerealia in the record (Dixit and Bera 2012a). During 10,810–7680 cal years B.P. Dixit and Bera (2012b) reported some cereal-type pollen, which might be an indication of enhanced human impact in the Chayagaon swamp, Kamrup District, western Brahmaputra flood plains of Assam. Based on the presence of anthropogenic indicator pollen taxa Chenopodiaceae, Caryophyllaceae, *Cannabis*,

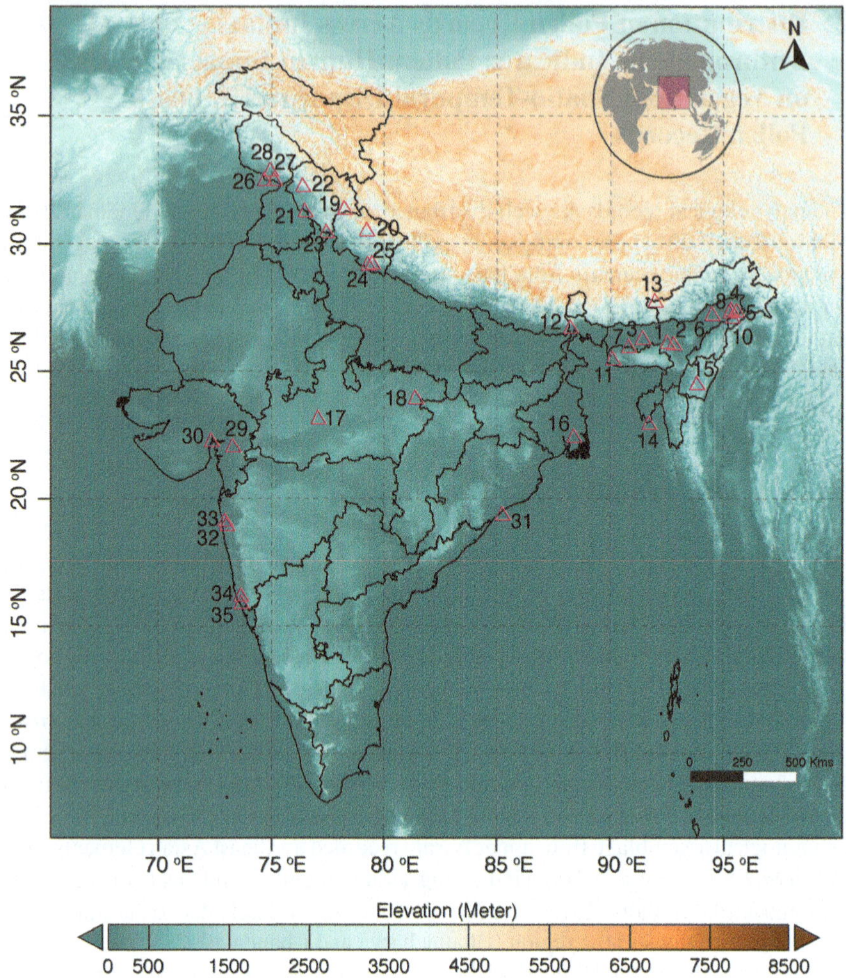

Fig. 1 Map of India showing palynological study sites which indicate anthropogenic activities in the fossil pollen records. (1) Mikir Hill, (Bera 2003); (2) Dabaka Swamp, (Dixit and Bera 2012a); (3) Chayagaon swamp, Kamrup, (Dixit and Bera 2012b); (4) Tinsukia Lake, (Bhattacharya and Chanda 1988); (5) Digboia (Bhattacharya and Chanda 1988); (6) Ledo-Matikhad (Bhattacharya and Chanda 1988); (7) Deosila Swamp, (Dixit and Bera 2011); (8) Merbeel Sawmp (Bera and Dixit 2011); (9) Jokai Forest (Bera et al. 2013); (10) Mothabeel reserve forest, (Bera et al. 2011); (11) Garobadha swamp, (Basumatary and Bera 2010); (12) Jore-Pokhari, (Chauhan and Sharma 1996); (13) PTTSO-Lake, (Mehrotra et al. 2019); (14) Srinagar, Tripura (Bhattacharyya et al. 2011); (15) Loktak Lake, (Nautiyal and Chauhan 2009); (16) Garia, Kolkata; (Das 2014); (17) Kachhar Lake, Sehore District, (Quamar and Chauhan 2011); (18) Kerha Swamp, Majhouli, (Chauhan 2005); (19) Rukti, Sangla, (Ranhotra et al. 2018); (20) Chopta Valley, (Misra et al. 2022); (21) Rewalsar Lake, (Sharma and Singh 1972a, b); (22) Khajiar Lake, (Sharma and Singh 1972a, b); (23) Parasram Lake, Sirmur District (Sharma 1985); (24) Sat Tal, (Gupta and Khandelwal 1982); (25) Naukuchia Tal (Dodia 1988); (26) Gharana Swamp (Quamar 2019); (27) Nanga (Quamar 2019); (28) Bajalta Lake, (Quamar 2022) (29) Wadhwana Lake, (Prasad et al.2014); (30) Pariyaj Lake, (Raj et al. 2015). (31) Chilka Lake, (Khandelwal et al. 2008); (32) Ariyoli,Thane Road, (33) Vashi, Palm beach, Navi Mumbai, (34) Hadi, Malvan-Achra Road, (35) Dhampur, Sindhudurg District (Limaye et al. 2014)

Artemisia, Cereal, *Xanthium* and *Plantago*, it was suggested that human activity persisted near the swamp since 12,450 cal years B.P.

Pastoral practice enhanced during 2970 to 1510 years B.P., as evidenced by an increase of Cerealia, *Plantago*, Amaranthaceae and Brassicaceae, near the Deosila swamp under Rangjuli Reserve forest of Assam (Dixit and Bera 2011). Pastoral activity during 2200–850 years B.P., was also reported based on the presence of Cerealia, *Xanthium, Celosia,* etc. in the pollen assemblage in the sediment profiles from Merbeel Swamp in Jeypore Reserve Forest, Dibrugarh District, Assam (Bera and Dixit 2011); Cerealia, *Xanthium, Plantago* pollen found in the sediments from Jokai reserve forests during 2800–2200 cal years B.P.(Bera et al. 2013); Cerealia, Caryophyllaceae, *Xanthium* and Amaranthaceae observed in sedimentary profile from Mothabeel and Dangrithan Reserve forest, Tinsukhia District, Assam indicated the presences of anthropogenic activities in the region during 1100 to 550 years B.P. (Bera et al. 2011).

In western Garo Hill, Meghalaya, palynological studies indicated deforestation due to anthropogenic activity in the region around 322 cal years B.P. onwards, as evidenced by Cerealia pollen along with other cultural pollen *(Xanthium,* Caryophyllaceae, etc.) observed in the assemblage (Basumatary and Bera 2010).The pollen assemblage from Jore-Pokhari, a temperate lake of Darjeeling, eastern Himalaya, included grasses together with other cultural pollen, viz., *Artemisia,* Caryophyllaceae, Asteraceae, Brassicaceae, Chenopodiaceae/ Amaranthaceae along with Cerealia type pollen indicating the expansion of agricultural practice during 1600–1000 cal years B.P. (Chauhan and Sharma 1996). Oak or *Quercus* spp. in the pollen was recorded from PT Tso Lake in Tawang, Arunachal Pradesh, eastern Himalaya, since 4195–2527 cal years B.P. Its presence was attributed to the utilization of the taxa by the humans or the local tribes (*Monpas* of *Tawang*) for various activities such as Yak breeding, utilizing leaves as manure in cultivation, also in exchange of Yak-based food products; firewood, etc. (Biswal 2006; Singh et al. 2015; Mehrotra et al. 2019).

Bhattacharyya et al. (2011) attributed the presence of >50 μm grass pollen from around 5700 years B.P. in Srinagar, Tripura, to the beginning of rice cultivation in the region. The northwest Tripura pollen record from Charilam also indicated the presence of larger (>50 μm) Poaceae pollen as evidence for human activities in the region since 4100 cal years B.P. to 1800 cal years B.P. (Mehrotra et al. 2022). In Manipur the cultural pollen taxa Cerealia, Chenopodiaceae/Amranthaceae, *Artemisia*, Caryophyllaceae, *Rumex*, Brassicaceae, etc., were recorded near Loktak lake site suggesting the presence of human activities in the vicinity of the lake from 1650 to 600 years B.P. (Chauhan and Nautiyal 2009).

2.2 Eastern India

The lower Bengal Basin region has pollen records indicating the existence of mangrove vegetation around 5000–6000 years B.P. in various records (Vishnu-Mittre and Gupta 1972; Mukherjee 1972) which are now restricted towards the Sunderbans regions in the present day. Cereal-type pollen, recorded together with *Plantago, Ammania, Capparis, Ocinum, Crotalaria, Tinospora* etc., reflect the possibility of early human settlement of primitive period. But due to the lack of additional evidence, such as macrofossils of seeds, carbonized food remains, charcoal, etc., the existence of primitive humans can be questionable (Chanda 1974). Recent pollen-based record of subsurface sediments from suburban region of Calcutta (Kolkata) at Garia metro railway excavation site showed the predominance of mangrove taxa in the three successive peat layers (Das 2014). This was followed by marshland (characterized by fern spore) and then successive transition towards terrestrial or inland vegetation. The enhancement of non-arboreal pollen throughout the successive peat layers from lower (7000 B.P.) to upper stratum (2600 B.P.) indicating the human influence in this area and gradual expansion of agricultural land (Das 2014). Thus, the overall impact of human activities transformed the existing mangrove vegetation and restricted it to a particular part of eastern India.

2.3 Central India

Humans are known to have a significant impact on the natural vegetation as it has been subjected to forest fires, deforested for agriculture and used as a fuel. The impact of these changes is well recorded in the pollen assemblages across the Core Monsoon Zone of India. The presence of Cerealia and ruderal plant taxa, such as *Cannabis sativa*, Chenopodiaceae/Amaranthaceae, Caryophyllaceae (around 2050 cal years B.P.) (Quamar and Chauhan 2011) and *Artemisia, Alternanthera* etc., signifies close proximity of cultivated land and human habitation (Quamar and Chauhan 2011) near Kachhar Lake, Sehore District (around 2050 cal years B.P.); Kerha Swamp, Sidhi (around 1600 cal years B.P.) (Chauhan 2005) of Madhya Pradesh. Besides this, Quamar and Chauhan (2011) have suggested *Madhuca indica, Butea monosperma, Melia azedarach, Bombax ceiba, Acacia nilotica Delile, Ficus tinctoria* and *F. benghalensis* as the indicator trees of human habitation. Substantial increase in Asteraceae (Tubulifloreae) has been ascribed to pastoral activities. It is to be mentioned that these studies do not show any direct or indirect evidence of deforestation in the palaeo-vegetation reconstructions.

2.4 Western India

The western Himalaya pollen records from Himachal Pradesh indicated the presence of anthropogenic activities, such as burning of wood around 5000 cal B.P. These were reported through charcoal layers found in Sangla town (Draganits et al. 2014; Ranhotra et al. 2018); additionally, through charcoal recorded in the sediments of 4300 to 1800 (uncalibrated radiocarbon ages) in Rukti valley (Chakraborty et al. 2006; Ranhotra et al. 2018). The vegetation records from Chopta Valley in western Himalaya, demarcated the existence of human activities in the pastoral activities due to the presence of Amaranthaceae, Brassicaceae, Caryophyllaceae, *Artemisia* and Polygonaceae, along with coprophilous fungi (dung-loving fungi), such as *Sporormiella, Sordaria, Podospora, Cercophora* and *Ascodesmis* in higher numbers in the region (Misra et al. 2022).

The lake deposits from Himachal Pradesh were investigated for the age determination of 'floating islands' at Khajiar and Rewalsar lakes dominated by the gregarious growth of *Phragmites communis* (Sharma 1970). The growth of the taxa at Khajiar Lake was closer to the lake center and lake margins, and the banks were devoid of it probably due to human clearance as a result of the village settlement nearby. But in case of the Rewalsar Lake these reeds are found both at the center and surroundings of the lake, as a consequence of the religious practices of the Buddhist communities near the lake conserving it. The Khajiar Lake deposits or floating islands were dated as 700 C.E., whereas the Rewalsar Lake deposits have *Phragmites* peat accumulations indicating much older existence of the taxa. Rewalsar Lake fossil pollen records observed an increase in Cerealia type pollen followed by a rise in Chenopodiaceae/ Amaranthaceae, *Artemisia* spp. and Compositae along with taxa such as Caryophyllaceae, Malvaceae, Umbelliferae, *Justicia* spp., *Peristrophe* spp., *Polygonum* and Labiatae. This rise in Cerealia type and other cultural pollen taxa was attributed to the forest clearance leading to intensive agriculture in the area around 569 cal years B.P. (Sharma and Singh 1972a, b). The Cerealia type, along with *Cannabis* spp. from the early part of the record (1737 cal years B.P.), from Khajiar Lake (at altitude 1950 m a.s.l.) indicated agricultural practices prevalent in the area resulting in forest clearance. Also, the advent of Buddhist cults in the region caused anthropogenic impact on the lake site, as inferred through the pollen records from the upper part of the lake. The Parasram Tal pollen records observed the presence of Cerealia type pollen since 3140 B.P. along with other cultural pollen indicating agricultural practices from the early part of the record, which enhanced manifolds around 680 $\pm$ 110 years B.P. (Sharma 1985).

Other records from the Kumaon Himalaya lakes, such as Bhim Tal, Sat Tal and Naukuchia Tal indicate the presence of anthropogenic indicator pollen. The pollen record from Sat Tal describes a 'land occupation phase' and a 'landnam phase' based on the occurrence and decline of marker pollen (Gupta and Khandelwal 1982). During the land occupation phase the occurrence of Cerealia, *Plantago*, Asteraceae, Chenopodiaceae/Amaranthaceae and Caryophyllaceae, considered as taxa favoring pastoral activities and indicating agricultural practices around 1400 $\pm$ 100 B.P. The landnam phase began around 815 $\pm$ 80 B.P., which recorded the reduction in all cultural pollen counts, apart from *Artemisia* and Asteraceae. These two taxa have a tendency to occupy vacant land areas, thus indicating reduction in agricultural activities around the lake, which was later occupied by arboreal taxa. Around Naukuchia Tal, Cerealia type pollen along with *Artemisia, Cannabis* and Brassicaceae pollen were found around 3170 $\pm$ 100 years B.P. in a lake sediment record dated up to 7150 $\pm$ 160 years B.P. (Dodia 1988). Pollen record from the lake sediments of Dewar Tal, in Garhwal Himalaya region indicated the existence of anthropogenic and agricultural practices, based on the presence of cultural pollen, such as Poaceae, Chenopodiaceae/Amaranthaceae, *Artemisia*, Caryophyllaceae and Urticaceae, since 2430 ± 100 years B.P. (Chauhan and Sharma 2000).

The post glacial fossil pollen records from the Kashmir Himalaya postulate the presence of cultural pollen or biotic changes in the vegetation, due to anthropogenic influences in the region (Vishnu-Mittre 1966).Gharana Wetland (GW) in Jammu District and Nanga Wetland (NW) in Samba District recorded Cerealia and cultural plant pollen, such as Amaranthaceae, Caryophyllaceae, *Artemisia* spp., *Alternanthera* spp., *Plantago* spp., Brassicaceae, Urticaceae, and *Cannabis sativa* between 8536 and 5296 cal years B.P. (at GW) and 2784 and 1584 cal years B.P. (at NW) (Quamar 2019).The existence of agricultural practices, in these regions of western Himalaya was validated through these records (Quamar 2019). Pollen records from Bajalta Lake in Jammu region, Lesser Himalaya, reported cultural pollen taxa: Cerealia, Amaranthaceae/Caryophyllaceae, *Artemisia* spp., *Alternanthera* spp., Brassicaceae, Urticaceae etc., which indicate that the lake surroundings witnessed some sort of agricultural practice and other human activity since 3205 cal years B.P. (Quamar 2022).

The pollen records in Gujarat, from Wadhwana Lake sediments included deciduous genera like *Madhuca*, *Terminalia*, *Mangifera*, high counts of *Chrozophora* and Cerealia pollen at the end of the 5th millennium B.P., thus indicating enhanced agricultural practices and arboriculture (Prasad et al. 2014). However, the archeological evidence is not compatible to these palynological findings, which do not particularly represent the actual palaeo-vegetation scenario. But the possibility of large-scale agricultural activities cannot be ruled out (Prasad et al. 2014). Raj et al. (2015) recorded Solanaceae, Brassicaceae, Asteraceae, Chenopodiaceae/Amaranthaceae, Poaceae, Caryophyllaceae pollen in a sediment core from Pariyaj Lake, Vatrak River basin, Gujarat. The lake sediments recorded a dry phase and low-lake stands during 5864–4680 cal years B.P., followed by a moist spell and higher lake stands after 4680 cal years B.P. The pollen taxa mentioned above flourished during the high-stand periods i.e., after 4680 cal B.P. and enhancement of other evidence of human influence, such as charcoal, burnt phytoliths were also recorded (Raj et al. 2015). This period (after 4680 cal years B.P.) is also known for the migration of Harapan population between 3200 and 1900 BCE (Clift et al. 2012) towards other favorable places. So, a lake body could have been a suitable spot for inhabitance and relocation. Thus, the authors of the present study comprehend that though these taxa were termed as indicators for winter precipitation and the commonalities of the anthropogenic indicator taxa mentioned in other pollen records cannot be overlooked.

3 Indian Coastal Records

Pollen records from the Indian coastline also attribute some of the vegetation changes, along the marginal areas, to human impact along with changing sea-level and variations in marine water influx towards the land areas. These regions essentially recorded mangrove vegetation along with associated flora and/ or back mangrove flora, in the fossil pollen records. In the east coast Chilka Lake records, the occurrence of Poaceae, Cyperaceae, Chenopodiaceae/ Amaranthaceae and Asteraceae recorded around 2000 cal years B.P. was attributed to human impact on the mangrove vegetation, dominant during the later part of the Holocene (Khandelwal et al. 2008). In the Cauvery River delta, pollen records from multiple core sites around Vellar and Thillaividangan discussed the presence of back mangrove taxa, such as *Avicennia* type mangroves, *Acacia* spp., and Euphorbiaceae and Solanaceae along with open ground herbaceous taxa *Derris* spp., and *Justicia* spp., thus indicating a rise in human impact on the coastal woodlands and lowlands due to pastoral activities, cutting and burning etc., (Srivastava and Farooqui 2017). The west coast mangrove records based on pollen data, mostly attribute the degradation of the mangroves from the Middle Holocene to the Late Holocene to hydrodynamical changes due to the changing strength of the Indian Summer Monsoon (ISM) and fresh water influx (Limaye et al. 2014).

4 Discussion

4.1 Initiation of Human Impact; Common Periods in the Pollen Records

The Late Holocene period is known for the climatic and environmental changes falling under the impact of anthropogenic impact. The term 'Anthropocene' was coined due to the spatio-temporal impact of human activities until the recent time on their surroundings globally (Crutzen and Stoermer 2000; Crutzen 2002, Smith and Zedar 2013). Climate change and vegetation fluctuations are now escalated by anthropogenic factors across India. The aforesaid pollen records are testament to its persistence with the advancement of human civilizations towards the Late Holocene. The common periods of these vegetation changes along with climatic influence began around 5000 cal years B.P. in the larger river basins and regions, where the human population has been documented to flourish in northern, central, western and eastern India; also, in most parts of northeastern India. Slowly, this impact moved towards the higher altitudes, especially around water bodies, such as lakes where anthropogenic indicator pollen was identified ca. 2000 cal years B.P. or beyond until present. The northeastern part of India has records dating back to 13,000 years B.P. (Bera 2003) of anthropogenic activities, along with agricultural practices in Mikir hills, Assam and pastoral activities in other parts of Assam in Early Holocene (~12,000 cal years B.P.); pollen records from Garia, Kolkata, West Bengal, recorded enhancement of agricultural lands around ~7000 cal years B.P. But the inclusive impact of reducing strength of the ISM and aridity across the Indian subcontinent enhanced the palaeo-vegetation changes and human adaptations, developments and modifications of human civilizations during the Late Holocene.

The pollen assemblages of the Late Holocene period, provide good evidence, when the agricultural practices escalated in most parts of India. A tentative outline of the periods in which agricultural practices and anthropogenic activities escalated across India, can be understood through these pollen-based records. Inferences drawn on the bases of fossil pollen records from similar taxa but the regional difference at the species level is inevitable specifically in the case of plants such as Poaceae.

Variations in the marker taxa are not very elaborate and common anthropogenic indicator plant taxa pollen recorded are Cerealia, *Plantago, Artemisia, Xanthium,* Caryophyllaceae, *Cannabis* spp., Chenopodiaceae/ Amaranthaceae, Brassicaceae, etc. These taxa have been mostly considered as indicators of human activities and agricultural practices in the various study areas in India. The pollen records have good amount of these taxa in the pollen assemblages across various forest and environmental regimes. Along with these taxa, some plant pollen indicators have been attributed to human induced changes in the natural vegetation related to pastoral activities. Thus, a number of factors contributed to changes in the natural vegetation in India during the Holocene, leading to cumulative anthropogenic impact alongside climatic changes. The adaptation to reduced strength of ISM, shrinking water bodies lead to the change in local vegetation, plantation activities and agricultural practices. Adaptations due to the migration of the human population dwelling around Lakes, river basins etc., due to paucity of resources and/or shifting cultivation practices, are also well-recorded in the pollen assemblages.

4.2 Tendencies in the Anthropogenic Pollen Preservation: The Indian Scenario

The role of Poaceae (Cerealia) pollen to decipher human impact has been crucial in the Indian scenario however, there is a school of thought with the justification that there are more than 2000 varieties of Poaceae pollen in the region with overlapping pollen characteristics, making the reliability of the anthropogenic nature of the preserved variety questionable (Vishnu-Mitre 1973). The extent of agriculture practices across the Indian subcontinent covered a vast region around subsequent floodplains and water bodies. These practices still persist and are the main occupation for a large pool of population in the region. Although the variety of cultural pollen recorded has been variable across most records, but then the composition of the assemblage indicating anthropogenic impact includes similar taxa. The evidence was conclusive for authors to estimate the beginning or influence of anthropogenic factors, and/ or biotic factors that interplay with the climatic factors to cause variations in the fossil pollen records. The advent of these taxa created reduction in the existing natural vegetation and prevailing taxa. Thus, sometimes it is complicated to understand if the replacement is natural or anthropogenic. But the reoccurrence of these particular set of taxa in a range of environments also signaling towards arid or dry conditions in most cases was a common factor in most of the records. The range of records from the high-altitude lakes to marginal marine environments to river delta all have witnessed this impact leading to parallel changes across the vegetation in these regions. The extent to which these anthropogenic marker taxa have been preserved as fossil pollen is noticeable in most records. The relative percentages of the marker pollen are adequate to draw subsequent inferences in most regions. But these records uphold an assemblage of these plant taxa which collectively indicate the advent, presence and persistence of human activities and induced vegetation changes. The beginning, extent and termination of such anthropogenic influence on the palaeo-vegetation at different sites however, were variable and also driven by coherent palaeo-environmental changes caused by climatic influences.

4.3 The Implications and Future Needs

To further understand these tendencies of anthropogenic indicator fossil pollen in an assemblage, more records with detailed assessment of human induced changes along with climate driven variations need to be investigated. Additional tools such as charcoal, DNA (human and plant), geo-chemical records and climate reconstructions can help understand the vegetation changes and the land-use patterns across India. This will further aid in understanding the past agricultural practices and human induced vegetation changes in India.

The advent of agriculture and pastoral activities set off an array of changes in the vegetation dynamics across India. The strength of palynological investigations in persistently recording these marker taxa provided conclusive evidence of anthropogenic impacts interlinked with environmental variations. Changing the overall vegetation scenario, the human influence during the Late Holocene is clearly evident in most of the pollen records in India. But higher resolution and well distributed records clubbed with multi-proxy and archeological evidence are further required. The identification of plant taxa indicating the presence and/or further absence of human impact from a region after a change in vegetation can be simply carried out by palynological studies. But the difficulties in species level identification and the existence of multiple species of taxa, such as Poaceae and Asteraceae can be difficult and raise questions about the palaeo-vegetation conditions. The key in palynology is to unlock the riddle of the past vegetative scenario and distinguish between the natural and anthropogenic factors causing the changes. High resolution microscopic studies such as Scanning Electron Microscopy, DNA-based evidence comparing modern and fossil pollen taxa, multi-proxy evidence such as isotope, mineral magnetism, geochemical data, along with archeological records can aid the palynological evidence specifically in the Indian scenario.

Acknowledgements The authors are grateful to the Director, Birbal Sahni Institute of Palaeosciences, Lucknow for the permission to publish this review work (BSIP/RDCC/Publication No: 7/2023-24). We sincerely thank the authors of all the above-mentioned publications for their lifetime of efforts and contribution to the science of Palynology. The authors and the future generations of palynologists shall always be indebted and benefited by this vast pool of palynological studies in India. These shall inspire to take up new approaches towards understanding the anthropogenic impacts on the environment through pollen-based evidences. The author NM is grateful to WISE KIRAN Division, Department of Science and Technology, Ministry of Science and Technology, Government of India, New Delhi, for the financial support given her under the Women Scientist project No. SR/WOS-A/EA-52/2018.

References

Basumatary SK, Bera SK (2010) Development of vegetation and climatic change in West Garo hills since late Holocene: pollen sequence and anthropogenic impact. J Indian Bot Soc 89:143–148

Behre KE (1981) The interpretation of anthropogenic indicators in Pollen diagrams. Pollen Spores 23:225–245

Behre KE (1988) The rôle of man in European vegetation history. In: Vegetation history. Handbook of vegetation science. Springer, Dordrecht, pp 633–672

Bera SK (2003) Early Holocene pollen data from Mikir Hills, Assam, India. The Palaeobotanist. 52:121–127. https://doi.org/10.54991/jop.2003.1760

Bera SK, Basumatary SK, Gogoi B (2011) Pollen analysis and its implications in climate change and vegetation succession in recent past: evidence from surface and subsurface fluvial sediment of Tropical forests of eastern Assam, North East India. J Frontline Res Arts Sci 1:62–74

Bera SK, Dixit S (2011) Pollen analysis of late Holocene lacustrine sediment from Joypore Reserve Forest, Dibrugarh, Assam. In: Singh DS, Chhabra NL (eds) Geological processes and climate change. Macmillan Publisher India Ltd, pp 85–94

Bera SK, Dixit S, Gupta K (2013) Late Holocene climate and vegetation succession as inferred from Jokai reserve forest, Dibrugarh, Assam: Pollen record and Anthropogenic impact. In: Hazarika M, Jamir T (eds) 51 Years after Daojali-Hading: emerging perspectives in the archaeology of Northeast India: Essays in Honour of Tarun Chandra Sharma. Research India Press, pp 72–78

Bhattacharya K, Chanda S (1988) Late-Quaternary vegetational history, palaeoecology and biostratigraphy of some deposits of Brahmaputra Basin, Upper Assam, India. J Palynol 23–24:225–237

Bhattacharya K, Chanda S (1992) Late-Quaternary vegetational history of Upper Assam, India. Rev Palaeobot Palynol 72:325–333. https://doi.org/10.1016/0034-6667(92)90035-F

Bhattacharyya A, Mehrotra N, Shah SK (2011) Holocene vegetation and climate of South Tripura based on Palynological analysis. J Geol Soc India 77:521–526. https://doi.org/10.1007/s12594-011-0056-x

Biswal A (2006) Mystic monpas of the Tawang Himalaya. Indus Publishing Company, New Delhi

Castri F, Hansen AJ, Debussche M (1990) Biological invasions in Europe and the Mediterranean Basin. Monogr Biol. Springer Netherlands

Chakraborty S, Bhattacharya SK, Ranhotra PS et al (2006) Palaeoclimatic scenario during Holocene around Sangla valley, Kinnaur Northwest Himalaya based on multi proxy records. Curr Sci 91:777–782

Chanda S (1974) Late Quaternary vegetational history of eastern region. In: Surange KR, Lakhanpal RN, Bharadwaj DC (eds) Aspects and appraisals of Indian Palaeobotany. Birbal Sahni Institute of Palaeobotany, Lucknow, India, pp 651–656

Chauhan MS (2005) Pollen record of vegetation and climatic changes in North Eastern Madhya Pradesh during last 1,600 years. Trop Ecol 46:265–271

Chauhan MS, Nautiyal CM (2009) Late Holocene vegetation and climate change in Loktak Lake region, Manipur, based on pollen and chemical evidence. Palaeobotanist 58:21–28. https://doi.org/10.54991/jop.2009.78

Chauhan MS, Sharma C (1996) Late Holocene vegetation of Darjeeling (Jore-Pokhari), eastern Himalaya. Palaeobotanist 45:125–129. https://doi.org/10.54991/jop.1996.1226

Chauhan MS, Sharma C (2000) Late Holocene vegetation and climate in Dewar Tal area, inner lesser Garhwal Himalaya. Palaeobotanist 49:509–514. https://doi.org/10.54991/jop.2000.164

Clift PD, Carter A, Giosan L et al (2012) U-Pb zircon dating evidence for a Pleistocene Sarasvati River and capture of the Yamuna River. Geology 40:211–214. https://doi.org/10.1130/G32840.1

Conedera M, Krebs P, Tinner W et al (2004) The cultivation of *Castanea sativa* (Mill.) in Europe, from its origin to its diffusion on a continental scale. Veg Hist Archaeobot 13:161–179. https://doi.org/10.1007/s00334-004-0038-7

Crutzen PJ, Stoermer EF (2000) The anthropocene. IGBP Newslett 41:12

Crutzen PJ (2002) Geology of mankind. Nature 415:23. https://doi.org/10.1038/415023a

Das S (2014) Palaeo-palynology of late Quaternary peat deposit from lower Bengal Basin, India: a palaeoecological approach. Quat Int 325:197–204. https://doi.org/10.1016/j.quaint.2013.07.042

Deza-Araujo M, Morales-Molino C, Tinner W et al (2020) A critical assessment of human-impact indices based on anthropogenic pollen indicators. Quat Sci Rev 236:106291. https://doi.org/10.1016/j.quascirev.2020.106291

Dikshit KR, Dikshit JK (2014) Natural vegetation: forests and grasslands of North-East India. In: North-East India: land, people and economy. Advances in Asian Human-Environmental Research. Springer Dordrecht, pp 213–255. https://doi.org/10.1007/978-94-007-7055-3_9

Dixit S, Bera SK (2011) Mid-Holocene Vegetation and climatic variability in tropical deciduous sal (*Shorea robusta*) forest of Lower Brahmaputra Valley, Assam. J Geol Soc India 77:419–432. https://doi.org/10.1007/s12594-011-0051-2

Dixit S, Bera SK (2012a) Holocene climatic fluctuations from Lower Brahmaputra flood plain of Assam, Northeast India. J Earth Syst Sci 121:135–147

Dixit S, Bera SK (2012b) Pollen-inferred vegetation vis-á-vis climate dynamics since Late Quaternary from Western Assam, Northeast India: signal of global climatic events. Quat Int 286:56–68. https://doi.org/10.1016/j.quaint.2012.06.010

Dodia R (1988) Holocene vegetation and climate change in Kumaun area based on palynological evidences. Curr Sci 57:71–73

Draganits E, Gier S, Hofmann CC et al (2014) Holocene versus modern catchment erosion rates at 300 MW Baspa II hydro-electricpower plant (India, NW Himalaya). J Asian Earth Sci 90:157–172. https://doi.org/10.1016/j.jseaes.2014.04.016

Fyfe R, Roberts N, Woodbridge J (2010) A pollen-based pseudobiomisation approach to anthropogenic land-cover change. Holocene 20:1165–1171. https://doi.org/10.1177/0959683610369509

Gupta HP, Khandelwal A (1982) Late Holocene palynology from the lake in Sat Tal valley, District Nainital, Kumaon Himalaya, U.P. Geophytology 12:313–321

Huntley BJ, Webb T (eds) (1988) Vegetation history. Kluwer Academic Publishers, Dordrecht

Iversen J (1944) Viscum, Hedera and Ilex as climate indicators: a contribution to the study of the post-glacial temperature climate. Geol Fören Stockh Förh 66:463–483. https://doi.org/10.1080/11035894409445689

Kar R, Quamar MF (2019) Pollen based Quaternary palaeoclimate studies in India: an overview of the recent advances. Palynology 43:76–93. https://doi.org/10.1080/01916122.2017.1410502

Khandelwal A, Mohanti M, Scharf BW et al (2008) Vegetation history and sea level variations during the last13,500 years inferred from a pollen record at Chilka Lake, Orissa, India. Veg Hist Archaeobot 17:335–344. https://doi.org/10.1007/s00334-007-0127-5

Krebs P, Pezzatti GB, Beffa G et al (2019) Revising the sweet chestnut (*Castanea sativa* Mill.) refugia history of the last glacial period with extended pollen and macrofossil evidence. Quat Sci Rev 206:111–128. https://doi.org/10.1016/j.quascirev.2019.01.002

Lang G (1994) Quatazre vegetationsgeschichte Europe: Methoden und Ergebnisse, Gustav Fischer, Jena/Stuttgart/New York

Limaye RB, Kumaran KPN, Padmalal D (2014) Mangrove habitat dynamics in response to Holocene Sea level and climate changes along southwest coast of India. Quat Int 325:116–125. https://doi.org/10.1016/j.quaint.2013.12.031

Mehrotra N, Shah SK, Bhattacharyya A (2014) Review of palaeoclimate records from Northeast India based on pollen proxy data of late Pleistocene-Holocene. Quat Int 325:41–54. https://doi.org/10.1016/j.quaint.2013.10.061

Mehrotra N, Shah SK, Basavaiah N et al (2019) Resonance of the '4.2ka event' and terminations of global civilizations during the Holocene, in the palaeoclimate records around PT Tso Lake, Eastern Himalaya. Quat Int 506:207–216. https://doi.org/10.1016/j.quaint.2018.09.027

Mehrotra N, Shah SK, Basavaiah N et al (2022) Middle to Late Holocene climate vegetation and sea-level changes in NW Tripura, Northeast India, based on palynological and mineral magnetic evidence. J Paleolimnol 68:297–313. https://doi.org/10.1007/s10933-022-00249-6

Misra AK, Mohanty RB, Ghosh R et al (2022) Modern pollen–vegetation relationships along an altitudinal transect in the Western Higher Himalaya, India: palaeoclimatic and anthropogenic implications. Holocene 32:835–852. https://doi.org/10.1177/09596836221096006

Mukherjee BB (1972) Pollen analysis of a few Quaternary deposits of Lower Bengal Basin. In: Ghosh AK, Chanda S, Ghosh TK et al (eds) Proceeding of seminar of palaeopalynology and Indian stratigraphy, Calcutta, 1971, pp 357–374

Prasad V, Farooqui A, Sharma A et al (2014) Mid-late Holocene monsoonal variations from mainland Gujarat, India: a multiproxy study for evaluation climate culture relationship. Palaeogeogr Palaeoclimatol Palaeoecol 397:38–51. https://doi.org/10.1016/j.palaeo.2013.05.025

Quamar MF (2019) Vegetation dynamics in response to climate change from the wetlands of Western Himalaya, India: Holocene Indian summer monsoon variability. Holocene 29:345–362. https://doi.org/10.1177/0959683618810401

Quamar MF (2022) Late Holocene vegetation dynamics and monsoonal climatic changes in Jammu, India. Acta Palaeobot 62:36–49. https://doi.org/10.35535/acpa-2022-0004

Quamar MF, Chauhan MS (2011) Late Holocene vegetation, climate change and human impact in southwestern Madhya Pradesh, India. Palaeobotanist 60:281–289. https://doi.org/10.54991/jop.2011.174

Raj R, Chamyal LS, Prasad V et al (2015) Holoceneclimatic fluctuations in the Gujarat Alluvial Plains based on multiproxy study of the Pariyaj Lake archive, Western India. Palaeogeogr Palaeoclimatol Palaeoecol 21:60–74. https://doi.org/10.1016/j.palaeo.2015.01.004

Ranhotra PS, Sharma J, Bhattacharyya A et al (2018) Late pleistocene-holocene vegetation and climate from the Palaeolake sediments, Rukti Valley, Kinnaur, Himachal Himalaya. Quat Int 479:79–89. https://doi.org/10.1016/j.quaint.2017.08.025

Ruddiman WF (2000) Earth's climate past and future, 1st edn. W.H. Freeman and Company, New York, pp 401–405

Sharma C (1970) Origin of 'floating island" in the lakes of Khajiar and Realsar in Himachal Pradesh. Palaeobotanist 19:270–276. https://doi.org/10.54991/jop.1970.876

Sharma C, Singh G (1972) Studies in the late quaternary vegetational history in Himachal Pradesh-1 Khajiar lake. Palaeobotanist 21:144–162. https://doi.org/10.54991/jop.1972.1477

Sharma C, Singh G (1972) Studies in the late quaternary vegetational history in Himachal Pradesh-2 Rewalsar lake. Palaeobotanist 21:321–338. https://doi.org/10.54991/jop.1972.1495

Sharma C (1985) On the late quaternary vegetational history in Himachal Pradesh-3 Parasram Tal. Geophytology 15:206–218

Singh RK, Singh A, Garnett ST et al (2015) Paisang (*Quercus griffithii*): a keystone tree species in sustainable agroecosystem management and livelihoods in Arunachal Pradesh, India. Environ Manage 55:187–204. https://doi.org/10.1007/s00267-014-0383-y

Smith BD, Zedar MA (2013) The onset of the anthropocene. Anthropocene 4:8–13. https://doi.org/10.1016/j.ancene.2013.05.001

Srivastava J, Farooqui A (2017) Holocene climate and relative sea level changes in Cauvery River delta, India based on pollen and sedimentary records. J Palaeontol Soc India 62:193–204

Srivastava G, Sub T, Mehrotra RC et al (2019) Bamboo fossils from Oligo-Pliocene sediments of northeast India with implications on their evolutionary ecology and biogeography in Asia. Rev Palaeobot Palynol 262:17–27. https://doi.org/10.1016/j.revpalbo.2018.12.002

Sugita S (2007) Theory of quantitative reconstruction of vegetation I: pollen fromlarge sites REVEALS regional vegetation composition. Holocene 17:229–241. https://doi.org/10.1177/0959683607075837

van Leeuwen JFN, Froyd CA, van der Knaap WO et al (2008) Fossil pollen as a guide to conservation in the Galápagos. Science 322:1206–1206. https://doi.org/10.1126/science.1163454

Vishnu-Mittre, Gupta HP (1972) Pollen analytical study of quaternary deposits in the Bengal basin. Palaeobotanist 19:297–306. https://doi.org/10.54991/jop.1970.880

Vishnu-Mittre (1966) Some aspects concerning pollen-analytical investigations in the Kashmir Valley. Palaeobotanist 15:157–175. https://doi.org/10.54991/jop.1966.763

Vishnu-Mittre (1973) Cereal versus non-cereal grass pollen and the inference of past agriculture. Pollen and spore morphology of the recent plants. In: Proceedings of the III international palynological conference Novosibirsk (U.S.S.R)

Modern Pollen Assemblage and Micro-morphometric Analysis of Arboreal and Non-arboreal Taxa from Lucknow District of Central Ganga Plain, India: A Window to Palaeoclimatic Studies

Rajeev Ranjan and Swati Tripathi

Abstract The present study was intended to establish a modern pollen analogue from the open-land areas of the Lucknow district as well as to investigate pollen morphological attributes from the same region to resolve and refine the taxonomic categorization and pollen identification. The flowers of 29 plant taxa belonging to 18 different families have been procured from the premises of BSIP and BBAU including the forest edge and open land localities of Lucknow district. Detailed pollen micro-morphological investigation of twenty nine plant taxa has been carried out under the Light Microscopy (LM). The pollen morphological characteristics like aperture number, position and character (NPC), pollen shape, size, exine sculpturing and thickness play a vital role in assigning different pollen taxa to genera, species and families. The palynological studies in the surface soil samples aids in tracing the marker pollen taxa which reflects the current vegetation pattern and climatic condition and hence have a plausible role in the palaeoclimatic interpretation. The pollen assemblage in the surface soil sample reflects the complete dominance of non-arboreal taxa (78.07%) over the arboreal taxa (9.99%). The prominent non-arboreal taxa are represented by Caryophyllaceae, Acanthaceae, Amaranthaceae, *Celosia argentea*, Lamiaceae, Brassicaceae and Asteroideae, followed by *Kydia* sp. The arboreal taxa are mainly represented by *Butea monosperma*, *Emblica officinalis*, *Alstonia scholaris*, *Holoptelea integrifolia*, *Lagerstroemia parviflora*, *Terminalia bellirica*, Meliaceae and *Syzygium cumini*. The distribution of marshy and aquatic taxa is less prominent. Thus, the overall palynological study would help in the precise identification of the pollen in the surface soil and the sedimentary sections for the formulation of the

R. Ranjan · S. Tripathi (✉)
Birbal Sahni Institute of Palaeosciences (BSIP), Uttar Pradesh, 53 University Road, Lucknow 226007, India
e-mail: swati.tripathi@bsip.res.in

R. Ranjan
Babasaheb Bhimrao Ambedkar University (A Central University) (BBAU), Uttar Pradesh, Vidya Vihar, Raebareli Road, Lucknow 226025, India

B. Samant and D. Thakre (eds.), *Applications of Palynology in Stratigraphy and Climate Studies*, Society of Earth Scientists Series, https://doi.org/10.1007/978-3-031-51877-5_9

modern pollen-vegetation relationship as well as palaeoecological reconstruction, respectively.

Keywords Palynology · Morphometric analysis · Modern pollen · Paleoecological reconstruction

1 Introduction

Northern India constitutes a large and rich phytogeographical province of our country and has not received adequate attention regarding botanical works, especially on the phenological and palynological points of view. In the present time, the most varied plant communities on the Earth are found in tropical forests. These forests are disappearing at alarming rates as a result of deforestation for the exploitation of timber and other forest products. The issue with chronic forest disturbance is that plants or ecosystems frequently do not have enough time to recuperate properly since human activity never ends (Singh 1998). Understanding the composition of a forest is necessary for its management, concerning other forests, surrounding forest edges and open-land areas. Moreover, to understand the effects of past impacts on the present status and the present relationship of the forest with surrounding land uses are always necessary for this direction (Geldenhuys and Murray 1993). Therefore, it is considered to be an urgent requirement to conserve these reserve forests and associated open-land areas to protect their unique biodiversity in the wake of future climate change. If properly managed, the forests are going to be a great source of wealth for this region leading to enrichment of the quality of its pristine environment. Tripathi et al. (2016) documented the modern pollen-vegetation relationship and its significance in decoding the past vegetation and climate of the Kukrail Reserve Forest and surrounding areas of the Lucknow district. In the present communication, our study would elaborate and provide an insight into the potential of generating pollen flora and formulating surface palynological studies exclusively from the forest edge and open-land areas of Lucknow to understand any discrepancies recorded in the pollen preservation and the surrounding vegetation (Fig. 1).

To date, pollen grains have been proven to be the best proxy for documenting the vegetation of a region. This is because pollen grains are produced in abundance and due to their minute size and resistant nature, are readily preserved in the sediments. Moreover, each plant produces pollen with unique morphology, which can be easily related to different living plant groups. Thus, pollen extracted from the sediments can provide a very good representation of vegetation growing in a region during a particular point in time, and since vegetation is controlled by climate, pollen assemblages can also aid in reconstructing the climate then prevalent (Tripathi et al. 2020). Pollen morphological examination is an extremely important proxy tool for decoding past environmental and climatic changes (Bradley 1985). Additionally, the pollen morphological characters of crop and tree pollen taxa of a region could also provide a way to trace the anthropogenic and arboriculture activities in the region

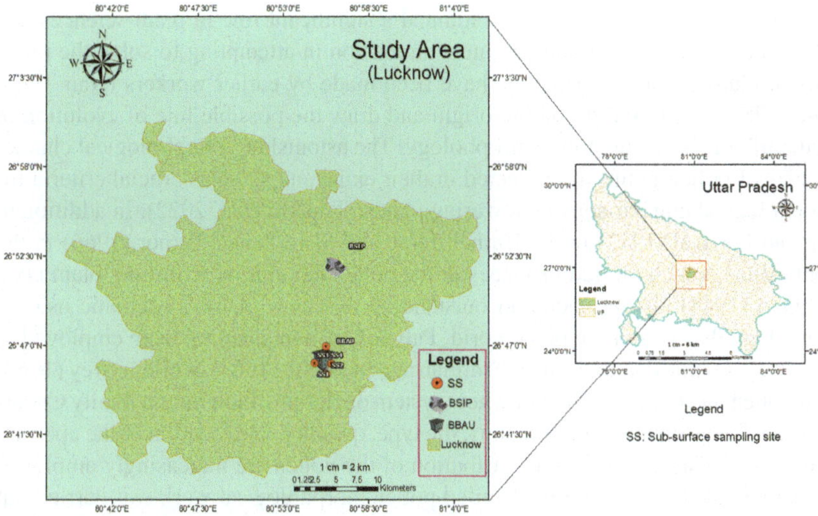

Fig. 1 Map indicating the location of sampling sites

during the past millennia (Tripathi et al. 2021). In this study, an effort has been made for the first time to gather pollen characters using LM (Light Microscopy) along with respective phenological data from the 29 selected plant taxa of the Lucknow district. Later on, these pollen micro-morphological studies were used to formulate modern pollen/vegetation relationships in the same area which could further aid in the reconstruction of past vegetation and climatic oscillation.

The studies dealing with the quantitative pollen-vegetation interaction in several vegetation types on the Indian subcontinent have been examined (Bera 2000; Singh et al. 2011; Basumatary et al. 2013; Chauhan et al. 2013; Tripathi et al. 2023). However, only a few investigations were concerned with the quantitative relationship between pollen and vegetation (Barboni and Bonnefille 2001). Earlier, detailed palyno-investigation on sedimentary soil sequences were carried out in northern India (Gupta 1978; Singh et al. 1990; Chauhan et al. 2004, 2009; Saxena et al. 2006; Chauhan and Chatterjee 2008; Ranhotra and Bhattacharyya 2010; Farooqui and Sekhar 2011; Trivedi et al. 2011, 2013); however, very few research works have been recorded on surface pollen samples (Sharma et al. 2006; Trivedi and Chauhan 2011; Tripathi et al. 2016; Bera et al. 2018). Therefore, the main objectives of our study are to examine the detailed pollen morphometric analysis of 29 different pollen taxa using a light microscope (LM) for the taxonomic characterization for palaeoecological studies and to explore the linear and non-linear relationship between the surface pollen and extant vegetation of the Lucknow district, Uttar Pradesh. This will provide us a starting point for understanding the palaeovegetation and past climate of the Central Ganga Plain. The modern pollen data collected from the forest edge and open-land areas of Lucknow could be compared to the other tropical regions at the global level (Tripathi et al. 2016).

In recent years, palynology has attained a significant role in plant sciences and pollen morphology has assumed a new dimension in attempting to solve the problems of plant evolution. Attempts have been made by earlier workers (Nair 1965; Feuer 1981) to throw light on the origin and draw the possible line of evolution in Thalamiflorae based on pollen morphology. The astonishing morphological characteristics of pollen grains are reflected in their exine and serve as crucial criteria for morphological comparisons in flowering plants (Tripathi et al. 2022). In addition to Cerceau-Larrival (1971) in the Umbelliferae and Skvarla and Turner (1966) in the Compositae, which offered instances of the pollen features in delimiting plant taxa, Erdtman (1943) also provided an outstanding overview of the systematic uses of palynology throughout the plant world. Palynology and anatomy were employed to solve a taxonomic conundrum by Erdtman and Metcalfe (1963), while Jeffrey (1964) used pollen morphology to create a new system of classification for the family Cucurbitaceae. In systematic investigations, the type, quantity, and location of the aperture as well as the size, shape, and distribution of the pollen are increasingly employed (Tripathi et al. 2019). In India, detailed pollen morphological work was carried out in North West Himalayas and Maharashtra (Gupta and Sharma 1986; Nayar 1990). However, scattered reports on pollen morphology of tropical regions of India have been published from time to time (Chauhan and Bera 1990; Dixit and Bera 2011; Tripathi et al. 2019).

1.1 Geology and Geography

The 2,40,928 square kilometres that make up the state of Uttar Pradesh are mostly made up of the middle Ganga drainage system's alluvial fill. Rocky uplands are prevalent in Uttar Pradesh's southern region. The geomorphic units like Tarai and Bhabhar that edge the Siwalik Foothills serve as a representation of the northern fringe region. The three mighty rivers—the Indus, the Ganga, and the Brahmaputra—cross the Indo-Gangetic foreland basin where Lucknow, the capital of Uttar Pradesh, is situated. The central section of the Indo-Gangetic foreland basin known as the Ganga plain is made up primarily of sandy silty deposits (Thakur et al. 2018). The geomorphology of the Ganga plain, which has a relatively low topography and a moderate regional slope, today is the consequence of a complex interplay between tectonic activity and climatic change throughout the Late Pleistocene-Holocene (Srivastava et al. 2003).

1.2 Soil and Climate

The Ganges River System's slow-moving rivers have spread a thick layer of alluvium over much of Uttar Pradesh. The sandy to clayey loam alluvial soils are incredibly fruitful. The soils in the state's southern region are typically a mixture of red and black or red to yellow.

Uttar Pradesh has a tropical monsoon climate with mild temperatures all year round. Between roughly 70 °F (20 °C) in January to more than 100 °F (38 °C) in May and June, Lucknow experiences its highest average high temperatures. Northwest of Faizabad, near Gonda, high temperatures of around 120 °F (50 °C) have been recorded. With a height of around 123 m above sea level, Lucknow City is situated geographically between 26.30° and 27.10° North and 80.30° and 81.13° East (Fig. 1). Due to the state's geographic location, the Indian Summer Monsoon (ISM), which is known for its moisture content, arrives from June to September. 75% of the precipitation throughout these four months may be attributed to the ISM. One of the nation's oldest meteorological observatories is located in the city. Lucknow receives 1010 mm of rain/year on average, according to meteorological data collected over the course of 49 years (IMD, Amausi station, Lucknow). Summer temperatures, which may easily reach 46 °C when combined with extreme heat waves, differ greatly from winter temperatures, which can only dip as low as 2 °C. The cold winds that are brought on by the western disturbances are another feature of the winters. In addition, mist and fog are notable winter features that might last for more than a fortnight.

The state has annual rainfall ranging from 24 to 40 inches (600–1,000 mm) in the west to 40–80 inches (1,000–2,000 mm) in the east. The southwest monsoon, which lasts from around June to September, brings about 90% of the rainfall. Floods are a persistent issue that may result in fatalities and significant agricultural and property damage, particularly in the eastern region of the state since the majority of the state's rainfall is concentrated during that four-month period. Drought conditions are caused by the monsoons failing on occasion.

1.3 Vegetation

Lucknow has a tropical mixed to dry deciduous vegetation (Tripathi et al. 2016) (Pl. 1(1)) which is gradually converting into open-land vegetation due to rise in human encroachment. Although poor, the natural vegetation is made up primarily of shrubs, grasses, and a few sparsely placed trees. As a result, the floral scenery has a scrubby woodland appearance. Patches of *Acacia arabica, Holoptelea integrifolia, Diospyros cordifolia, Cordia dichotoma, Syzygium cumini, Capparis decidua, Butea monosperma, Mimosa* sp., *Albizia lebbeck, Flacourtia indica, Ziziphus mauritiana, Carissa spinarum, Adhatoda vasica, Nyctanthes arbor-tristis* and *Lagerstroemia speciosa* despite being dispersed, represent the basic components of tree and shrubby taxa. In openland areas, there are also a few stands of *Holoptelea integrifolia, Syzygium cumini, Dalbergia sissoo, Bombax ceiba, and Acacia catechu.*

Along the banks of Gomti River, ravine thorn forest, consisting of *Capparis decidua, Aegle marmelos, Ziziphus mauritiana, Carissa spinarum, Calotropis gigantea* and *Adhatoda vasica* exist. Recently plantations of *Eucalyptus* sp., *Terminalia arjuna, Pongamia pinnata, Melia azedarach, Prosopis juliflora, Syzygium cumini, Acacia arabica, Parkinsonia* sp., *Gardenia, Kalancho epinnata* etc., have been raised. The herbaceous elements of terrestrial nature include *Ageratum conyzoides, Euphorbia hirta, Euphorbia thymifolia, Oxalis acetosella, Chenopodium album, Chenopodium murale, Mazus japonicus, Melilotus alba, Justicia simplex, Justicia diffusa, Solanum nigrum, Solanum xanthocarpum, Heliotropium strigosum, Launaea nudicaulis, Portulaca oleracea*, etc., whereas along the margins of lakes, ditches, and in wet places *Polygonum plebeium, Polygonum glabrum, Rotalasp., Ammania baccifera, Alternanthera sessilis, Hydrocotyle* sp., *Eriocaulon quinquangularis*, are quite common. There is abundant aquatic vegetation (Pl. 1 (2)) in the lakes, ponds, streams, and rivers. *Eichhornia crassipes* (water hyacinth) frequently obstructs the Gomti River by forming a dense mat on the water surface. *Lemna polyrrhiza, Nymphaea* sp., *Nymphoides cristatum, Potamogeton indicus, Jussiaea repens, Ipomoea aquatica, Wolfia arrhiza, Vallisnaria spiralis, Hydrilla verticillata* etc. are some of the frequent lake water plants (Chauhan and Bera 1990).

2 Material and Methods

2.1 Extraction of the Pollen/Spores from the Living Polliniferous Samples for the Pollen Morphological Studies

The materials for the pollen morphological study include fresh flowers of 29 plant taxa that were collected near BBAU, BSIP and Lucknow University (Pl. 1 (3 and 4)). The acetolysis process for the extraction of pollen/spores from the flowers was carried out using Erdtman (1952) and provided below:

1. Remove all the petals and sepals from the flower.
2. After that, keep the anthers in a fresh vial tube.
3. Then, gently tease the samples with the help of a glass rod by adding some amount of water.
4. After gently teasing and crushing, the material was then sieved through a 150 μm-mesh sieve to remove the residue materials.
5. After the sieving process, the left filtrate was transferred to the clean 15 ml centrifuge tube.
6. The material was then dehydrated using centrifugation and glacial acetic acid (GAA) treatment.
7. After that, the material was treated in a water bath for 5 min, or until the water begins to boil, with an acetolysing solution (9:1 ratio of acetic anhydride to concentrated sulphuric acid). This method is performed for the removal of cellulose and the clear visibility of pollen morphology.

8. The sample was then centrifuged and decanted off to remove the acid.
9. Again, the material was treated with GAA and washed 2–3 times with distilled water.
10. To prevent the samples from microbial decomposition, 50% glycerin and a few drops of phenol were then added to the residue.
11. Permanent pollen slides were made with the help of DPX and polyvinyl for the preservation, identification, and studies of pollen morphology under LM.

2.2 *Extraction of the Pollen/Spores from the Surface Soil Samples for the Modern Pollen-Vegetation Relationship*

The materials for the present study include four surface samples (surface soil) that were procured in a transect from the forest edge to open-land areas near BBAU of Lucknow (Pl. 1 (5 and 6)) and the chemical procedure adopted was stepwise given below:

1. To deflocculate pollen from the soil and dissolve humic acid, 20 g of soil was heated for 5 min in a 10% potassium hydroxide (KOH) solution.
2. Using the decantation method, samples were washed three times in distilled water to eliminate alkalis.
3. To remove the coarse debris, the materials were sieved using a 150 μm-mesh sieve after the KOH treatment.
4. If carbonates were present, they were then removed from the samples using 10% hydrochloric acid (HCL), and the samples were then rinsed twice with distilled water to remove the acid using decantation. Then the material was treated with 40% hydrofluoric acid (HF) in plastic jars and kept for 2–5 days, to remove the silicates.
5. To rid the samples of silica and excess HF, they were once more rinsed twice with distilled water.
6. The left filtrate was treated with glacial acetic acid (GAA) and centrifuged 2 times in a dehydration process.
7. After that, the sample was treated in a water bath for five minutes, or until the water began to boil, with an acetolyzing solution (9:1 ratio of acetic anhydride to concentrated sulphuric acid). Samples were then centrifuged and decanted to remove the acid.
8. The remaining material was treated with GAA once more and cleaned with distilled water three or four times.
9. To prevent the samples from microbial decomposition, 50% glycerin and a few drops of phenol were then added to the residue.
10. In order to identify and count pollen and non-pollen palynomorphs (fungal remains), temporary pollen slides were prepared.

3 Pollen Morphology

The pollen morphological study presented here is based upon detailed pollen morphometrical analysis of 29 plant taxa belonging to 28 genera and 18 families distributed in the region under study. The chief objective of preparing this pollen morphological data are to identify modern and fossil pollen grains preserved in the surface soil and sediment core, respectively. This section includes the major pollen characters which have been made sufficiently exhaustive to meet all requirements in identification (Pl. 2). In preparing this pollen morphological data, primarily importance is given to kind, number and distribution of aperture and sexine ornamentation. Terminology adopted is after Erdtman (1952), Nayar (1990), Faegri and Iverson (1989) and Punt et al. (2007). This is strictly a regional key applicable to the region under investigation.

3.1 Results of Pollen Morphometry of Selected Plant Taxa

1. ***Rosa alba* L.** Pl. 2 (1–2) (Rosaceae)

Pollen 3-colporate, prolate, 33.42 × 18.66 μm. Colpi length 29.95 μm. Os lalongate. Exine 1.74 μm, sexine pattern striate. Flowering: January to April.

2. ***Tectona grandis* L.f.** Pl. 2 (3) (Verbenaceae)

Pollen 3-colpate, perprolate, 55.2 × 23.1 μm. Colpi long, broad, apices obtuse, margin thick. Exine 3.2 μm, crassisexinous, obscure, tegillate. Flowering: March to April.

3. ***Cassia fistula* L.** Pl. 2 (4) (Fabaceae)

Pollen 3-colporate, prolate-spheroidal, 34.12 × 33.48 μm. Colpi long, broad. Os lalongate. Exine 2.1 μm, sexine as thick as nexine, ±obscure, tegillate. Flowering: March to April.

4. ***Kydia calycina* Roxb.** Pl. 2 (5) (Malvaceae)

Pollen panporate, spheroidal, 86 μm (excluding spines). Pore circular, 4.9 μm, membrane thickened, interpolar distance, 5.45 μm. Exine 4.49 μm (excluding spines), sexine as thick as nexine, spines 14.9 μm long, 6μm broad at base, granulate, tegillate. Flowering: whole year.

5. ***Dimorphotheca ecklonis* DC.** Pl. 2 (6) (Asteraceae)

Pollen 3-colporate, oblate-spheroidal, 25.63 × 25.81 μm (with spine). Exine 3.68 μm thick, echinate, sexine as thick as nexine, ±obscure. Flowering: February to April.

6. ***Albizia julibrissin* Durazz., 1772 non sensu Baker, 1876** Pl. 2 (7) (Fabaceae)

Polyads, 16 celled, 72.9 μm, monads loosely fitted. Exine 2.12 μm, sexine slightly thicker than nexine, finely granulate, tegillate. Flowering: March to April.

7. ***Tabernaemontana divaricata* (L.) R.Br. ex Roem. & Schult**. Pl. 2 (8–9) (Apocynaceae)

Pollen 3-4-colpate, prolate, 32.92 × 24.38μm. Colpi 26.34 μm long, tenuimarginate. Exine 1.48 μm, sexine thicker than nexine, psilate with punctae distributed all over the surface. Flowering: Throughout the year.

8. ***Acacia oxyphylla* Graham ex Benth**. (Fabaceae). Pl. 2 (10)

Polyad, 24 celled, size 50.11 × 48.05μm. Exine 2.06 μm thick, sexine as thick as nexine, finely granulate. Flowering: November to December.

9. ***Acacia pennata* Willd**. (Fabaceae)

Polyad, 24 celled, size 48.9 × 47.1 μm. Exine 2.48 μm thick, sexine much thicker than nexine, coarsely granulate. Flowering: November to December.

10. ***Lannea coromandelica* (Houtt.) Merr**. Pl. 2 (11) (Anacardiaceae)

Pollen 3-colporoidate, sub-porolate, 37.38 × 26.81 μm. Colpi 32.52 μm long, crassimarginate, constricted at the middle. Exine 3.1 μm thick, sexine thicker than nexine, striate, tegillate. Flowering: February to March.

11. ***Lagerstroemia parviflora* Roxb**. Pl. 2 (12) (Lythraceae)

Pollen 3-colporate, prolate-spheroidal, 40 × 36 μm. Colpi long, margin thick and slightly undulated. Os lalongate. Exine 3.4 μm thick, sexine thicker than nexine and much thicker at the poles, reticulate, heterobrochate, tegillate. Flowering: March to April.

12. ***Barringtonia acutangula* Gaertn**. Pl. 2 (13) (Lecythidaceae)

Pollen 3-syncolpate, prolate, 42.15 × 32.19 μm. Colpi crassimarginate. Exine 3.06 μm thicker at aperture, sexine thicker than nexine, granulate tegillate. Flowering: September to October.

13. ***Bombax Ceiba* L**. Pl. 2 (14) (Malvaceae)

Pollen 3-colporate, planaperturate. Amb triangular, 65.08 × 25.06 μm. Exine 3.61 μm thick, sexine thicker than nexine, reticulate, distinctly heterobrochate, brochi broad at the centre and finer towards angle, integillate. Flowering: February to March.

14. ***Euphorbia milii* Des Moul**. Pl. 2 (15–16) (Euphorbiaceae)

Pollen 3-colporate, shape prolate, 29.73 × 46.25 μm. Colpi length 39.48μm. Os lalongate, diameter 8.95 μm. Exine 1.22 μm thick, sexine thicker than nexine, sexine pattern reticulate. Flowering: Throughout the year.

15. ***Emblica officinalis* Gaertn.** (Euphorbiaceae)

Pollen 4-colporate, oblate-spheroidal, 20 × 20 μm. Colpi tenuimarginate. Os circular. Exine 2.02 μm thick, sexine thicker than nexine, distinctly reticulate, tegillate. Flowering: March to April.

16. ***Syzygium cumini* (L.) Skeels.** (Myrtaceae)

Pollen 3-colporate, parasyncolpate, oblate, 11.1 × 17.2 μm. Amb triangular. Colpi long, Os lalongate, 1.62 × 1.93 μm. Exine 2.62 μm thick, sexine thicker than nexine, ±obscure. Flowering: March to April.

17. ***Terminalia bellirica* (Gaert.) Roxb.** (Combretaceae)

Pollen 3-colporate, heterocolpate, distinct pseudo-colpi visible, prolate, 19.74 × 14.62 μm. Colpi 13.89 μm long. Os circular—slightly lalongate, 2.51 × 1.82 μm, margin indistinct. Exine 1.51 μm thick, sexine thinner than nexine, ±obscure. Flowering: March to April.

18. ***Madhuca indica* J.F.Gmel.** (Sapotaceae)

Pollen large 3-4-colporate, prolate-spheroidal, 54.2 × 48.4 μm. Colpi narrow, long, 39.8 μm. Os large, lalongate, 5.5 × 4.1 μm, margin distinct. Exine 2.3 μm thick, sexine thicker than nexine, ±obscure. Flowering: April to May.

19. ***Sapindus rarak* DC.** (Sapindaceae)

Pollen 3-colpate, sub-spheroidal, 18.6 μm. Colpi long, broad, tenuimarginate. Exine 1.52 μm, sexine slightly thicker than nexine, finely reticulate, tegillate. Flowering: June to July.

20. ***Ziziphus mauritiana* Lam.** (Rhamnaceae)

Pollen 3-colporate, sub-oblate, 20.2 × 25.3 μm. Amb triangular. Colpi long, narrow, apices obtuse. Os circular, 3.01 μm, sharply delimited. Exine 2.02 μm thick, sexine slightly thicker than nexine, ±obscure, tegillate.

21. ***Bauhinia purpurea* L.** (Fabaceae)

Pollen large 3-4-colporoidate, sub-prolate, 79.22 × 51.24 μm. Colpi 60.02 μm long, broad, crassi-marginate. Os slightly lalongate, 6.1 μm. Exine 3.8 μm thick, sexine thicker than nexine, pseudo-reticulate. Flowering: February to March.

22. ***Erythrina stricta* Roxb.** (Fabaceae)

Pollen 3-porate, sub-oblate, 24.3 × 37.2 μm. Amb subtriangular. Pore circular. Exine 2.51 μm, sexine as thick as nexine, finely reticulate, tegillate. Flowering: February to March.

23. ***Mallotus philippensis* (Lam.) Müll.Arg.** (Euphorbiaceae)

Pollen 3-colporate, prolate-spheroidal, 22.3 × 21.1μm. Colpi short, broad. Os lalongate, 2.52 μm crassimarginate. Exine 1.64 μm thick, sexine as thick as nexine, psilate. Flowering: February to March.

24. ***Nerium oleander* L.** (Apocynaceae)

Pollen 3-5-porate, spheroidal, 39.52 μm. Pore circular, 4.5μm, membrane pronounced. Exine 2.09 μm thick, sexine as thick as nexine, finely reticulate, tegillate. Flowering: April–June.

25. ***Butea monosperma* (Lam.) Kuntze** (Fabaceae)

Pollen 3-colporate, sub-prolate, 39.3 × 34.9 μm. Colpi long, broad, tenuimarginate. Oslalongate, 7.51 μm, lalongate. Grain protruded at the equator. Exine 1.54 μm thick, sexine slightly thicker than nexine, pseudo-reticulate. Flowering: April–May.

26. ***Artocarpus chaplasha* Buch.-Ham.** (Moraceae)

Pollen 3–4-porate, subprolate, 20.3 × 17.4 μm. Pore circular, 4 μm, tenuimarginate. Exine 1.53 μm thick, sexine slightly thicker than nexine, granulate-foveolate, tegillate. Flowering: March to April.

27. ***Azadirachta indica* A. Juss.** (Meliaceae)

Pollen 4-5-colporate, subprolate, 34.3 × 28.1 μm. Colpi long, margin thick. Os circular, 2.53 μm with central portion tapering into lateral appendages. Exine 2.08 μm thick, sexine as thick as nexine, ±obscure, tegillate. Flowering: February to March.

28. ***Adina cordifolia* (Rox.) Hook.f. & Benth.** (Rubiaceae)

Pollen 3-colporate, sub-oblate, 16.19 × 12.51μm. Colpi long, tenuimarginate. Os circular, margin distinct. Exine 1.81 μm thick, sexine slightly thicker than nexine, finely reticulate, tegillate. Flowering: February to March.

29. ***Alstonia scholaris* (L.) R. Br.** (Apocynaceae)

Pollen 3-colporate, spheroidal, 37.2 μm. colpi short, broad, membrane well-defined. Os lalongate, 3.2 × 4.1 μm. Exine 1.21 μm thick, sexine slightly thicker than nexine, psilate, tegillate. Flowering: February to April.

4 Results of Modern Pollen Rain

4.1 General Account of Pollen/Vegetation Relationship

This phenomenon is a physical process of both vertical and horizontal transport of tiny pollen released, from anther. After its release enormous amounts of pollen grain have been sifted all over the atmosphere and under favourable conditions, few become airborne for a long time whereas, the rest gets deposited at places through various mediums. As the dispersal patterns may vary in a particular area, the study should be conducted in a strategic manner i.e., (a) pollen dispersal within the forested region, (b) pollen dispersal in the forest edge and (c) pollen dispersal across open land to understand and formulate a pollen/vegetation relationship model which in turn can help in interpreting the pollen diagrams from respective areas.

Before the investigation of the sediment cores/profiles from the investigation areas, the study of the modern pollen/vegetation relationship is indispensable. To achieve the target a series of surface samples both moss cushions and subsurface soil has to be analyzed from the various forest stands, edge of forests and adjoining open areas which incorporate wetlands also, to understand the respective representation of the forest constituents in the pollen rain as well as their dispersal efficiency and preservation in the sediment in relation to their factual composition in the regional forest types. This comparative database generated from pollen rain and the extant vegetation through the analysis of subsurface soil is taken as an analogue for the appropriate evaluation of the fossil pollen spectra in terms of past vegetation and climate in and around tropical dry deciduous forests of the Lucknow district.

The sample collection follows a survey of sampling sites and procurement of samples in a transect from the edge of the forest and across the open land. The study areas selected for the purpose were in and around the scattered forests near BBAU, Lucknow (Fig. 1).

4.2 Pollen Spectra (Forest Edge: Fig. 2)

Sample nos-1 & 2 (Surface soil): Palynological studies of subsurface soil demonstrate how ferns (8.61%), highland taxa (5.44%), and arboreals (18.26%) dominate the average value of non-arboreal taxa (67.69%). Other terrestrial herbs like Asteroideae (9.76%), Chenopodiaceae (5.48%), *Sonchus* (4.43%), *Celosia argentea* (4.27%), *Silene* (3.82%), *Heracleum* (2.45%), *Artemisia* (2.29%), and *Xanthium* (2.14%) are followed by grasses with non-cereal (15.65%) and cereal (10.91%) values, totalling 26.56% among non-arboreals. Additionally, Brassicaceae and Apiaceae are rather frequent. Ranunculaceae (3.20%) and Onagraceae (2.44%) are two marshy taxa that have moderate values, whilst Cyperaceae, *Polygonum serrulatum*, and *Rumex nepalensis* (1.82% each) have low values. Monolete and trilete spores are frequently seen in the pollen assemblage.

Among arboreals, tree species include *Butea monosperma*, Meliaceae, *Emblica officinalis*, *Alstonia scholaris* and *Holoptelea* in trace values but slightly improved as compared to other deciduous taxa like *Lagerstroemia parviflora*, *Terminalia bellirica*, *Syzygium cumini*, *Madhuca indica*, *Bombax ceiba*, Sapotaceae, *Adina cordifolia*, *Flacourtia* and *Acacia*. *Clerodendrum*, *Melastoma malabathricum* (1.69%), *Ligustrum robustum* (1.38%), and *Jasminum sambac* (1.37%) are significant shrubs that are present in lesser values. *Ziziphus mauritiana* and *Bauhinia purpurea*, on the other hand, are infrequent. *Pinus* (2.67%), *Alnus* and *Betula* (1.69% each), and *Corylus* (1.39%) are the only highland taxa with moderate values, whereas *Abies* is occasional. Moderate values of fungal remains are also found.

4.3 Pollen Spectra (Open Land: Fig. 2)

Sample nos-3 and 4 (Surface soil): Palynological studies of subsurface soil demonstrate that non-arboreal taxa (78.07%) have a higher average value than arboreal taxa (9.99%), ferns (6.01%), and highland taxa (5.93%). Other terrestrial herbs such as Asteroideae (11.86%), Chenopodiaceae (6.17%), Sonchus (4.79%), *Celosia argentea* (4.60%), *Silene* (4.23%), *Heracleum* (2.58%), *Xanthium* (2.39%), and *Artemisia* (2.30%) are also found in high concentrations among non-arboreals. Grasses, which include both non-cereal (18.37%) and cereal (15.63%) with a combined value of 34%. Ranunculaceae (3.49%), *Impatiens balsamina* (2.30%), and Cyperaceae (1.11%) are among the marshy taxa that are present in moderate values. Other marshy taxa like *Rumex nepalensis,* Onagraceae, and *Polygonum serrulatum* are sporadic. The representation of aquatic taxa including *Myriophyllum indicum*, *Typha latifolia* and *Potamogeton pectinatus* (1.01% each) are observed. *Nymphoides*, however, is infrequent. Both monolete and trilete spores are found in reasonably good numbers in ferns.

Among arboreals, tree taxa like *Butea monosperma*, *Emblica officinalis*, *Alstonia scholaris*, *Holoptelea integrifolia*, *Lagerstroemia parviflora*, *Terminalia bellirica*, Meliaceae, *Syzygium cumini*, *Bombax ceiba*, Sapotaceae, *Adina cordifolia*, Rutaceae, *Palaquium* and *Acacia* are poorly represented along with trace occurrence of other scattered arboreals. *Melastoma malabathricum* and *Clerodendrum* are the only shrubs represented within 1.11%. However, *Ligustrum robustum*, *Jasminum sambac*, *Ziziphus mauritiana* and *Bauhinia purpurea* mark their sporadic presence. The primary highland taxa are represented by *Pinus* (2.12%), which has a moderate value, followed by *Alnus* (1.36%) and *Betula* (1.27%), which have low values. Rarely are *Abies* and *Corylus* shown. Fungal remains are encountered in relatively low value as compare to forest edge assemblage.

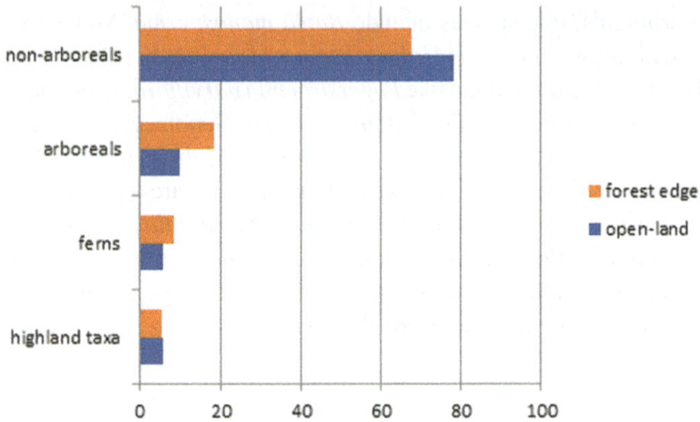

Fig. 2 Composite Pollen spectra from the forest margin and open-land areas near BBAU, Lucknow

5 Discussion and Conclusions

The pollen characteristics, such as form, size, apertures, polarity, and tectum types, in the current pollen morphological research demonstrate tremendous variation. The most important pollen characteristics, however, on which diverse pollen types may be identified geographically, are the exine pattern and apertural kinds. Particularly in entomophily and anemophily, a clear link between pollen characteristics and pollen types is demonstrated. Compound apertures in the pollen grains of entomophilous species are their defining feature i.e., 3–6 colporate, prolate–spheroidal shape, large size, thickened wall, dark colorations with thick reticulate tectum while pollen of anemophilous taxa are of simple apertures, viz., colpate-colporoidate, oblate spheroidal shape, small size, thinned wall with psilate-scabrate-finely reticulate tectum.

In the present scenario, it is found that the volume of vegetation has been minimized due to encroachment of forestland by road construction, building development programs, cattle ranchers, loggers, and slash and burn farmers in and around the study area. Many important vascular plant taxa especially medicinal plants are found damaged due to massive soil erosion. As a result, rolling grasslands became evident in the place of natural vegetation. Therefore, pollen morphological and phenological data especially the flowering period may play an important role in the sustenance of this scattered vegetation in and around the main Lucknow city.

5.1 *Overall Modern Pollen Deposition and Its Relationship with Extant Vegetation*

Under the current climatic circumstances, the overall palynoassemblage recovered from the examined surface samples indicates the presence of tropical mixed deciduous to open-land vegetation. The pollen spectra revealed a predominance of non-arboreals (terrestrial and marshy/aquatic herbs) and comparatively lower frequencies of arboreals (trees and shrubs), but the arboreals are significantly more prevalent than non-arboreals in the actual composition of the forest edge. As a result, the palynoassemblage and the local vegetation of today are not entirely consistent. In terms of arboreals, some of the key taxa are underrepresented and some are overrepresented, which may be because various plant species and climatic circumstances affect pollen distribution and preservation differently.

Although Cyperaceae pollen is typically found in the palynoassemblage in soil samples from open-land and forest edges, its thinner exine renders it more susceptible to crumpling or corroding. Since most Cyperaceae pollen grains are lost as soil column depth increases (Chauhan et al. 1991), Cyperaceae pollen percentages and concentrations correspondingly decline. Cyperaceae are the most abundant pollen taxa in the marsh, hence their destruction lowers the total pollen concentration. In our study region, the structural and ornamentation variations of exines can have an impact on pollen retention in the soil. In the deep layers of the soils (sedimentary profile), there is a general tendency towards larger percentages of Chenopodiaceae and Asteraceae, but in the surface levels, there is a tendency towards higher percentages of Poaceae (cereal and non-cereal).

While the exines of Poaceae are thin and smooth, those of Chenopodiaceae and Asteraceae are thick and densely ornamented. *Acacia, Butea, Dillenia, Emblica, Lagerstroemia*, Meliaceae, Moraceae, *Bombax, Syzygium*, and *Ziziphus* are the principal arboreal species retrieved from the surface samples. The remaining forest species, such as *Erythrina, Flacourtia, Madhuca, Mitragyna, Pongamia*, and *Trewia*, are either underrepresented in the various pollen substrates or not present at all. The cause might be due to the entomophilous nature of pollination as well as recent monsoon irregularities, which affect the taxa's flowering season. It is also impossible to completely exclude out the low pollen dispersal efficiency, poor pollen preservation in the sediments, and differential identification of poorly preserved grains (Quamar et al. 2017).

5.2 *Pollen Rain Along the Forest Edge*

The study was conducted on two subsurface soil samples procured from the outskirt (forest edge) of the scattered forest stands near BBAU. In the forest margin areas, the study reflects low frequencies of arboreals (18–19%). Shrubs like *Melastoma malabathricum*, *Bauhinia purpurea*, Oleaceae, *Clerodendrum* and *Ziziphus mauritiana* are represented without any significant changes. However, the non-arboreal taxa (67–68%) are recorded in dominant frequency uniformly. Prominent ground flora is represented by herbaceous taxa like Caryophyllaceae, Acanthaceae, Amaranthaceae (*Celosia argentea*), Lamiaceae, Brassicaceae, Asteroideae, Cichorioideae, *Artemisia*, *Plantago*, *Heracleum*, *Sonchus*, *Silene* and *Xanthium*. The occurrence of highland taxa (5–6%) like *Pinus*, *Betula*, *Alnus*, *Corylus* and *Abies* in moderate frequencies might be linked to the changing nature of the forest's periphery, where powerful winds were seen transporting thermophilic species from local or distant upper elevations that eventually mixed with sediment. Ferns (8–10%) are consistent in their values and are represented by monolete and trilete spores. The occurrence of moderate frequency of fungal elements in forest edge is due to the adjacent crop fields where generally air gets charged with pathogens and finally incorporated in the sediment.

5.3 Pollen Rain Across the Open Land

The study was conducted on two surface soil samples procured from the open land area near BBAU, Lucknow. The significant feature of palynoassemblage in open-land surface soil is the gradual loss of tree taxa that is directly correlated with the distance travelled from the forest margin to the open-land. The reasonable improvement of non-arboreal taxa (78–79%) is the second characteristic (Fig. 2). In the ultimate case the samples procured further off the forest stand, the tree pollen becomes less or sporadic in frequency. However, there is no evident change in the frequency of a few terrestrial herbaceous taxa in this case. Most of the marshy as well as aquatic taxa, are well represented because of their local origin as flourished in and around the wetland areas.

Thus, in a broader sense, there is a non-linear relationship between pollen rain and vegetation in forested areas of Lucknow district (Fig. 2). The abnormal representation of many pollen taxa which are frequently growing along the forest margin could be attributed to their poor distribution, reduced pollen production due to entomophily, and pollen biodegradation in sediments. Sometimes locally abundant pollen taxa also create problems and consequently serve a distorted picture due to over-representation. Hence, while translating the fossil pollen spectra in terms of past vegetation and climate, all these factors and databases should be considered with great care.

Plate 1 Field photographs: 1. A view showing Tropical Deciduous forest in and around BBAU, Lucknow, 2. Swampy area near the forest margin, 3& 4. Collection of fresh polliniferous samples for the pollen morphological studies, 5& 6. Collection of surface soil samples for the modern pollen rain

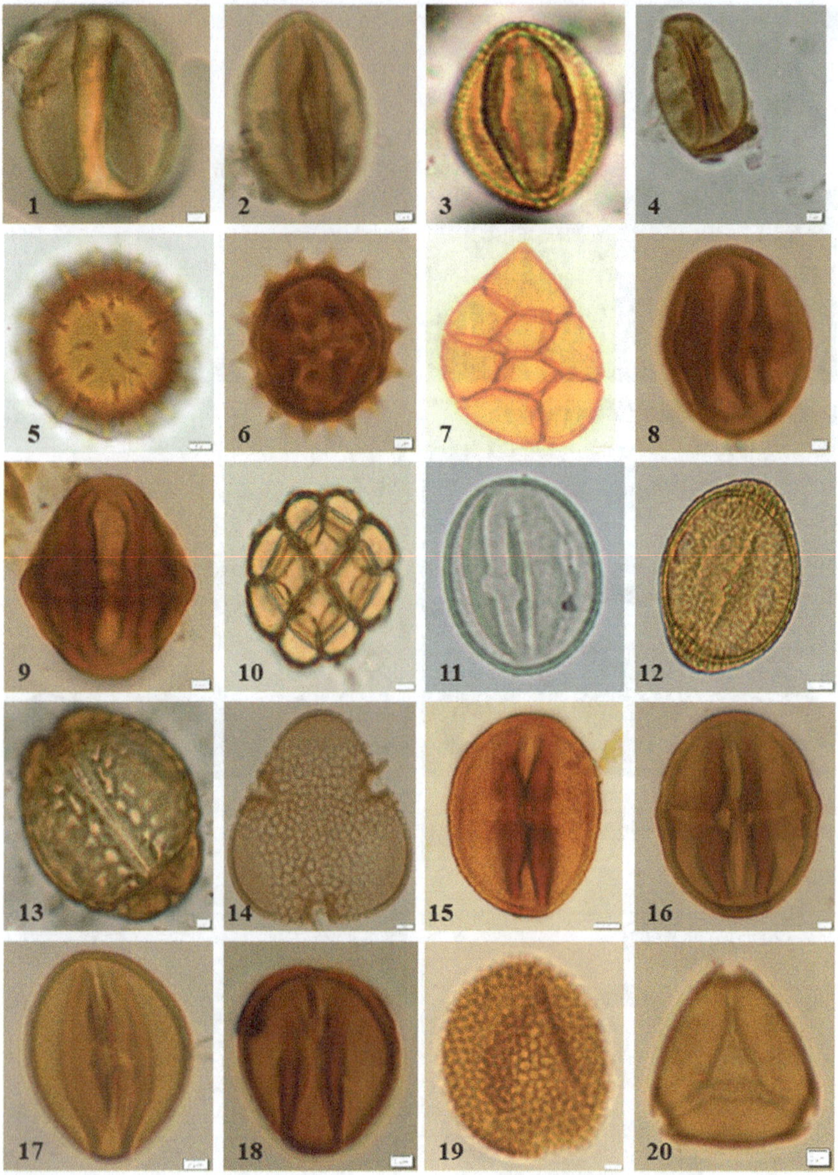

Plate 2 LM micrographs of recovered pollen taxa from the procured flowers. **Description of Plate 2, 1–2**. *Rosa alba*; 3. *Tectona grandis,* 4. *Cassia fistula,* 5. *Kydia calycina*. 6. *Dimorphotheca* sp., 7..*Albizia julibrissin*;; 8–9. *Tabernaemontana divaricata*; 10. *Acacia oxyphylla*, 11. *Lannea coromandelica*, 12. *Lagerstroemia parviflora*, 13. *Barringtonia acutangula*; 14. *Bombax ceiba*, 15–16. *Euphorbia milii*, 17–18. *Petunia* sp., 19. *Jatropha integerrima*, 20. *Callistemon lanceolatus*

Acknowledgements We are highly thankful to Professor Mahesh G. Thakkar, Director, BSIP, Lucknow for permission to carry out the research. This research article is a part of the M. Sc. Dissertation of the first author (RR). RR thanks Dr. Narendra Kumar, Head of Department, Dept. of Geology, BBAU, Lucknow for his encouragement and permission to pursue such interesting research topic. Thanks are also due to Ms. Arya Pandey and Mr. Jagdeesh for their help.

References

Barboni D, Bonnefille R (2001) Precipitation signal in modern pollen rain from tropical forests of south India. Rev Palaeobot Palynol 114:239–258

Basumatary SK, Dixit S, Bera SK, Mehrotra RC (2013) Modern pollen assemblages of surface samples from Cherrapunjee and its adjoining areas, Meghalaya, northeast India. Quat Int 298:68–79

Bera SK, Tripathi S, Gupta S, Roy S (2018) Pollen and spores in yellow rain from Lucknow, northern India. Palynol 42(4):504–515

Bera SK (2000) Modern pollen deposition in Mikir Hills. Assam. Palaeobot. 49:325–328

Cerceau-Larrival MT (1971) Morphologiepollinique et correlations phylogénétiqueschez les Umbelliferous. In: Heywood VH (ed) The biology and chemistry of the umbelliferae. Academic press, London, pp 109–155

Chauhan MS, Bera SK (1990) Pollen morphology of some important plants of tropical deciduous Sal forest, District Sidhi, Madhya Pradesh. Geophytol 20:30–36

Chauhan MS, Chatterjee S (2008) Holocene vegetation, climate and human habitation in the Central Ganga plain, based on pollen records from the lake deposits. Palaeobot 57:265–275

Chauhan MS, Khandelwal A, Bera SK, Gupta HP (1991) Palynology of Kathauta Tal, Chinhat, Lucknow. Geophytol 21:191–194

Chauhan MS, Kumar K, Quamar MF, Sharma A (2013) Correlation of data on loss-on-ignition and palynology for late quaternary climate change in southwestern Madhya Pradesh, India. Curr Sci 104(3):299–301

Chauhan MS, Pokharia A, Singh IB (2009) Pollen record of Holocene vegetation, climate change and human habitation from Lahuradewa Lake, Sant Kabir Nagar District, Uttar Pradesh, India. Man Environ 34:88–100

Chauhan MS, Sharma C, Singh IB, Sharma S (2004) Proxy records of late Holocene vegetation and climate changes from Basaha Jheel, Central Ganga plain. J Palaeontol Soc Ind 49:27–34

Dixit S, Bera SK (2011) Mid-Holocene vegetation and climatic variability in tropical deciduous Sal (*Shorea robusta*) forest of Lower Brahmaputra valley, Assam, Northeast India. J Geo Soc Ind 77:419–432

Erdtman G (1943) An introduction to pollen analysis. The Chronic Botanica Co, Waltham Mass (MA)

Erdtman G (1952) Pollen morphology and plant taxonomy. The Chronica Botanica Co., Waltham, Mass, USA

Erdtman G, Metcalfe CR (1963) Affinities of certain genera incertae sedis suggested by pollen morphology and vegetative anatomy. Kew Bull 17:249–256

Fægri K, Iversen J (1989) In: Fægri K, Kaland PE, Krzywinski K (eds) Textbook of pollen analysis, 4th. Wiley, Chichester, New York, Brisbane, Toronto, Singapore

Farooqui A, Sekhar B (2011) Climate change and vegetation succession in Lalitpur area, Uttar Pradesh (India) during late Holocene. Trop Eco 52:69–77

Feuer SM (1981) Pollen morphology and relationships of the Misodendraceae (Santalales). Nord J Bot 1:731–734

Geldenhuys CJ, Murray B (1993) Floristic and structural composition of Hanglip forest in the South pansberg, Nothern Transvaal. South Africa Forest J 165:9–20

Gupta HP (1978) Holocene palynology from Meander Lake in the Ganga Valley, district Pratapgarh. Up. Palaeobot. 25:109–119

Gupta HP, Sharma C (1986) Pollen flora of North-West Himalaya. Indian Association of Palynostratigraphers, Lucknow (India)

Jeffrey C (1964) A note on pollen morphology in Cucurbitaceae. Kew Bull 17:473–476

Nair PKK (1965) Trends in the morphological evolution of pollen and spores. J Ind Bot Soc 44(4):468–478

Nayar TS (1990) Pollen flora of Maharashtra State, India. Today's and Tomorrow's Publishers and Printers, New Delhi (India)

Punt W, Hoen PP, Blackmore S, Nilsson S, Le Thomas A (2007) Glossary of pollen and spore terminology. Rev Palaeobot Palynol 143(1–2):1–81

Quamar MF, Ali SN, Morthekai P, Singh V (2017) Confocal (CLSM) and light (LM) photomicrographs of different plant pollen taxa from Lucknow, India: Implications of pollen morphology for systematics, phylogeny and preservation. Rev Palaeobot Palynol 247:105–119

Ranhotra PS, Bhattacharyya A (2010) Holocene Palaeoclimate and glacier fluctuations within Baspa valley, Kinnaur, Himachal Pradesh. J Geo Soc Ind 75:527–532

Saxena A, Prasad V, Singh IB, Chauhan MS, Hasan R (2006) On the Holocene record of wild and cultivated rice from Ganga Plain: evidence for rice based agriculture. Curr Sci 90:1547–1551

Sharma S, Joachimski MM, Tobschall HJ, Singh IB, Sharma C, Chauhan MS (2006) Correlative evidences of monsoon variability, vegetation change and human inhabitation in Sanai Lake deposit: Ganga Plain. India Curr Sci 90:973–978

Singh G, Wasson RJ, Agarwal DP (1990) Vegetational and seasonal climatic changes since the last full glacial in the Thar Desert, north-western India. Rev Palaeobot Palynol 64:351–358

Singh S, Scharf BW, Khandelwal A, Mohanti M (2011) Modern pollen-vegetation relationship as an adjunct in the interpretation of fossil pollen records in the Chilka Lagoon, Odisha, India. Palaeobot 60:265–271

Singh SP (1998) Chronic disturbances, a principal cause of environmental degradation in developing countries (Editorial). Environ Conserv 25:1–2

Skvarla JJ, Turner BL (1966) Systematic implications from electron microscope studies of Compositae pollen—a review. Ann Miss Bot Gard 53:220–256

Srivastava P, Singh IB, Sharma M, Singhvi AK (2003) Luminescence chronology and late quaternary geomorphic history of Ganga Plain, India. Palaeogeog Palaeoclimat Palaeoecol 197:15–41

Thakur B, Saxena A, Singh IB (2018) Paddy cultivation during early Holocene: evidence from diatoms in Lahuradewa lake sediments. Ganga Plain Curr Sci 114(10):2106–2115

Tripathi S, Arya A, Basumatary SK, Bera SK (2016) Modern pollen and its ecological relationships with the tropical deciduous forests of central Uttar Pradesh, India. Palynology 40(2):264–279

Tripathi S, Basumatary SK, Pandey A, Khan S, Tewari P, Thakur B (2021) Palaeoecological changes from 580 to 1220 CE from the Indo-Burma region: a biotic assessment from the Barak Valley of Assam, northeast India. CATENA 206:105487

Tripathi S, Farooqui A, Singh VK, Singh S, Roy RK (2019) Morphometric analysis of *Ceiba* Mill. (Bombacoideae, Malvaceae) pollen: a sacred plant of the Mayan (Mesoamerican) civilization. Palynology 43:551–573

Tripathi S, Garg A, Shukla AK, Farooqui A, Pandey A, Tripathi T, Singh VK (2022) Pollen micromorphometry of two endangered species of *Rauvolfia* L. (Apocynaceae) from the Indo-Gangetic Plains of Central India using LM, CLSM and FESEM. Palynology 46(4):1–15

Tripathi S, Thakur B, Nautiyal CM, Bera SK (2020) Floristic and Climatic reconstruction in Indo-Burma region for the last 13000 cal. yr: a palynological interpretation from endangered wetlands of Assam, northeast India. Holocene 30(2):315–331.

Tripathi S, Thakur B, Sharma A, Phartiyal B, Basumatary SK, Ghosh R, Kumar K, MC, Manoj, Agrawal S, Farooqui A, Tiwari P, Saikia K, Tiwari A, Pandey A, Ali N, Agnihotri R, Prasanna K, Morthekai P, Ranhotra PS, Pandey S, Bose T (2023) Modern biotic and abiotic analogues from the surface soil of Ganga-Ghaghara-Gandak interfluves of the Central Ganga Plain (CGP), India: implications for the palaeoecological reconstruction. Catena 224:106975

Trivedi A, Chauhan MS (2011) Modern pollen rain-vegetation relationship study in Jalesar, Unnao district, Uttar Pradesh. J Palynol 47:11–21
Trivedi A, Chauhan MS, Sharma A, Nautiyal CM, Tiwari DP (2013) Record of vegetation and climate during late Pleistocene-Holocene in Central Ganga Plain, based on multiproxy data from Jalesar Lake, Uttar Pradesh, India. Quat Int 306:97–106
Trivedi A, Singh DS, Chauhan MS, Arya A, Bhardwaj V, Awasthi A (2011) Vegetation and climate change around Ropan Chhapra tal in Deoria District, the Central Ganga plain during the last 1350 yrs. J Palaeontol Soc Ind 56:39–43

Decoding Ancient Vegetation and Environment Using Potsherd Palynology: A Case Study from an Archaeological Site of Deltaic West Bengal, India

Oindrila Biswas, Dipak Kumar Paruya, Binod Saradar, Krishna Kali Mandal, and Subir Bera

Abstract Archaeobotany deals with the interpretation of preserved plant remains from the archaeological sites in order to reconstruct the past vegetation and climate change and past human-plant relationship. The archaeobotanical remains found in the archaeological sites are classified into two groups; macro and microbotanical remains. Usually, the microbotanical investigations includes the extraction of spores/pollen, phytoliths, diatoms, fungal spores and starch grains from the soils/sediments of the archaeological sites. But sometimes archaeological artefacts including potsherds may also provide us with valuable information. Palaeobotanical investigations and palaeoclimate reconstruction from the archaeological potsherds are rare especially from the coastal regions of the globe. In the present study two earthen vessels associated potsherds collected from the Tilpi region of Sundarbans, deltaic West Bengal were studied in a multi-proxy approach. The age of the earthen pots and potsherds found to be 2630 ± 25 years BP. The pollen, phytolith and non-pollen palynomorph data of the potsherds revealed that a non-littoral forest was present in the area under tropical warm and humid climate condition. However, occurrence of pollen grains and phytoliths of palms, and *Concentricystes* sp. indicate that the coast line was not far away from the area of deposition (Rao 1990, Ramanujam et al. 1998).

O. Biswas · D. K. Paruya · B. Saradar · S. Bera (✉)
Centre of Advanced Study, Palaeobotany-Palynology Laboratory, Department of Botany, University of Calcutta, 35, Ballygunge Circular Road, Kolkata 700019, India
e-mail: sbbot@caluniv.ac.in

O. Biswas
e-mail: oindrilabiswas.m@srmus.edu.in

O. Biswas
Department of Botany, School of Basic Sciences, SRM University Sikkim, Gangtok 737102, India

K. K. Mandal (Deceased)
Dakshin Chobbish Pargana Protna Itihas Gobeshona Kendra, South 24 Parganas, Dhopagachi, Baruipur, West Bengal, India

B. Samant and D. Thakre (eds.), *Applications of Palynology in Stratigraphy and Climate Studies*, Society of Earth Scientists Series,
https://doi.org/10.1007/978-3-031-51877-5_10

Keywords Potsherds · Palynology · Phytolith · Climate reconstruction · Deltaic West Bengal

1 Introduction

The analysis of prehistoric sediments as well as artefacts from archaeological sites have gained importance in the recent times in order to solve problems related to origin, evolution and dispersal of agricultural system (Colledge et al. 2005; Li et al. 2007 and references there in; Fuller and Qin 2009; Zhou et al. 2020; Wang et al. 2021), availability and usage of wild plants (Colledge and Conolly 2014; Long et al. 2017; Mooney and Seijo 2021), chronology of domesticated crops (Piperno 2011; Prasad et al. 2011; Gremillion 2014), human-plant relationship and socio-economic development (Ghosh et al. 2006a; Lu et al. 2014; Masojć et al. 2019; Kolobova et al. 2020; Gao et al. 2020). Earlier, the palynomorphs and other flora and faunal remains were studied from the sub-surface sediments recovered from the archaeological sites and this was considered as the conventional method (Long et al. 1999; Ghosh et al. 2005; 2006a, b; Hu et al. 2007; Backels 2020; Mckey et al. 2020 and many more). But in modern times emphasis are given in multiproxy approaches such as study of stone tools, dental remains, fabric composition of the ceramics, starch grains and other geochemical analyses along with the conventional analysis (Barton 1997; Piperno et al. 2000; Haslam 2004; Prasad et al. 2005; Piperno 2006; Dumpe and Stivrins 2015; Forshaw 2015; Brown et al. 2016; Ma et al. 2017; Roffet-Salque et al. 2017; Min-Kyu 2017; Luncz et al. 2019; Sarkar et al. 2020; Das et al. 2021).

The use of palynology in the archaeological studies has been started since early twentieth century (Von Post 1916). Through pollen analysis direct quantification of the changing vegetation and environment is possible and thus, this proxy is conventionally used in archaeological and palaeoecological studies. However, preservation of pollen grains in archaeological sites should also be taken into consideration. There are several complex factors that govern the preservation of spores/pollens (Bryant and Holloway 1978) and sometimes preservation of pollen grains are poor due to mechanical degradation, chemical and biological factors. Thus, now a day besides spores/pollens, the non-pollen palynomorphs such as, phytoliths, fungal spores, and diatoms are also being studied from the same archaeological sediments. Phytoliths are opaline, biogenic silica bodies which are observed under the microscope and are resistant to the forces of taphonomic decay. They can survive in well preserved state over a long period of time compared to other plant remains. Therefore, they are considered as an excellent proxy in archaeological studies especially when recovered palynomorphs are poorly preserved (Piperno 2006). Archaeobotanists working in several regions of the world have advanced this proxy well beyond its formative stage, allowing for precise identification of important crop plants such as, maize, bananas, cucurbits, taro in ancient sediments (Bozarth 1987; Mbida et al. 2000; Mindzie et al. 2001; Iriarte 2003; Pearsall et al. 2003; Piperno and Stothert 2003; Denham et al. 2003), documenting the early domestication of wheat, barley, millet and rice as well in distinguishing between these domestic crops and their wild ancestors (Rosen and Weiner 1994; Pearsall et al. 1995; Zhao and Piperno 2000; Saxena et al. 2006; Lu et al. 2009; Pamirsky et al. 2017).

Among the different archaeological artefacts, potsherds have been proved to be a reliable proxy of prehistoric sediments for past climate reconstruction. The potters mostly use clayey soil for making the earthenware from a suitable source such as, river banks, pond or lake around their habitation. Spores, pollen grains, leaves, fruits and food grains of agricultural fields in the vicinity can get deposited and mixed with the clay layers and thus, surface sediments used for pottery making may reflect the vegetation growing around the habitation site. Earlier many researchers have reported palynomorphs as well as non-pollen palynomorphs from ancient potsherds. Udatsu et al. (1999), by studying the potsherd phytolith, established that rice cultivation expanded 5000–6000 years ago in north Jiangsu province of China. Later, Li et al. (2021) supported the idea of Udatsu et al. (1999) and reconstructed the trajectory of the rice cultivation during 6000–8500 BP. Ghosh et al. (2006a) studied the palynomorphs recovered from ancient partially fired potsherd from an excavation site (ca. 3320–2080 BP) at Gangetic Bengal and inferred tropical, moist deciduous vegetation was present at that time. On the basis of palynomorph analysis Hu et al. (2007) resolved the origin of 2200 years old terracotta figurines of the Qin Empire presently situated in Lintong District, Xi'an Shaanxi province, China. Later, Yao et al. (2012) studied the phytoliths along with pollens from ancient potsherds collected from an excavation site of Turpan, North-west China (ca. 2700 BP) and established a temperate and arid climate prevailed in the area during deposition. Study of phytolith from pottery also revealed the uses of spices like garlic and mustard in European prehistoric cuisine around ca. 6100 BP ago (Saul et al. 2013). Here we present another case study on the reconstruction of the past vegetation and climate of deltaic West Bengal with the help of spores, pollen and phytoliths recovered from the pottery remains excavated from the Tilpi region of Sundarbans.

2 Methodology

2.1 *Sampling Site*

The area under investigation is an archaeologically rich small place named Tilpi (22°14′13.8″N; 88°33′40.67″E) situated at the deltaic West Bengal. It falls under Joynagar I Police Station of South, 24 Parganas (Fig. 1). It is referred in various literatures that Tilpi and its adjacent area (Chandaneswar) in ancient times was an important and busiest port city beside the magnificent river Piyali (Roy 2006). At present this river confluences with the Matla River near the delta (Mandal 2002). This area is inhabited by people who mainly depend on agriculture as their livelihood. The modern climatic condition of the area is tropical and warm humid supporting typical mesophytic vegetation. The pottery vessels and their associated remains that were investigated for the present study were collected from Chandaneswar village within Tilpi which is 400 m away from the previous excavation site. Out of the two pottery remains collected, one was AMS C^{14} dated from Silesian University of Technology, Poland (Fig. 2). The absolute date of the sample was 2630 ± 25 year BP (Lab no. GdA-4740).

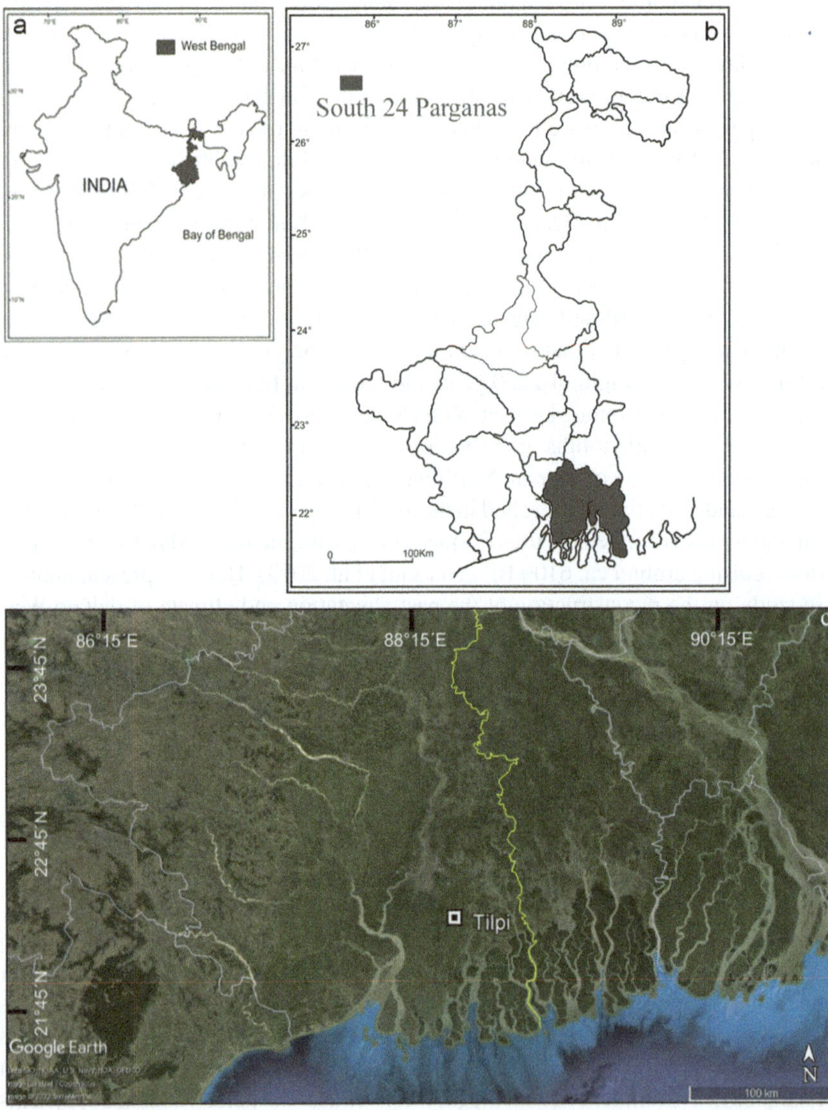

Fig. 1 **a** Map showing the location of West Bengal in India, **b** The dark region marks the position of deltaic West Bengal, India; **c** Magnified view of the archaeological site Tilpi from where the samples were collected

Fig. 2 **a–c** Pottery vessels recovered from the excavation site of Tilpi; **d** Magnified view of the collection site (curtesy, Mr. Krishna Kali Mandal)

2.2 Extraction of Microbotanical Remains

Out of two pottery vessels one was not-painted having a small nozzle (Potsherd I) and the other one was painted with black color (Potsherd II). To avoid contamination the potsherds were washed with distilled water several times to remove the surface contaminants followed by washing with pure alcohol and air dried. After that small broken part of the vessels were taken and crushed with the help of mortar and pestle. The non-pollen palynomorphs were retrieved from the potsherds following the conventional method of Faegri and Iversen (1950). Subsequently, standard acetolysis method of Erdtman (1954) was followed for clearing the cellulosic content of the pollen wall. The crushed potsherds were boiled with a 10% potassium hydroxide (KOH) solution, followed by treatment with hydrofluoric acid (HF) and thorough washing. At final step acetolysis was done with a mixture of acetic anhydride and concentrated sulphuric acid (H_2SO_4) (9:1 ratio). The extracted palynomorphs were stored in water and studied under the light microscope (Zeiss Axioskop 2).

Phytoliths were extracted following the protocol of Pearsall (2000). The potsherd samples were taken, dried, homogenized, sieved and then treated with strong acids (conc. nitric acid and hydrochloric acid in a ratio 1:1 v/v) in a hot water bath until the reaction stops followed by thorough washing with distilled water. Then the samples were treated with hydrogen peroxide for removal of the organic matter, followed again by thorough washing. Density separation was done using cadmium iodide and potassium iodide (sp. gr 2.3) followed by centrifugation. Phytoliths float up into the supernatant and the heavy liquid was washed to recover the phytoliths. 250–300 phytoliths were observed and counted for each potsherd samples. Phytoliths were observed under light microscope and photographed using Zeiss Axioskop 2. The phytoliths were identified and classified following the classification of International Code of Phytolith Nomenclature (ICPN) 1.0 and 2.0.

3 Results

3.1 Palynomorph and Non-pollen Palynomorph Analyses

Palynomorphs recovered from the ancient, burnt potsherd may have been destroyed due to high temperature during firing of the same. Therefore, a small number of deformed pollen grains were recorded from the potsherds which survived the firing process. The yield of potsherd II was extremely poor and thus, we have only considered the non-pollen palynomorph data for this sample.

A mixture of pteridophyte and angiosperm palynoflora were retrieved from potsherd I. The identifiable palynomorphs are presented in Table 1 along with their family, habit and climatic regime to which they grow. Thirteen palynomorphs were recovered from the potsherds, with three palynomorph types belonging to three families of pteridophytes and the remaining ten genera belonging to nine groups that represented the angiosperm flora. (Table1, Fig. 3). In comparison to pteridophytes (14.28%), angiosperm pollen grains were found to be more abundant (47.61%). In addition to this, *Concentricystes* sp., *Aulacoseira*a long with a good number of other types of algal cysts and fungal spores also retrieved from both the potsherds (Fig. 3).

Table 1 Palynotaxa recovered from the archaeological site along with its family, habitat and ecology

Palynotaxa	Family	Habit	Ecology
Pteridophytes			
Pteris sp.	Pteridaceae	Herb	Cosmopolitan
Ceratopteris sp.	Parkeriaceae	Herb	Tropical, aquatic
cf. Polypodiaceae	Polypodiaceae	Herb	Tropical to subtropical
Angiosperms			
cf. Poaceae	Poaceae	Herb	Cosmopolitan
Borassus sp.	Palmae	Woody tree	Tropical coastal
cf. Palmae	Palmae	Woody tree	Tropical coastal
Datura sp.	Solanaceae	Shrub	Tropical to subtropical
Hygrophila sp.	Acanthaceae	Herb	Tropical swampy
Nymphaea sp.	Nymphaeaceae	Perennial Herb	Tropical aquatic
Polygonum sp.	Polygonaceae	Herb	Cosmopolitan
Jasminum sp.	Oleaceae	Herb	Tropical to subtropical
Mimusops sp.	Sapotaceae	Small woody tree	Tropical evergreen
Cucurbita sp.	Cucurbitaceae	Herb	Tropical
Algal component			
Algal cyst Type I	–	Thallophyta	Fresh water, swampy
Concentricystes sp. *Aulacoseira* sp.	Zygnemataceae Aulacoseiraceae	Thallophyta	Brackish water, swampy Fresh water
Fugal component			
Alternaria type *Podospora* type	Deuteromycota Ascomycota	Thallophyta	Tropical, humid
Others			
Aroid starch grains	Araceae	Herbs	Cosmopolitan, Swampy

3.1.1 Phytolith Analyses

A significant yield of various phytolith morphotypes were observed in both the potsherds and thus, the phytolith data were briefly discussed. 18 phytolith morphotypes recovered from the two ancient potsherd among which 11 were important types having diagnostic significance (Fig. 4). Potsherd I yielded abundance of BILOBATE phytoliths followed by SADDLE, CROSS and SADDLE PLATEAUED. These are diagnostic grass phytoliths and other types of grass phytoliths recovered were POLYLOBATE, SADDLE TALL, SADDLE COLLAPSED, RONDEL, TOWER AND BULLIFORM FLABELLATE. Among the non-grass types, SPHEROID PSILATE, SPHEROID ORNATE and SPHEROID ECHINATE recovered significantly. Rest were all non-diagnostic types such as, BLOCKY, ACUTE BULBOSUS, TRACHEARY and ELONGATE types (Fig. 5).

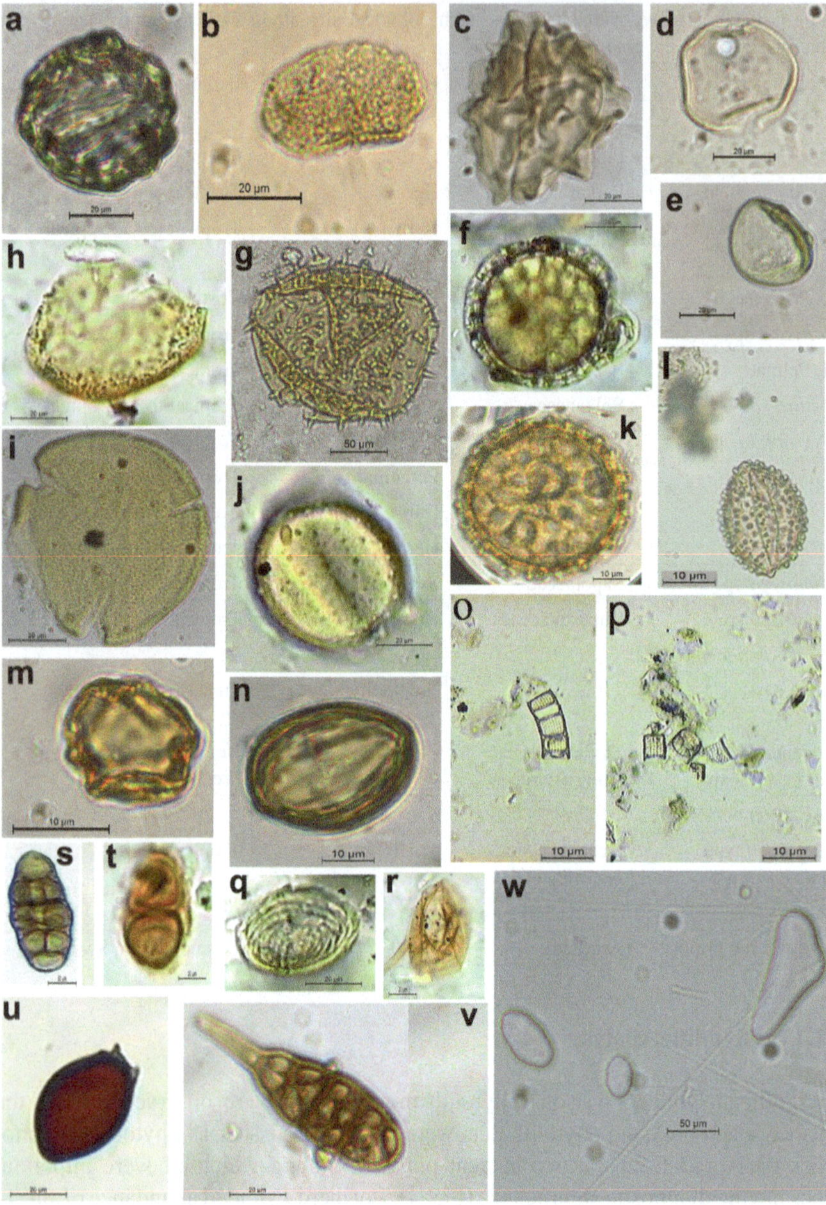

Fig. 3 Light microscopic images of palynomorphs recovered from the potsherd I. **a** *Ceratopteris* sp.; **b** cf. Polypodiaceae; **c** *Pteris* sp., **d–e**. Cf. Poaceae, **f** *Jasminum* sp., **g** *Cucurbita* sp., **h** cf. *Palmae*, **i** *Datura* sp., **j** *Nymphaea* sp., **k** *Polygonum* sp., **l** *Borassus* sp., **m** *Mimusops* sp., **n** *Hygrophila* sp., **o–p**. *Aulacoseira* sp., **q** *Concentricystes* sp., **r** Algal cyst, **s–t**. Fungal spores, **u** *Podospora* type (fungal spore), **v** *Alternaria* type (fungal spore), **w** Aroid starch grains. (scale = 10 μm)

The phytolith assemblage in potsherd II was also similar to potsherd I, BILOBATE being recovered in highest frequency followed by SADDLE, CROSS AND SADDLE PLATEAUED. POLYLOBATE and SADDLE COLLAPSED were not recovered from potsherd II and Saddle tall recovered in negligible amount. The frequencies of BLOCKY and BULLIFORM FLABELLATE were comparatively higher in potsherd II and rest of the phytolith assemblage were more or less similer to potsherd I (Fig. 5).

BILOBATE, POLYLOBATE AND CROSS are considered to be diagnostic for the subfamily Panicoideae and SADDLE for Chloridoideae grasses. Both of these are C4 subfamily growing in warm and humid to dry condition (Bremond et al. 2008; Biswas et al. 2016). SADDLE PLATEAUED are dominantly found in *Phragmites* sp. (Arundinoideae grass) growing in swamy or marshy conditions, Rondels are the indicators of Pooideae subfamily, SADDLE TALL and SADDLE COLLAPSED are commonly found in Bambusoideae plants. All of these are C3 indicating morphotypes (Bremond et al. 2008; Biswas et al. 2016). SPHEROID ORNATE AND SPHEROID PSILATE are the indicator of woody dicots and SPHEROID ECHINATE are considered diagnostic for Palmae family (Neumann et al. 2019 and references therein). BULLIFORM FLABELLATE considered as an indicator type for water stress condition of an area (Bremond et al. 2008). Overall, in the both samples grass phytoliths dominate over the non-grass ones.

4 Discussion

4.1 Microbotanical Remains from Tilpi Interpreting the Palaeovegetation and Palaeoclimate

Pollen can be retrieved from archaeological sites, either clinging to the inner side of pots, on the edges of stone tools or within archaeological features such as storage pits or living floors. These palynomorphs reflect what people used to feed, cultivate and feed their animals or what materials they used to build their homes with as well as gives us an insight into the local vegetation and climate change. The palyno-assemblage recovered from the potsherd sample I show the prevalence of angiosperms taxa followed by pteridophytes. Most of these plants grow in the present day tropical, moist and humid climatic condition. Occurrence of *Mimusops, Datura, Cucurbita, Jasminum, Polygonum, Hygrophila* with a moderate number of ferns under the family Pteridaceae and Polypodiaceae indicate tropical, non-littoral and mesophytic vegetation in the area of archaeological investigation (Naskar et al. 2021). Presence of palm pollen i.e., *Borassus* and other deformed palm pollen grains indicate the position of the pre-historic settlement in a near shore area. This observation matches with the literature (Roy 2006) about ancient Tilpi where it has been mentioned that Tilpi and its adjoining area were an ancient port city and thus settlement was very close to the shores.

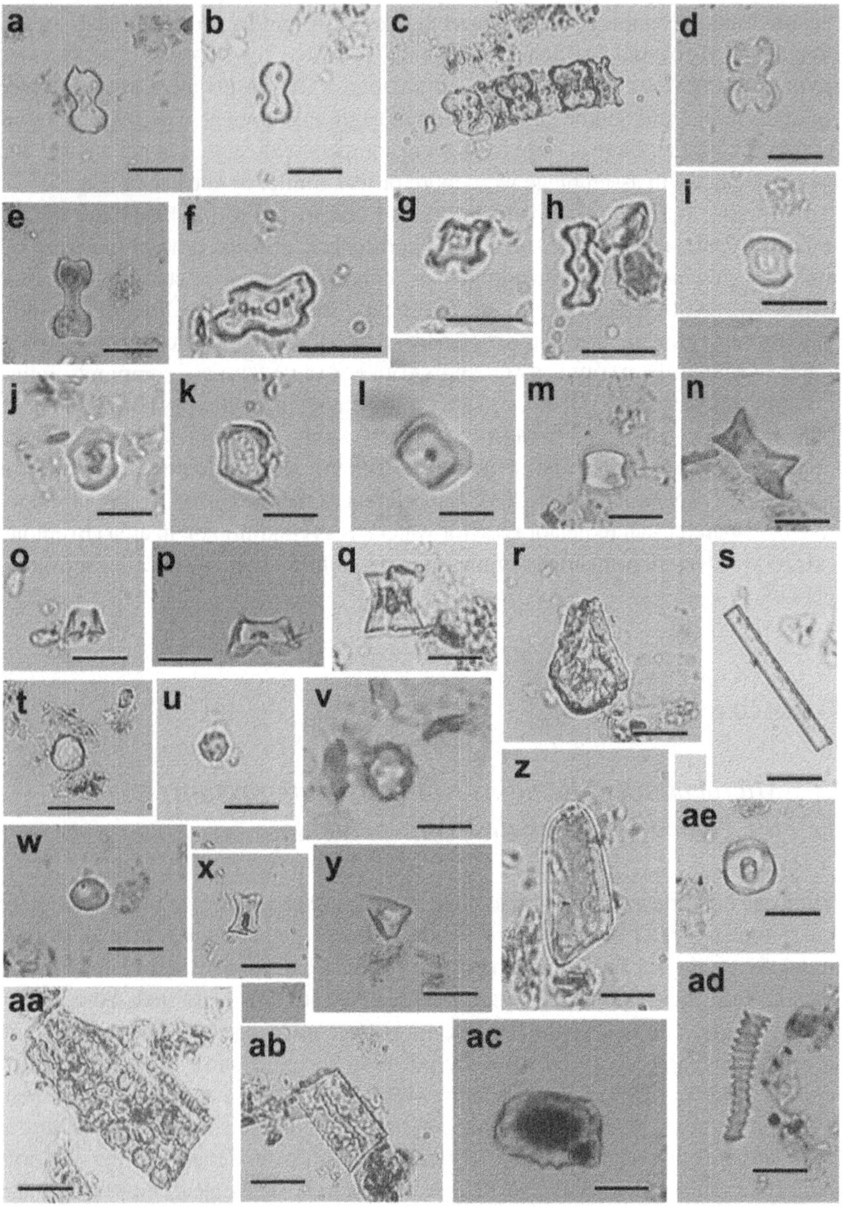

Fig. 4 Light microscopic images of phytoliths recovered from the potsherds. **a–e**. Different types of BILOBATE; **f–g**. CROSS; **h** POLYLOBATE; **i–k** SADDLE; **l–m** SADDLE PLATEAUED; **n** SADDLE COLLAPSED; **o–p** RONDEL; **q**, **x** TOWER; **r** BULLIFORM FLABELLATE; **s** ELONGATE ENTIRE; **t**, **w** SPHEROID PSILATE; **u** SPHEROID ORNATE; **v** SPHEROID ECHINATE; **y–z** ACUTE BULBOSUS; aa. Epidermal plate with SPHEROID ORNATE morphotypes; ab-ac. BLOCKY; ad. TRACHEARY; ae. Unidentified. (scale = 10 μm)

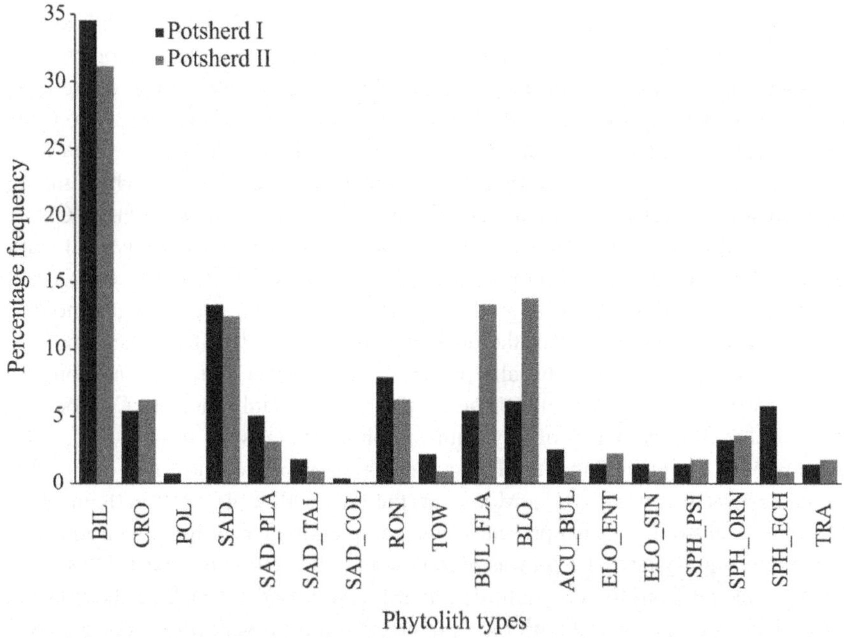

Fig. 5 The abundance frequencies of phytolith morphotypes recovered from the two potsherds. BIL = Bilobate; CRO = Cross; POL = Polylobate; SAD = Saddle; SAD_PLA = Saddle plateaued; SAD_TAL = Saddle tall; SAD_COL = Saddle collapsed; RON = Rondel; TOW = Tower; BUL_FLA = Bulliform flabellate; BLO = Blocky; ACU_BUL = Acute bulbosus; ELO_ENT = Elongate entire; ELO_SIN = Elongate sinuate; SPH_PLI = Spheroid psilate; SPH_ORN = Spheroid ornate; SPH_ECH = Spheroid echinate; TRA = Tracheary

Moreover, occurrence of certain algal forms *Concentricystes, Aulacoseira* and different types of algal cysts in low frequency suggests that freshwater bodies were present near the vicinity of the settlement (Tang et al. 2013). Along with this a good number of dispersed fungal spores also indicate tropical and humid climatic condition with high rainfall (Das et al. 2021). The record of *Nymphaea* pollen taxa, fresh water loving fern taxa *Ceratopteris* and aroid starch grains indicates that the source of soil for making the potsherd was from a fresh water body or from a swampy area present nearby the settlement.

Phytolith analysis has been applied to answer several distinct questions in the field of archaeology, such as whether certain cultigens are present at an archaeological site (Wilson 1985; Bozarth 1987, 1990; Pearsall 2000), what kind of vegetation grew in an area (Alexandre et al. 1997, 1999), and what climate prevailed in a region (Fredlund and Tieszen 1997; Alexandre et al. 1999). In India, Chowdhury and Ghosh (1954) were the first to identify *Saccharum* phytoliths from the archaeological site of Hastinapura (age c. 1100 BC to c. 800 BC). Subsequently, Kajale et al. (1995) recognized the potential of phytoliths as an important proxy in archaeological studies from the Indian subcontinent. Kajale and Eksambekar (1997, 2001) by studying the grass phytoliths recovered from the archaeological sites of Budihal, south India and Balathal region, Rajasthan could establish the occupations and farming practices of the settlements of that area during the Chalcolithic and Neolithic time respectively. In the present study the phytolith data matches with the pollen data. Among the different phyto-morphotypes, BILOBATE recovered in highest percentage. BILOBATE AND CROSS are the diagnostic morphotypes of Panicoideae grasses which mainly grow in warm and humid climatic conditions. Along with this, SADDLE of Chloridoideae grasses was also recovered in moderate frequency indicating typical warm and tropical climate. The Bambusoideae phytolith type i.e., SADDLE COLLAPSED and SADDLE PLATEAUED found predominantly in *Phragmites* sp. of Arundinoideae subfamily were retrieved from the ancient potsherd in a low frequency and both of this subfamily of grasses usually prefers warm and wet climatic condition. The reason behind the retrieval of RONDEL types which mainly indicate C3 Pooideae may be due to redundancy of phytolith morphotypes (Biswas et al. 2021). It has already been established that the Panicoideae grasses produce RONDEL at very low frequency (Biswas et al. 2021). Moreover, the type SPHEROID ECHINATE, indicator of Palmae family were also present in the sample along with the pollen grains of this family signifying that Tilpi was present not far away from the coast. Thus, it can be summarised from the phytolith data that the soil used for making the potsherd was taken from a wet area with mesophytic, non-littoral vegetation where grasses mainly dominate and a warm and humid tropical climatic condition prevailed at that time. The observation of the present study corroborated with the data of Naskar et al. (2021) where they have observed the dominance of non-littoral vegetation between c. 3.9–2.2 ka in deltaic West Bengal.

Farooqui et al. (2013) explored the vegetation-climate change of the Harappan period from archaeological sites of Gujarat by studying the pollen and phytolith of the recovered potsherds and found an arid/dry climate prevailed since 2230 cal BC. Sarkar et al. (2020) also recorded a hyper-arid condition after 1600-year BC by studying isotopic and microbotanical remains of the potsherds found in the archaeological sites of Karim Shahi and Vigakot, Great Rann of Kachchh, Thar Desert which led to migration of the settlement. Furthermore, monsoonal variations during the late Holocene were recorded from the archaeological site of Gujarat (Singh et al. 2006; Prasad et al. 2014) and Kerala (Veena et al. 2014) with the help of grass phytoliths and it was observed that a humid or moist climate prevailed during that time. Our observation is further supported by the reported prevalence of tropical humid climate at protohistoric-historic site at Erenda situated in the eastern coastal region of India (Das et al. 2021), at the Chalcolithic site of Pakhanna in Bankura District of West Bengal (Chattopadhyaya et al. 2005; Ghosh et al. 2005, 2006b, 2013) and Lahuradewa archaeological site of Ganga Plain (Saxena et al. 2013).

4.2 *Possible Age of the Pre-historic Settlement of Tilpi*

The Directorate of Archaeology and Museums, Government of West Bengal conducted an archaeological excavation during 2005–2006 in Mollapara within Tilpi village located on the banks of river Piyali, in order to explore the cultural sequences of the site, the nature of occupation and causes behind the abandonment of the settlement. The authorities excavated a trench of 1.88 m and encountered 5 successive deposits. A large number of antiquities and potsherds were recovered from all the layers. The potsherds were further classified according to colour such as red, grey, dark-grey, dull-red, grey with red patches etc. According to the archaeological report, on the basis of the nature of deposition at the excavation site and nature of excavated antiquities and potteries Tilpi is thought to be contemporary to pre-Gupta dynasty i.e., third century BC to 1st centuryAD (Roy 2006). In our present study it was observed that the absolute age of the pottery collected from Tilpi was much older than pre-Gupta dynasty (2630 $\pm$ 25 year BP). Thus, it can be said that the inhabitants of Tilpi are much ancient than our traditional belief based on archaeological data. Subsequently, it can be concluded that beside environment and ecology the age of the sampling site also holds a significant importance, as for the first time we can have an accurate idea on the how old was the settlement of deltaic Sundarbans.

5 Conclusion

Microbotanical analysis from the archaeological site of Tilpi revealed that a warm and humid climate prevailed during approx. 2600 years ago with predominance of C4 grasses along with other mesophytic, non-littoral vegetation which was growing near the coast. The present-day vegetation of Tilpi village and adjacent areas may be described as typical mesophytic with open grass land and agricultural field with frequent fresh water bodies like ponds, road side ditches etc. Therefore, the general climate and vegetation had not suffered much change in the present area of investigation. This data is important as it reflects the ecological and environmental condition that influenced human settlement at the site. Moreover, archaeological data from the coastal India is meagre thus, the present study might be considered as a baseline data in the future.

Acknowledgements The authors pay tribute to Late Dr. Ruby Ghosh (Scientist E, Birbal Sahni of Palaeosciences), who started the archaeobotanical research in the Palaeobotany and Palynology laboratory, Department of Botany, University of Calcutta under the supervision of Late Prof. Manju Banerjee and the corresponding author of this work. Authors acknowledge the infrastructural support provided by the Centre of Advance Study (Phase-VII grant), Department of Botany, University of Calcutta. Authors are also thankful to the authorities of the Silesian University of Technology, Poland for providing us with the AMS radiocarbon dates of the archaeological sample.

References

Alexandre A, Meunier JD, Lézine AM et al (1997) Phytoliths indicators of grasslands dynamics during the late Holocene in intertropical Africa. Palaeogeogr Palaeoclimatol Palaeoecol 136:213–219

Alexandre A, Meunier JD, Mariotti A et al (1999) Late Holocene phytolith and carbon-isotope record from a latosol at Salitre, South-Central Brazil. Quatern Res 51:187–194

Backels C (2020) Pollen and Archaeology. In: Henry A (ed) Handbook for the analysis of micro-particles in archaeological samples. Interdisciplinary contributions to archaeology, Springer Cham, https://doi.org/10.1007/978-3-030-42622-4

Barton CM (1997) Stone tools, style, and social identity: an evolutionary perspective on the archaeological records. Archaeol Pap Am Anthropol Assoc 7(1):141–156

Biswas O, Ghosh R, Paruya DK et al (2016) Can grass phytoliths and indices be relied on during vegetation and climate interpretations in the eastern Himalayas? Studies from Darjeeling and Arunachal Pradesh, India. Quatern Sci Rev 134:114–132

Biswas O, Ghosh R, Agrawal S et al (2021) A comprehensive calibrated phytolith based climatic index from the Himalaya and its application in palaeotemperature reconstruction. Sci Total Environ 750:142280

Bozarth SR (1987) Diagnostic opal morphotypes from rinds of selected Cucurbita species. Am Antiq 52(3):607–615

Bozarth SR (1990) Diagnostic opal phytolith from Pods of selected verities of common beans (*Phaseolus vulgaris*). Antiquity 55:98–104

Bremond L, Alexandre A, Wooller MJ et al (2008) Phytolith indices as proxies of grass subfamilies on East African tropical mountains. Global Planetary Change 61:209–224

Brown AG, Petit F, James LA (2016) Archaeology and human artefacts. In: Kondolf MG, Piégay H (eds) Tools in fluvial geomorphology. Chichester, Wiley

Bryant VM, Holloway RG (1978) The role of palynology in archaeology. Adv Archeol Method Theory 6:191–224

Chattopadhyaya BD, Sengupta G, Chakrabarty S (2005) An annotated archaeological atlas of West Bengal Volume I: pre and protohistory. Manohar in Association with Centre for Archaeological Studies and Training, Eastern India, New Delhi, pp 83–102

Chowdhury KA, Ghosh SS (1954) Plant Remains from Hastinapura. Ancient India. 10:120–137

Colledge S, Conolly J, Shennan S (2005) The evolution of Neolithic farming from SW Asian Origins to NW European limits. Eur J Archaeol 8:137

Colledge, S, Conolly J (2014) Wild plant use in European Neolithic subsistence economies: A formal assessment of preservation bias in archaeobotanical assemblages and the implications for understanding changes in plant diet breadth. *Quat Sci Rev* 101. https://doi.org/10.1016/j.quascirev.2014.07.013

Das SK, Gangopadhyay K, Ghosh A et al (2021) Organic geochemical and palaeobotanical reconstruction of a Late Holocene archaeological settlement in coastal eastern India. The Holocene 31(10):1511–1524

Denham T, Haberle SG, Lentfer C et al (2003) Origins of agriculture at Kuk Swamp in the Highlands of New Guinea. Sciences 301:189–193

Dumpe B, Stivrins N (2015) Organic inclusions in Middle and Late Iron Age (5th–12th Century) hand-built pottery in present-day Latvia. J Archaeol Sci 57:239–247

Erdtman G (1954) An introduction to pollen analysis, 2nd edn. Chronica Botanica Company, Waltham, Massachusetts

Faegri K, Iversen J (1950) Textbook of modern pollen analysis. Ejnar Munksgaard, Copenhagen

Farooqui A, Gaur AS, Prasad V (2013) Climate vegetation and ecology during Harappan period: excavations at Kanjetar and Kaj, mid-Saurashtra coast, Gujarat. J Archaeol Sci 40:2631–2647

Forshaw R (2015) Unlocking the past: the role of dental analysis in archaeology. Dent Hist 60(2):51–62

Fredlund GG, Tieszen LT (1997) Calibrating grass phytolith assemblages in climatic terms: Application to late Pleistocene assemblages from Kansas and Nebraska. Palaeogeogr Palaeoclimatol Palaeoecol 136(1–4):199–211

Fuller DQ, Qin L (2009) Water management and labour in the origins and dispersal of Asian rice. World Archaeol 41(1):88–111

Gao Y, Yang J, Ma Z et al (2020) New evidence from the Qugong site in the central Tibetan Plateau for the prehistoric Highland Silk Road. Holocene 1–10. https://doi.org/10.1177/0959683620941144j

Ghosh R, Bera S, Chakraborty S et al (2005) Significance of study of Phytoliths in understanding vegetational pattern in an archaeological site of West Bengal, India. Phytomorphology 55(3&4):221–232

Ghosh R, D'Rozario A, Bera S (2006a) Can palynomorphs occur in burnt ancient potsherds? An experimental proof. J Archaeol Sci 33:1445–1451

Ghosh R, Bera S, D' Rozario A et al (2006b) Plant remains from an archaeological site as indicators of vegetation and agricultural practice between (3320 ± 400) and (2080 ± 80) yr BP in Gangetic West Bengal, India. J Integr Biol 48(6):628–641

Ghosh R, Bera S, Banerjee M (2013) Man-environment interaction and its impact since ca. 3600 years BP in West Bengal, India: a multiproxy approach. Chin Sci Bull 58(S1):9–28

Gremillion K (2014) Plant domestication and cultivation in archaeology. In: Smith C (ed) Encyclopedia of global archaeology. Springer, New York. https://doi.org/10.1007/978-1-4419-0465-2_70

Haslam M (2004) The decomposition of starch grains in soils: implications for archaeological residue analyses. J Archaeol Sci 31(12):1715–1734

Hu YQ, Zhang ZL, Bera S et al (2007) What can pollen grains from the Terracotta Army tell us? J Archaeol Sci 34:1153–1157

Iriarte J (2003) Assessing the feasibility of identifying maize through the analysis of cross-shaped size and three-dimensional morphology of phytoliths in the grasslands of southeastern south America. J Archaeol Sci 30:1085–1094

Kajale MD, Eksambekar SP (1997) Application of Phytolith Analyses to a Neolithic site at Budihal, district Gulbarga, South India. Centro De Ciencias Medioambientales, CSIC, Monografias, Madrid 4:219–229

Kajale MD, Eksambekar SP (2001) Phytolith approach for Investigating ancient occupations at Balathal, Rajasthan, India, Part 1: Evidence of crops exploited by initial farmers, Part 2: Late chalcolithic to earlihistoric stratigraphic reconstruction. In: Meunier JE, Colin F (eds) Phytoliths: application in earth science and human history. AA Balkema Publishers, Lisse, pp 199–212

Kajale MD, Mulholland S, Rapp G Jr (1995) Applications of Phytolith analysis: a potential tool for quaternary environmental and archaeological investigations in the Indian subcontinent. Mem Geol Soc India 32:530–544

Kolobova KA, Roberts RG, Chabai VP et al (2020) Archaeological evidence for two separate dispersals of Neanderthals into southern Siberia. Proc Natl Acad Sci 117(6):2879–2885. https://doi.org/10.1073/pnas.1918047117

Li X, Dodson J, Zhou X et al (2007) Early cultivated wheat and broadening of agriculture in Neolithic China. Holocene 17(5):555–560

Li H, Liu Z, James N et al (2021) Agricultural transformations and their influential factors revealed by archaeobotanical evidence in Holocene Jiangsu province, Eastern China. Front Earth Sci 9:661684. https://doi.org/10.3389/feart.2021.661684

Long DJ, Milburn P, Bunting MJ et al (1999) Black Henbane (*Hyoscyamus niger* L.) in the Scottish Neolithic: a re-evaluation of Palynological findings from grooved ware pottery at Balfarg riding school and Henge, Fife. J Archaeol Sci 26:45–52

Long T, Wagner M, Demske D et al (2017) Cannabis in Eurasia: origin of human use and Bronze Age trans-continental connections. Veg Hist Archaeobotany 26:245–258

Lu H, Li Y, Zhang J et al (2014) Component and simulation of the 4,000-year-old noodles excavated from the archaeological site of Lajia in Qinghai, China. Chin Sci Bull 59(35):5136–5152

Lu H, Zhang J, Wu N et al (2009) Phytoliths analysis for the discrimination of foxtail millet (*Setariaitalica*) and common millet (*Panicum miliaceum*). PLoS ONE 4(2):e4448. https://doi.org/10.1371/journal.pone.0004448

Luncz LV, Gill M, Proffitt T et al (2019) Group-specific archaeological signatures of stone tool use in wild macaques. eLife 8:e46961. https://doi.org/10.7554/eLife.46961

Ma Z, Zhang C, Li Q et al (2017) Understanding the possible contamination of ancient starch residues by adjacent sediments and modern plants in Northern China. Sustainability 9(5):752. https://doi.org/10.3390/su9050752

Madella, M, Alexandre A, Terry B (2005) International code for phytolith nomenclature 1.0. Ann Bot 96(2):253–260. https://doi:https://doi.org/10.1093/aob/mci172

Mandal KK (2002) A documentation of artefacts and a study of archaeological, historical and socio-religious features of some new archaeological spots of coastal West Bengal. Dakshin Banger Notun Pratnasthal, Nilachal Press, Kolkata, India

Masojć M, Nassr A, Kim JY et al (2019) Saharan green corridors and Middle Pleistocene hominin dispersals across the Eastern Desert, Sudan. J Hum Evol 130:141–150

Mbida CM, Van Neer W, Doutrelepont H et al (2000) Evidence for Banana cultivation and animal husbandry during the first Millennium BC in the forest of southern Cameroon. J Archaeol Sci 27:151–162

Mindzie C, Doutrelepont H, Vrydaghs L et al (2001) First archaeological evidence of banana cultivation in central Africa during the third millennium before present. Veg Hist Archaeobotany 10:1–6. https://doi.org/10.1007/PL00013367

McKey D, Renard D, Blatrix R (2020) The multiple roles of soil animals in the interpretation of archaeological soils and sediments in lowland tropical South America. In: Odonne G, Molino JF (eds) Methods in historical ecology. Routledg, London, p 9

Min-Kyu P (2017) A method for studying human teeth excavated in archaeological sites: a focus on recent research sites. Anthropol Notebs 23(2):5–19

Mooney DE, Martín-Seijo M (2021) Archaeobotany in the wider landscape. Environ Archaeol 26(2):115–121

Naskar M, Ghosh R, Das S (2021) Grass phytoliths in surface sediments of the Sunderbans, India and their implications in reconstructing past deltaic environmental changes. Holocene 1–16. https://doi.org/10.1177/09596836211041736

Neumann K, Strömberg CAE, Ball T et al (2019) International code for phytolith nomenclature (ICPN) 2.0. Ann Bot 124:189–199

Pamirsky IE, Klykov AG, Murugova GA et al (2017) Silica biominerals (Phytolith) compound of cultured barley plants (*Hordeum vulgare* L.). Mater Sci Eng 225:012238. https://doi.org/10.1088/1757-899X/225/1/012238

Pearsall DM (2000) Paleoethnobotany: a handbook of procedures, 2nd edn. Academic Press, New York

Pearsall DM, Chandler-Ezell K, Chandler-Ezell A (2003) Identifying maize in neotropical sediments and soils using cob phytoliths. J Archaeol Sci 30:611–627

Pearsall DM, Piperno DR, Dinan EH et al (1995) Distinguishing rice (*Oryza sativa* Poaceae) from wild *Oryza* species through phytolith analysis: results of preliminary research. Econ Bot 49:183–196

Piperno DR (2006) Phytoliths: a comprehensive guide to archaeologists and paleoecologists. Altamira Press, Lanham, MD

Piperno DR (2011) The origins of plant cultivation and domestication in the new world tropics. Curr Anthropol 52(4):453–470

Piperno DR, Ranere AJ, Holst I et al (2000) Starch grains reveal early root crop horticulture in the Panamanian tropical forest. Nature 407:894–897

Piperno DR, Stothert KE (2003) Phytolith evidences for early Holocene Cucurbita domestication in south-west Ecuador. Science 299:1054–1057

Prasad V, Farooqui A, Sharma A et al (2014) Mid–late Holocene monsoonal variations from mainland Gujarat, India: a multi-proxy study for evaluating climate culture relationship. Palaeogeogr Palaeoclimatol Palaeoecol 397:38–51

Prasad V, Strömberg CAE, Leaché AD, Samant B et al (2011) Late Cretaceous origin of the rice tribe provides evidence for early diversification in Poaceae. Nat Commun 2:480. https://doi.org/10.1038/ncomms1482

Prasad V, Strömberg CAE, Alimohammadian H et al (2005) Dinosaur coprolites and the early evolution of grasses and grazers. Science 310:1177. https://doi.org/10.1126/science.1118806

Ramanujam CGK, Reddy PR, Ramakrishna H (1998) Pollen types of arecaceae (Palmae) from the subsurface Miocene sediments of Krishna-Godavari Basin. APJ Swamy Botanical Club 15:55–57

Rao MR (1990) Palynological investigation of Arthungal bore-hole, Alleppey District, Kerala, In: Jain PK, Tiwari RS (ed) Proc symp, vista in Indian Palaeobotany, Palaeobotanist, vol 38. pp 243–255

Roffet-Salque M, Dunne J, Altoft D et al (2017) From the inside out: upscaling organic residue analyses of archaeological ceramics. J Archaeol Sci Rep 16:627–640

Rosen AM, Weiner S (1994) Identifying ancient irrigation: a new method using opaline phytoliths from Emmer wheat. J Archaeol Sci 21:125–132

Roy A (2006) Early historic settlement of the Piyali valley. A preliminary report of the excavation at Dhosa and Tilpi. In: The Nimna-Gangeya Sunderban Sanskrit patra, pp 7–8

Sarkar A, Deshpande Mukherjee A, Sharma S et al (2020) New evidence of early Iron Age to Medieval settlements from the southern fringe of Thar Desert (western Great Rann of Kachchh), India: Implications to climate-culture co-evolution. Archaeol Res Asia 21:100163. https://doi.org/10.1016/j.ara.2019.100163

Saul H, Madella M, Fischer A et al (2013) Phytoliths in pottery reveal the use of spices in European Prehistoric Cuisine. PLoS ONE 8(8):e70583. https://doi.org/10.1371/journal.pone.0070583

Saxena A, Prasad V, Singh IB (2013) Holocene palaeoclimate reconstruction from the phytoliths of the lake-fill sequence of Ganga Plain. Curr Sci 104(8):1054–1062

Saxena A, Prasad V, Singh IB et al (2006) On the Holocene record of phytoliths of wild and cultivated rice from Ganga Plain: evidence for rice-based agriculture. Curr Sci 90(11):1547–1552

Singh V, Prasad V, Chakraborty S (2006) Phytoliths as indicators of monsoonal variability during mid–late Holocene in mainland Gujarat, western India. Curr Sci 92(12):1754–1759

Tang L, Mao L, Lü X et al (2013) Palaeoecological and palaeoenvironmental significance of some important spores and micro-algae in Quaternary deposits. Chin Sci Bull 58:3125–3139. https://doi.org/10.1007/s11434-013-5747-9

Udatsu T, Houben Z, Fujiwara H et al (1999) Analysis of excavated phytoliths in Neolithic pottery from Jiangsu province. Agric Archaeol 1:36–45

Veena MP, Achyuthan H, Eastoe C et al (2014) A multi-proxy reconstruction of monsoon variability in the late Holocene, South India. Quatern Int 325:63–73

Von Post L (1916) Einigesüdschwedischen Quellmoore. Bull Geol Inst Univ Uppsala 15:219–277

Wang J, Sun L, Fang Q et al (2021) Phytolith evidence for early agriculture in the East Liao River Basin, Northeast China. Archaeol Anthropol Sci 13:156

Wilson S (1985) Phytolith analysis at Kuk, an early agricultural site in Papua New Guinea. Archaeol Ocean 20:90–97

Yao YF, Li X, Jiang HE et al (2012) Pollen and phytolith from fired ancient potsherds as potential indicators for deciphering past vegetation and climate in Turpan, Xinjiang, NW China. PLoS ONE 7(6):e39780. https://doi.org/10.1371/journal.pone.0039780

Zhao Z, Piperno DR (2000) Late Pleistocene/Holocene Environments in the Middle Yangtze River Valley, China and rice (*Oryza sativa* L.) domestication: the phytolith evidences. Geoarchaeology 15:134–145

Zhou X, Yu J, Spengler RN (2020) 5,200-year-old cereal grains from the eastern Altai Mountains redate the trans-Eurasian crop exchange. Nat Plants 6:78–87

Hydrocarbon Biomarkers in Elucidating Source Organic Matter and Palaeodepositional Environment: A Case Study from Cambay Basin, India

Sumit Kumar, Deepti Niyolia, and Suryendu Dutta

Abstract A detailed biomarker characterisation has been carried out from the Early Palaeogene samples from the Cambay Basin. The present study examines the biomarker applications in interpreting organic matter sources and palaeodepositional environment from the Palaeogene shales of the Cambay Basin, India. The detailed saturated and aromatic biomarkers identified from the mass chromatograms of GC-MS and GC × GC-TOF-MS, such as sesquiterpanes, tricyclic, tetracyclic, and pentacyclic triterpanes including hopanoids and methylhopanes, oleananes, bicadinanes, secobicadinanes, C_{30} tetrapolyprenoid, C_{29} 28-nor-spergulane, aromatic sesquiterpanes, diaromatic tricyclic terpane, aromatic non-hopanoid triterpenoids, aromatic diterpenoids, diaryl isoprenoids, and acyclic isoprenoids efficiently discriminate between the organic matter sources, compositions, and palaeodepositional environment in north and south blocks of the Cambay Basin. Low values of pristane/phytane, along with the extended series of tricyclic terpanes, 3β-methylated hopanes, diaryl isoprenoids, C_{30} tetrapolyprenoid and C_{29} 28-norspergulane in the northern block samples, indicate that the organic matter was deposited in the photic zone under suboxic to anoxic settings, most likely in a restricted marine/lacustrine environment. Similarly, the occurrence of oleanane, bicadinanes, secobicadinanes and higher plant aromatic triterpenoids in the southern block samples suggests dominant angiospermous organic matter input deposited in an oxic fluvio-deltaic environment. Moreover, the presence of bicadinanes and secobicadinanes clearly indicates the advent of evergreen tropical rainforests in the Early Eocene.

Keywords Biomarker · Early Eocene · Cambay Basin · Depositional environment · Bicadinanes

S. Kumar (✉) · D. Niyolia · S. Dutta (✉)
Department of Earth Sciences, Indian Institute of Technology Bombay, Mumbai 400076, India
e-mail: sumit.kumar@iitb.ac.in

S. Dutta
e-mail: s.dutta@iitb.ac.in

D. Niyolia
e-mail: deepti.niyolia@iitb.ac.in

B. Samant and D. Thakre (eds.), *Applications of Palynology in Stratigraphy and Climate Studies*, Society of Earth Scientists Series,
https://doi.org/10.1007/978-3-031-51877-5_11

1 Introduction

In hydrocarbon exploration and exploitation, it is crucial to decipher and reconstruct the sedimentary depositional environment since it provides the essential framework for delineating and predicting the petroleum system components for a basin, especially in frontier petroliferous basins. In these petroleum systems, source rocks form an integral part, the compositions and characteristics of which are substantially influenced by the depositional setting (Magoon and Dow 1994; Law 1999; Peters et al. 2005). The depositional environment of organic matter constituting the source rocks can be portrayed using an array of geochemical, geophysical, and petrological proxies. Among them, advanced and applied organic geochemistry is a widely used tool for understanding the organic matter sources and its depositional conditions, like, Eh, pH, hydrology, etc. Molecular fossils (biomarkers) in crude oil and source rock extracts offer a more precise and better understanding of organic matter source and depositional environment as these are generally more resistant to diagenetic elimination and other secondary processes (Peters et al. 2005). Moreover, a comprehensive geochemical assessment of organic matter and its responses to depositional conditions allows both qualitative and quantitative assessment of source rocks (Tissot and Welte 1984; Peters and Cassa 1994), which may guide to build strategies for economic hydrocarbon exploration.

Biomarkers are chemical fossils of complex molecular structures produced by once-living organisms and are adequately stable to be recognised in crude oil or the organic matter of ancient sediments (Hunt 1995). Biological marker assemblages divulge the particular source of organic matter and restructure the palaeoenvironment where the source material was deposited (Moldowan et al. 1986). Moreover, the biomarkers studies can elucidate specific age, thermal maturity, and degree of organic matter biodegradation in source rocks and crude oils (Peters et al. 2005). Hence, considering the relevance of source rock reconstruction and oil-source rock correlation, precise and comprehensive documentation of biomarker assemblages in the crude oil and bitumen fraction of source rocks is essential and imperative for any petroliferous basin.

Cambay Basin is an intracratonic rift basin located in the western onshore region of India, in the state of Gujarat, and covers an area of approximately 53,500 km^2 (Chowdhary 2005). The basin is classified as Category I based on its commercial hydrocarbon productivity. This makes it one of the significant onshore petroliferous basins of India. The detailed geology and stratigraphy of this basin are elaborated in Bhandari and Choudhary (1975), Biswas (1982), Kundu et al. (1997), and Raju (1969). The Cambay oil/gas reservoirs and potential source rocks are distributed mainly in the Palaeocene–Eocene sequences of the Olpad, Cambay Shale, and Kalol formations (Yalcin et al. 1988; Banerjee and Rao 1993). Previous investigations reported that the oil/gaseous hydrocarbons were derived from similar organic matter sources (Raju and Srinivasan 1993; Banerjee et al. 2002; Sivan et al. 2008). However, Kumar et al. (2021) recently analysed crude oil samples from the Cambay Basin and differentiated the source organic matter of north and south blocks based on the biomarker distributions. Moreover, they also concluded that the depositional environment and thermal maturity of organic matter in the north blocks and south blocks of the basin are distinct, regardless of having similar ages and geographical locations.

Although the source rock characteristics of the Cambay Basin were previously studied, comprehensive research based on the variations of saturated and aromatic biomarkers in the different source rock sequences for elucidating the palaeodepositional environment has been hardly documented hitherto. In the present study, we have evaluated and compared the organic matter source and depositional environment using extractable hydrocarbon biomarkers from the different shale formations belonging to the Ahmedabad-Mehsana and Jambusar-Broach blocks of the Cambay Basin. This detailed application of molecular fossils in palaeoenvironmental interpretations is beneficial in researching oil-source rock correlation and its responses to palaeoclimatic paradigms.

2 General Geology

The Cambay Basin (21°–25° N and 71°30′–79°30′ E; Chowdhary 2005) extends from Sanchor in the North to Surat in the South. The basin has rifted along the NNW-SSE trend. It is an elongated narrow graben, which is flanked by the Aravalli Range, Saurashtra Uplift, Deccan Plateau, and Radhanpur-Barmer High. The Cambay rift is surrounded by well-divided basin margin faults. This basin is divided into five tectonic blocks (from north to south) based on the cross trends (Kundu et al. 1993; Chandra et al. 1994) (Fig. 1). Deccan basalt progressively subsided the rift graben and acted as a platform for the post-Deccan Tertiary rift sedimentation. The Olpad Formation, consisting mainly of trap-derived clastics, was deposited over the Deccan volcanic basement during the Early Palaeocene under a fluvio-deltaic environment (Choudhary 2005). During the Late Palaeocene to Early Eocene, a major marine transgression deposited organic-rich Cambay Shale under a restricted marine environment. During the Middle Eocene, the regressive phase led to the deposition of the Kalol Formation as a deltaic sequence. Another cycle of transgression during the Middle Eocene to Early Oligocene deposited the Tarapur Shale, which acts as a regional hydrocarbon seal. Subsequently, after a period of non-deposition/erosion, the Neogene sediments were continuously deposited throughout the Cambay Basin. The Cambay Basin form two significant petroleum system (henceforth PS) namely, the Cambay-Kadi/ Kalol PS in the north part and the Cambay-Hazad PS in the south part (Fig. 2).

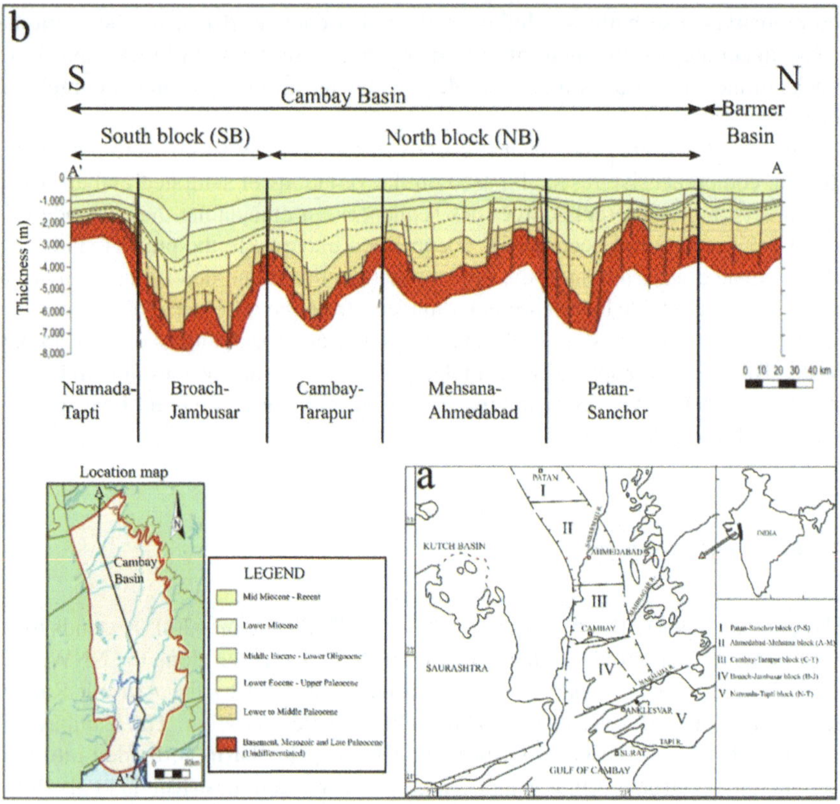

Fig. 1 **a** Tectonic map of Cambay Basin (modified after Chandra et al. 1994); **b** Regional cross-section of the Cambay basin showing the thickness of sediments in different depocenters (modified after Ariketi et al. 2017). Reprinted from Kumar et al. 2021, copyright (2021), with permission from Elsevier (License Number: 5417670440638)

3 Samples and Methodology

For the present study, ten (10) core samples were collected from the Cambay Basin belonging to the Ahmedabad-Mehsana block (henceforth AM) and Jambusar-Broach block (henceforth JB). Among these, five samples belong to the AM, including one sample from the Younger Cambay Shale Formation (YCS), three from the Older Cambay Shale Formation (OCS), and the remaining one from the Olpad Formation. Another set of 5 samples from the JB, comprised two samples from the YCS, another two from the OCS, and the rest one from the Olpad Formation. The core samples were powdered before soluble organic matter extraction. The soluble organic matter or bitumen was extracted five times with a solution of dichloromethane: methanol in a ratio of 9:1 (v/v) through ultrasonication. This total lipid extract was separated into saturated and aromatic hydrocarbon fractions by eluting them with

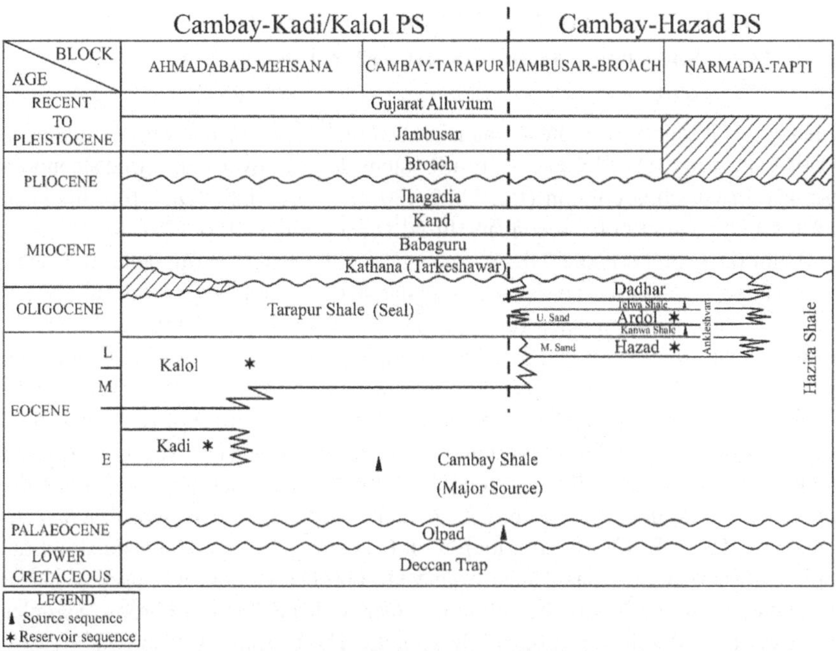

Fig. 2 Generalised stratigraphic of Cambay Basin, showing the major petroleum system (modified after Choudhary 2005 and reference therein). Reprinted from Kumar et al. 2021, copyright (2021), with permission from Elsevier (License Number: 5417670440638)

n-hexane and a mix solution of *n*-hexane and dichloromethane (4:1), respectively, through a liquid chromatography using activated silica gel of 100–200 mesh size. The collected saturated and aromatic fractions were investigated using gas chromatography–mass spectrometry and two-dimensional gas chromatography time-of-flight mass spectrometry.

3.1 Gas Chromatography-Mass Spectrometry (GC–MS)

The saturated and aromatic hydrocarbon fractions were examined using an Agilent 7890 GC coupled with the Agilent 5975 MS. The GC was configured with a HP-5MS capillary column, (30 m × 0.25 mm ID × 0.25 μm film thickness). The GC oven temperature program began at 40 °C (held for 5 min), then raised at a ramp of 4 °C/min to 310 °C for 5.5 min. The carrier gas helium (He) was used with a constant flow rate of 1 ml/min. Mass spectrometry operating parameters, such as ion source and quadrupole temperature, were kept at 250 °C and 280 °C, respectively, and electron ionisation took place at 70 eV. The Agilent ChemStation software package was employed for the data acquisition, and interpretation.

3.2 Two-Dimensional Gas Chromatography Time-of-Flight Mass Spectrometry (GC × GC-TOFMS)

The GC × GC analysis of the aliphatic and aromatic fractions was performed using a LECO Pegasus 4D GC × GC-TOFMS system. The GC oven comprised a nonpolar primary fused silica column (Rxi-5MS) (30 m × 0.18 mm ID × film thickness 0.2 μm) and a mid-polar secondary fused silica column (Rxi-17Sil MS) (2 m × 0.10 mm ID × film thickness 0.1 μm) integrated through a consumable free modulator. The secondary column was accommodated in a separate oven inside the primary GC oven. The initial temperature of the main oven was kept at 50 °C for 1 min with a subsequent increase up to 120 °C at a ramp of 10 °C/min. The final temperature of GC rose to 300 °C at a ramp of 2 °C/min, which was maintained isothermally for 10 min. The secondary oven and modulator were kept at an offset of 10 °C and 15 °C, respectively, corresponding to the main oven. The modulation period was 5.5 s, where the hot pulse lasted for 0.9 s between the cold pulse. The operating temperatures of ion source and transfer line were 250 °C and 280 °C, respectively, with a detector voltage of 1500, 70 eV electron ionisation (EI) energy, and a data acquisition rate took at 100 Hz with a mass range from 50 to 600 Da. Helium was used as carrier gas with a constant flow of 1.5 ml/min. A Gerstel MPS-2 (MultiPurpose Sampler) was used for autoinjection in a splitless mode. The ChromaTOF software (LECO Corporation) was used for the data acquisition, processing and interpretation.

4 Results

4.1 n-alkanes and Acyclic Isoprenoids

The *n*-alkanes are the predominate components in the saturated fractions of all the samples. The distributions of *n*-alkanes and acyclic isoprenoids are substantial for evaluating the origin and deposition of organic matter as they are widespread in the biosphere. The chromatograms of the saturated fraction of the crude oils reveal a partial or total absence of the short chain *n*-alkanes (up to $C_{13/14}$). Meanwhile, longer chains with higher molecular weight are detected in all the samples (up to $C_{36/37}$).

The samples of AM are characterised by unimodal distributions of *n*-alkanes (maximum at $C_{16/17}$) (Fig. 3) with low odd-over-even predominance (OEP), which is related to aquatic macrophyte origin (Ficken et al. 2000). However, samples like AM-2 with a unimodal distribution at C_{31} and AM-5 with a bimodal distribution (with peak maximum at C_{16} and C_{27}) (Figs. 3a, b) represent significant terrigenous organic matter input (Eglinton et al. 1962). Thus, *n*-alkane profiles propose that in the AM, the organic matter was substantially derived from the aquatic macrophytes with a moderate contribution from terrestrial plants (Ficken et al. 2000).

Further, the *n*-alkane profiles in the YCS samples (JB-1 and JB-2) exhibit slight OEP with a bimodal distribution (with maxima in the vicinity of $C_{22/23}$ and C_{27}

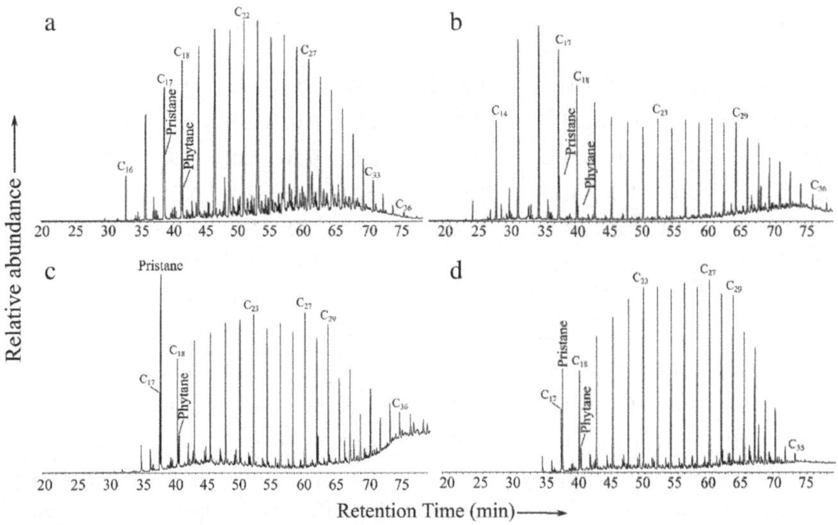

Fig. 3 GC–MS total ion chromatograms of the saturated hydrocarbon fraction of **a** AM-4, **b** AM-5, **c** JB-1, **d** JB-2

(Figs. 3c, d), suggesting a dominant terrigenous higher plant input with subordinate algae/microbial contribution to the sedimentary organic matter (Tissot and Welte 1984; Peters et al. 2005). Additionally, samples belonging to OCS and Olpad Formations, including JB-3, JB-4, and JB-5, depict unimodal *n*-alkane profiles with a peak maximum at $C_{17/23}$, representing the dominant contribution of aquatic macrophytes to the kerogen.

Despite variations in *n*-alkane profiles in the saturated fractions, major acyclic isoprenoids (pristane (Pr) and phytane (Ph)) exhibit less variabilities. The content of Pr and Ph are used to deduce the redox conditions of the depositional environment. The measured values of the Pr/Ph ratio range between 1.06 and 3.08, with an average of 1.92 for the AM samples, and for the JB samples vary between 1.44 and 12.26, with an average value of 5.70 (Table 1), which indicate suboxic to oxic depositional conditions. Another common parameter Pr/nC_{17} describes the depositional environment. The $Pr/nC_{17} < 0.5$ reflects a swampy environment, whereas a value greater than 1 suggests an open water environment (Lijmbach, 1975). The range of this ratio in the AM samples varies in the range of 0.08–1.52, with only one sample (AM-1) displaying value greater than 1 (Table 1). Therefore, the ratio may indicate that the depositional environment for the AM samples was possibly dominated by the inland swamp environment over open water conditions. Similarly, the ratio for the JB samples ranges from 0.14 to 4.27 (Table 1), with only two samples having values less than 0.5, suggesting that an open water environment more likely dominated during the deposition of the JB samples.

Table 1 Ratios of parameters calculated from the saturated and aromatic fractions of the Early Eocene crude oils from Cambay Basin

Sample name	CPI	OEP	Pr/nC_{17}	Ph/nC_{18}	Pr/Ph	OI	BI	DSR	Fla/(Fla + Py)
AM-1	1.02	0.95	1.52	3.45	1.06	0.07	0.01	ND	0.48
AM-2	1.12	1.05	0.45	0.18	2.12	0.10	0.06	ND	0.46
AM-3	1.17	1.00	0.08	0.06	1.43	0.08	0.06	ND	0.51
AM-4	1.04	0.99	0.19	0.08	1.93	ND	ND	ND	0.31
AM-5	1.05	1.05	0.23	0.11	3.08	ND	ND	ND	0.40
JB-1	1.19	1.09	3.10	0.33	7.07	0.10	0.05	0.91	0.36
JB-2	1.22	1.17	2.88	0.22	12.26	0.17	0.12	0.81	0.57
JB-3	1.13	1.07	0.40	0.12	4.53	0.15	0.03	0.93	0.44
JB-4	1.07	1.01	0.14	0.05	3.18	0.37	0.27	ND	0.57
JB-5	0.76	0.85	4.27	7.97	1.44	0.10	0.87	ND	0.89

CPI = [(nC23 + nC25 + nC27) + (nC25 + nC27 + nC29)]/2 × (nC24 + nC26 + nC28); OEP = (nC21 + 6 × nC23 + nC25)/4 × (nC22 + nC24); Pr/Ph = pristane/phtane; OI = oleanane/ (oleanane + C_{30} hopane); BI = T-bicadinane/(T-bicadinane + C_{30} hopane); DSR = D2/(D2 + D6) Diaromatic secobicadinane ratio; Fla/(Fla + Py) = fluoranthene/ (fluoranthene + pyrene)

4.2 Saturated Hydrocarbon Biomarkers

Petroleum biomarkers detected in the GC × GC analysis of the saturated fractions, like sesquiterpanes and triterpenoids, including tricyclic terpanes, tetracyclic terpanes, and pentacyclic triterpenoids, can also be employed to explicate the organic matter source and palaeodepositional environment (Table 2, Figs. 4 and 5).

4.2.1 Sesquiterpenoids

Several bicyclic sesquiterpenoids ($C_{14/15}$–$C_{20/21}$) were identified in the *m/z* 123 mass chromatogram (Philp et al. 1981) (Fig. 4). Among these compounds, the C_{15} drimane and C_{16} homodrimane dominated all the samples. The drimane and homodrimanes are supposed to have been generated from multiple sources, such as from the diagenetic transformation of tricyclic terpanes and hopanes and/or unknown precursors (Alexander et al. 1983). 4β(H)- Eudesmane (*m/z* 109) is derived from β-eudesmol, a higher plant constituent (Alexander et al. 1983). Cadinanes (*m/z* 109) is related explicitly to the resinous input from higher plants (van Aarssen and de Leeuw 1992).

4.2.2 Triterpenoids

Triterpenoids in the studied samples are represented by tricyclic terpanes, tetracyclic terpanes, and pentacyclic terpanes. The distributions of all triterpanes, including

Table 2 List of abbreviation for the compounds given in Figs. 4 and 5

Ts	18α-22,29,30-trisnorneohopane
Tm	17α-22,29,30-trisnorhopane
βTm	17β-22,29,30-trisnorhopane
MeTs	Methyl-18α-22,29,30-trisnorneohopane
MeTm	Methyl-18α-22,29,30-trisnorhopane
H29	17α(H),21β(H)-30-norhopane
M29	17β(H),21α(H)-30-norhopane
2αMeH30	17α(H),21β(H)-2α-methylhopane
H30	17α(H),21β(H)-30-hopane
M30	17β(H),21α(H)-30-hopane
3βMeH31	17α(H),21β(H)-3β-methylhopane (S + R)
HH31	17α(H),21β(H)-30-homohopane (22S + 22R)
HH32	17α(H),21β(H)-30,31-bishomohopane (22S + 22R)
HH33	17α(H),21β(H)-30,31,32-trishomohopane
HH34	17α(H),21β(H)-30,31,32,33-tetrakishomohopane (22S + 22R)
HH35	17α(H),21β(H)-30,31,32,33,34-pentakishomohopane (22S + 22R)
HH36	17α(H),21β(H)-30,31,32,33,34,35-hexakishomohopane (22S + 22R)
2αMeH31	17α(H),21β(H)-2α-methylhomohopane
2αMeH32	17α(H),21β(H)-2α-methylbishomohopane (S + R)
2αMeH33	17α(H),21β(H)-2α-methyltrishomohopane (S + R)
2αMeH34	17α(H),21β(H)-2α-methyltetrakishomohopane (S + R)
2αMeH35	17α(H),21β(H)-2α-methylpentakishomohopane (S + R)
2αMeH36	17α(H),21β(H)-2α-methylhexakishomohopane (S + R)
MH31	17α(H),21β(H)-methylhopane
3βMeH32	17α(H),21β(H)-3β-methylbishomohopane (S + R)
2αMeH33	17α(H),21β(H)-3β-methyltrishomohopane (S + R)
2αMeH34	17α(H),21β(H)-3β-methyltetrakishomohopane (S + R)
2αMeH35	17α(H),21β(H)-3β-methylpentakishomohopane (S + R)
TT	Tricyclic terpane
TET	Tetracyclic terpane
S	Sterane
O	Oleanane
SH	8,14-secohopane
B_W	cis-cis–trans bicadinane
B_T	Trans–trans-trans bicadinane
MB_W	Methylated cis-cis–trans bicadinane
MB_T	Methylated trans–trans-trans bicadinane
C_{30}TPP	C_{30} tetrapolyprenoid
NSP	C_{29} 28-nor-spergulane

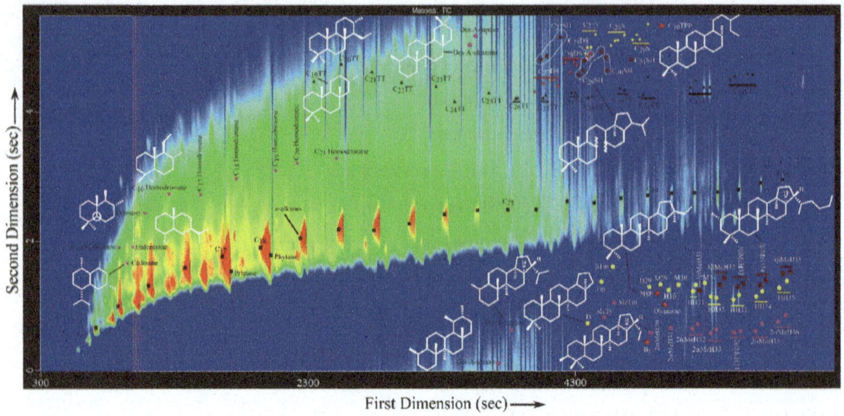

Fig. 4 GC × GC total ion chromatogram of the saturated fraction of the sample AM-1 belonging to AM (see Table 2 for details)

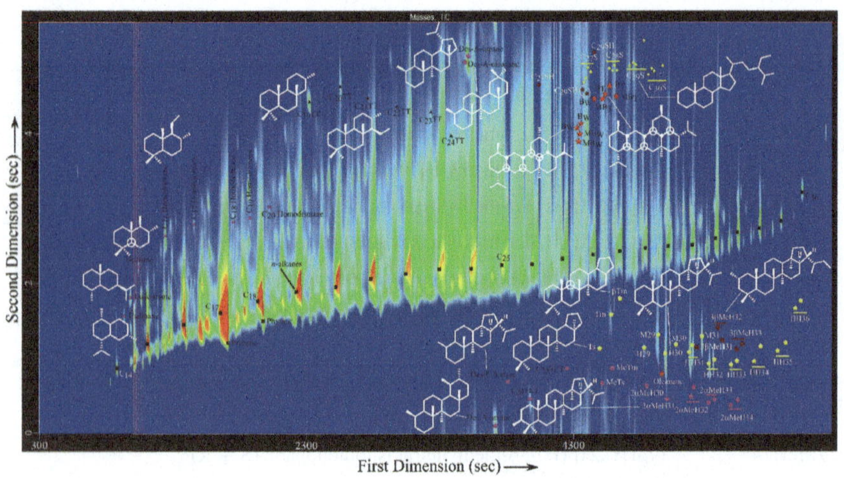

Fig. 5 GC × GC total ion chromatogram of the saturated fraction of the sample JB-2 belonging to JB (see Table 2 for details)

hopanes and most of the non-hopanoids triterpanes, can be obtained from *m/z* 191 mass chromatogram (Seifert and Moldowan 1979).

Tricyclic Terpanes

A homologous series of tricyclic terpanes (TT) were identified in the partial mass chromatogram of *m/z* 191 in all the samples. A homologous tricyclic terpane (C_{19}-$C_{30/34}$) series was detected in all the AM samples, with a low abundance of C_{32}

isomers (Fig. 4). This low abundance of C_{32} isomers may arise due to insufficient cleavage of two carbon–carbon bonds in the precursor compounds to form these homologs (Peters et al. 2005). In contrast, the JB samples reveal the presence of low and uniform short chain TT (C_{19}-$C_{21/24}$) in GC × GC chromatograms (Fig. 5).

High concentrations of tricyclic terpanes in the samples are the typical characteristics of marine (Philp and Gilbert 1986) and lacustrine organic matter (De Grande et al. 1993) and indicate a possible algal organic matter input (Volkman et al. 1989; Azevedo et al. 1992). Moreover, the low abundance of C_{19}-C_{24} tricyclic terpanes in the JB samples suggests oil generation linked to terrigenous source materials (Peters et al. 2005).

Tetracyclic Terpanes

Distribution of ring degraded tetracyclic triterpenoids, des-A-oleanane (*m/z* 191), des-A-lupane (*m/z* 287), des-A-ursane (*m/z* 191), and des-E-hopane (*m/z* 191) were identified in all the studied samples (Woolhouse et al. 1992) (Figs. 4 and 5). These compounds were derived from pentacyclic oleanane, lupane, and ursane-type terpenoids of angiosperms (Woolhouse et al. 1992; Chattopadhyay and Dutta 2014), implying angiospermous organic matter contribution to the depositional environment.

Pentacyclic Terpenoids (Hopanes and Related Compounds)

Hopanes are the pentacyclic terpenoids which are abundant in the biosphere and geosphere. They originate from the membrane lipids of prokaryotes, viz. bacteria, cyanobacteria, and some other prokaryotic primitive organisms, where they provide rigidity and strength. A high abundance of (C_{27-35}) regular hopanoid series was detected in all the samples, dominated by C_{30} hopane (Figs. 4 and 5). C_{29}–C_{30} 17β(H), 21α(H)- Moretanes were noticed in relatively low amounts in all the samples. 17α(H), 21β(H)-Hopanes (C_{30}H) are thermally more than moretanes (C_{30}M) and, thus, C_{30}H increases more rapidly in their relative abundance with increasing maturity (Kvenvolden and Simoneit 1990).

Along with the regular hopanes series, unusual methylated hopanes (3β- and 2α-methylhopanes) were identified in the samples because of the higher resolution and sensitivity of GC × GC. Methyl substituted hopanes were monitored by *m/z* 205 partial mass chromatogram. The 2α- methylhopanes were detected in a range of C_{28}–C_{36} in the AM samples. It could be a reliable biomarker for cyanobacteria (Ricci et al. 2014). 3β-Methylhopanes were identified in greater abundance and range in the AM samples than the JB (Fig. 4). Neunlist and Rohmer (1985) and Zundel and Rohmer (1985) reported that various microorganisms, including methanotrophic bacteria, produce 3β-methyl hopanoids. A higher abundance of 3β-methylhopanes is related to lacustrine depositional environments (Farrimond et al. 2004; Eigenbrode et al. 2008).

4.3 Steranes

Steranes in the crude oil or source rocks are derived from complex mixtures of sterols found in plants and animals. Most of the sterane homologs can be identified from *m/z* 217 and 218 mass fragmentograms (Seifert and Moldowan 1979). Steranes are extremely useful for correlation purposes, and sometimes individual steranes indicate the organic matter source and palaeoenvironment of deposition. Regular steranes (C_{27}–C_{29}) were found in all the samples, where C_{29} steranes distribution was the most prominent compared to C_{27} and C_{28} homologs (Figs. 4 and 5). High C_{29} sterane predominance indicates the terrestrial organic matter input to the deposition center (Pande et al. 1994).

4.4 Unusual Saturated Hydrocarbon Biomarkers

Oleananes are produced by the defunctionalisation of higher plant-derived natural precursors, in particular amyrins, in an acid-induced environment (Ames et al. 1954, ten Haven and Rullkötter 1988). A high abundance of oleanane commonly relates to a large amount of angiospermous derived organic matter input to the Tertiary source rocks (Henderson et al. 1968). The partial mass chromatogram corresponding to *m/z* 191 revealed the presence of both 18α(H)- and 18β(H)- oleanane in all the saturated fractions, and they elute just before 17α(H), 21β(H)-hopane. Oleanane is used to calculate the oleanane index (OI), a crucial parameter that determines the extent of contribution from terrestrial higher plants. The OI ranged from 0.10 to 0.37 in the JB samples, whereas a low OI was encountered in the AM samples (0.05 on average) (Table 1), suggesting significant contribution of angiosperm derived organic matter to the source of the JB.

Bicadinanes are the pentacyclic non-hopanoid triterpenoids derived from dammar resin related to the Dipterocarpaceae family of tropical lowland rainforest trees (Grantham et al. 1983). In the subsurface, under thermal stress, the polycadinene macromolecule transforms to form bicadinanes and other cadinanes and polycadinanes through depolymerisation (van Aarssen et al. 1990). Two groups of bicadinane were detected and reported (e.g., Alam and Pearson 1990; Pearson and Alam 1993; Murray et al. 1994). These groups are cis-cis-trans-bicadinane (V, W, W1, W2) and trans-trans-trans-bicadinane (T, T1), which were detected in the JB samples. Only T-bicadinane was encountered in the AM samples except in the deeper AM-4 and AM-5 samples, which barely reveal any signature of bicadinanes. Bicadinane index (BI), calculated from the bicadinane distributions, determines the extent of angiospermous (higher plants) organic matter input in the deposition center along with the thermal maturity of the source rocks. BI ranged between 0.02 and 0.87 for the JB samples, with an average of 0.20. The average value of BI in the AM samples is 0.03 (Table 1).

C_{30} tetrapolyprenoid and C_{29} 28-nor-spergulane are the other unusual biomarkers that were detected in the AM samples of the YCS Formation (Fig. 4). However, no such compounds were observed in the JB samples. The presence of tetrapolyprenoid in crude oil and/or sediments is a diagnostic marker of the hyposaline lacustrine environment (Holba et al. 2003). C_{29} 28-Nor-spergulane designates lacustrine input as it was identified in several lacustrine crude oils from south China (Huang et al. 2003).

4.5 Aromatic Hydrocarbons

The major compounds in the total ion chromatogram (TIC) obtained from the GC × GC-TOFMS of aromatic fractions of the AM and JB samples yield similar distributions. The TIC of the aromatic fractions of AM-1 and JB-1 are shown in Figs. 6 and 7, respectively. The chromatograms are dominated by the isomers of alkylated naphthalenes and alkylated phenanthrenes, cadalene, isocadalene, diaromatic tricyclic terpane, aromatic non-hopanoid triterpenoids, benzohopanes, and other polycyclic aromatic hydrocarbons (PAHs).

In sesquiterpenoids, ionene, methyl ionene, cadalene, isocadalene, and 5, 6, 7, 8-tetrahydrocadalene were identified in all the samples. Aromatisation of these sesquiterpenoids occurs without the loss of structural integrity. Cadalane and isocadalene source from cadinene or cadinol type precursors found in higher plants (Simoneit 1986; Bordoloi et al. 1989). Isohexylalkylaromatic compounds are considered to originate from higher plant terpenoids (Peters et al. 2005). These compounds are considered to originate through the aromatisation and rearrangement mechanism in the sediment that leads to the opening of the A-ring in the terpenoid. The presence

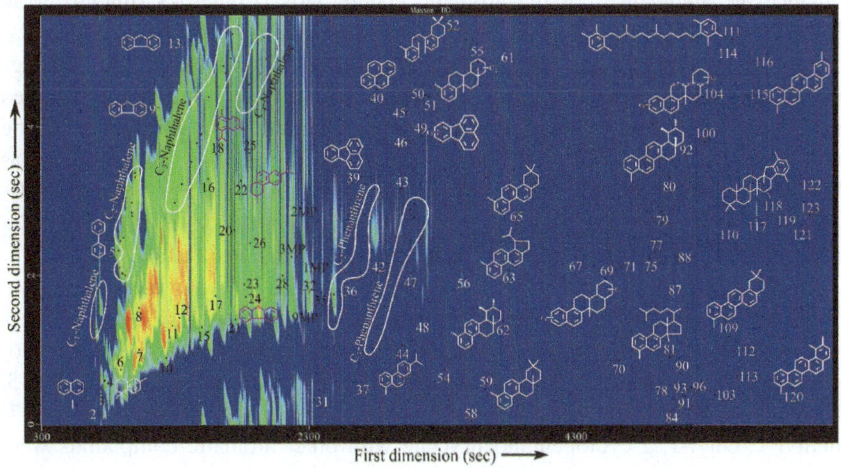

Fig. 6 GC × GC total ion chromatogram of the aromatic fraction of the sample AM-1 belonging to AM (peak table is given in Table 3)

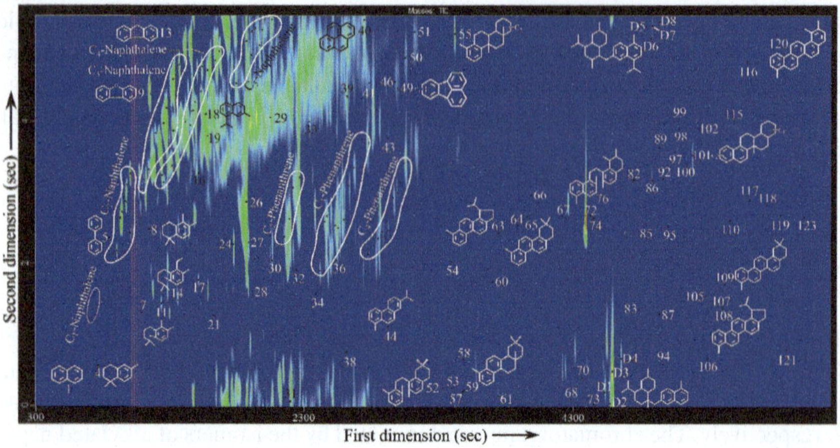

Fig. 7 GC × GC total ion chromatogram of the aromatic fraction of the sample JB-1 belonging to JB (peak table is given in Table 3)

of these compounds in petroleum products can be attributed to non-specific higher plant input (Ellis et al. 1996).

Although the samples lack the presence of saturated diterpenoids, various aromatic diterpenoids were ascertained in aromatic fractions, including bisnorsimonellite, tetrahydroretene, retene, methyl retene, and simonellite. So, this contrasting phenomenon may result from the aromatisation of the saturated compounds induced by thermal maturity. Additionally, clay minerals also felicitate catalysed molecular rearrangement reactions and aromatisation of saturated diterpenoids and thereby enhance the abundances of the aromatic counterparts (Romero-Sarmiento et al. 2010).

Moreover, tricyclic diaromatic terpanes (C_{17-19}), in which B and C rings are aromatised, were detected in the samples. These compounds are considered to source from tricyclic monoaromatic terpene (Azevedo et al. 1992). C-ring cleaved des-A-compounds with oleanane, and lupane structure were detected as diaromatic compounds. Triaromatic compounds, characterised by tetracyclic C-ring cleaved triterpenoids (C_{27} triaromatic-8, 14-seco-pentacyclo-oleanane and C_{27} triaromatic-8,14-seco-pentacyclo-oleanane), mostly dominated the JB samples.

Besides, many aromatised triterpenoids were identified in the samples (Figs. 6 and 7). Monoaromatic triterpenoids are described by dinoroleana (ursa)-1, 3, 5(10), 12-tetraene, and unknown A-ring aromatised compound (*m/z* 169). Moreover, 24,25,26-trisnor-lupa-1, 3, 5 (10)-tetraene, des-A-26, 27, 28-trisnor-oleana-5, 7, 9, 11, 13, 15, 17-heptaene, 24, 25, 26, 27-tetranor-lupa-1, 3, 5 (10), 6, 11, 13-heptaene, and 7-methyl-1'-ethyl-1, 2-cyclopentanochrycene with other aromatic compounds were noticed in both the AM and JB samples. These aromatic triterpenoids are related to oleanane, lupane, or ursane skeletons (Stout 1992; Jacob et al. 2007) abundant in terrestrial higher plants.

Diaromatic pentacyclic terpane (M+ 374) with aromatic rings at A and B (Hazai et al. 1986; Jiang and George 2019) was also observed. This compound would have derived from the euphane or lanostane structures (Jacob et al. 2007). Both euphane and lanostane are considered to originate from two different floral groups. Euphane is considered to be an angiosperm input (Anderson and Botto 1993; Killops et al. 1995; Wang et al. 2003), while Otto and Wilde (2001) explained lanostane as a triterpenoid and associated to conifers.

Further, polycadalene-derived C_{27} diaromatic des-isopropyl secobicadinanes (*m/z* 156; D1-D4) and C_{30} diaromatic secobicadinanes (*m/z* 198; D5-D8), observed only in JM source rock samples, were derived from the polycadinene dammar resins, produced by the Dipterocarpaceae tree of the tropical angiosperm family (van Aarssen et al. 1990). The occurrence of diaromatic secobicadinanes is indicative of the organic matter entering the early catagenetic zone (Chattopadhyay and Dutta 2014). The diaromatic secobicadiane ratios (DSR) are calculated and tabulated in Table 1.

Low abundance of benzohopanes (C_{31}–$C_{32/34}$), with the base peak at *m/z* 191, was detected in the studied samples. Other compounds noticed were arborane/fernane series aromatic triterpenes including 25-norarbora-5, 7, 9-triene and its demethylated substitute, which were originated from the bacterial hopenes after the rearrangement and aromatisation (Hauke et al. 1992). Another series of monoaromatic C_{15}–C_{35} aryl isoprenoids (*m/z* 133) represented by doublets was detected in the AM samples (Fig. 6). Various 2, 4, 6/3, 4, 5-trimethyl substituted diaryl isoprenoids were also recognised in the AM samples (Fig. 6). However, in the JB samples, low abundances of C_{15}–C_{17} aryl isoprenoids (*m/z* 133) were observed (Table 3, Fig. 7).

5 Discussion

5.1 Source Organic Matter

The cross-plot between Pr/nC_{17} and Ph/nC_{18} provides information on kerogen type, maturity trends, and redox conditions. Figure 8 illustrates that the AM samples contain mixed type III/II organic matter, whereas the JB samples are dominated by type III kerogen. Drimane, eudesmane, cadalene, and isohexyl alkylaromatic in the samples support the contributions from non-specific terrestrial vascular plants to the organic matter. Moreover, the occurrence of des-A compounds (des-A-oleanane, des-A-lupane, and des-A-ursane) in the samples specify the angiosperm input. The occurrence of oleanane along with aromatised triterpenoids (e.g., B-ring aromatic des-A-triterpenoid, 8,14-des-A-triterpenoid, 3, 3, 7, 12a-tetramethyl-1, 2, 3, 4, 4a, 12, 12a-octahydrochrysene, C_{21} triaromatic-des-A-lupane, C_{21} triaromatic-des-A-ursane, 24, 25, 26, 27-tetra-oleana-1, 3, 5(10), 6, 11, 13-heptaene, C_{25} tetraromatic lupane, C_{25} tetraaromatic oleanane etc.) in all the samples marks significant proportion of angiosperm input to the depositional centre. The presence of

Table 3 Tentative identification of aromatic compounds for Figs. 6 and 7

S. No	Compounds	Molecular ion
1	Naphthalene	128
2	C_{11} Aryl isoprenoid	148
3	C_{12} Aryl isoprenoid	162
4	Ionene	174
5	Biphenyl	154
6	C_{13} Aryl isoprenoid	176
7	C_{14} Aryl isoprenoid	190
8	Methyl ionene	188
9	Dibenzofuran	168
10	C_{16} Aryl isoprenoid	218
11	C_{15} Isohexyl alkyl aromatic	204
12	C_{15} Aryl isoprenoid	204
13	Fluorene	166
14	C_{16} Isohexyl alkyl aromatic	212
15	C_{17} Aryl isoprenoid	133
16	16, 17-Bisnorsimonellite	224
17	C_{16} Aryl isoprenoid	218
18	Cadalene	198
19	Isocadalene	198
20	Bisnordehydeoabietane (?)	242
21	C_{18} Aryl isoprenoid	246
22	Tetrahydroretene	238
23	C_{17} Aryl isoprenoid	232
24	Dibenzothiophene	184
25	Isomer of tetrahydroretene	238
26	Phenanthrene	178
27	Anthracene	178
28	C_{18} Aryl isoprenoid	246
29	C_{17} Isohexyl alkyl aromatic	226
30	C_{18} Aryl isoprenoid	246
31	Bisnorsimonellite	224
32	C_{19} Aryl isoprenoid	260
33	C_{18} Isohexyl alkyl aromatic	240
34	C_{21} Aryl isoprenoid	288
35	C_{21} Triaromatic des-A-oleanane	218
36	C_{20} Aryl isoprenoid	274
37	Isomer of bisnorsimonellite	224

(continued)

Table 3 (continued)

S. No	Compounds	Molecular ion
38	Tetrahydroretene	238
39	Fluoranthrene	202
40	Pyrene	202
41	C_{19} Isohexyl alkyl aromatic	254
42	C_{21} Aryl isoprenoid	228
43	C_1-Fluoranthene	216
44	Retene	234
45	Monoaromatic des-A-triterpenoids	296
46	C_1-Pyrene	216
47	C_{22} Aryl isoprenoid	302
48	Isomer of retene	230
49	C_1-Fluoranthene	216
50	C_1-Pyrene	216
51	C_1-Pyrene	216
52	C-ring cleaved 8, 14-seco des-A-triterpenoid	306
53	3, 3, 7, 12a-Tetramethyl-1, 2, 3, 4, 4a, 12, 12a-octahydrochrysene (oleanane series)	292
54	C_1-Retene	248
55	B-ring aromatised des-A-triterpenoid	310
56	C_{25} Aryl isoprenoid	330
57	Isomer of 3, 3, 7, 12a-tetramethyl-1, 2, 3, 4, 4a, 12, 12a-octahydrochrysene (oleanane series)	292
58	Monoaromatic triterpenoid	324
59	3, 3, 7, 12a-Tetramethyl-1, 2, 3, 4, 4a, 12, 12a-octahydrochrysene (oleanane series)	292
60	Isomer of 3, 3, 7, 12a-tetramethyl-1, 2, 3, 4, 4a, 12, 12a-octahydrochrysene (ursane series)	292
61	Phenyl anthracene	254
62	3, 3, 7, 12a-Tetramethyl-1, 2, 3, 4, 4a, 12, 12a-octahydrochrysene (ursane series)	292
63	C_{21} Triaromatic-des-A-lupane	274
64	C_{21} Triaromatic des-A-ursane	274
65	3, 4, 7-Trimetyl-tetrahydrochrysene	274
66	3, 3, 7-Trimethyl-1,2,3,4-tetrahydrochrysene	274
67	3, 4, 7, (?)-Tetramethyl-1,2,3,4-tetrahydrochrysene	288
68	Dinor-oleana (ursa)-1, 3, 5(10)-triene	378
69	Tetramethyl diaromatic pentacyclic triterpenoids	360
70	Demethylated 25-norarbora-5, 7, 9-triene	378

(continued)

Table 3 (continued)

S. No	Compounds	Molecular ion
71	C_{27} Aryl isoprenoid	414
72	Dimethyl 3, 4, 7-trimetthyl-1, 2, 3, 4-tetrahydrochrysene	302
73	Dis-oleana (lupa)-1, 3, 5(10)-triene	378
74	C_{27} Triaromatic 8, 14-seco-pentacyclo-oleanane	356
75	C_{28} Aryl isoprenoid	428
76	C_{27} Triaromatic 8, 14-seco-pentacyclo-ursane	356
77	Tetramethyl diaromatic pentacyclic triterpenoids	360
78	Lanosta(eupha)pentaene	376
79	Tetramethyl diaromatic pentacyclic triterpenoids	360
80	Unknown triaromatic triterpenoid	340
81	Lanosta(eupha)pentaene	386
82	Tetramethyl triaromatic 8, 14 secotriterpenoid	370
83	Isomer of demethylated 25-norarbora-5, 7, 9-triene	378
84	Isomer of unknown triaromatic triterpenoid	340
85	Tetramethyl diaromatic triterpenoid	360
86	Isomer of C4-tetramethyl diaromatic triterpenoid	368
87	25-Norarbora-5, 7, 9-triene	392
88	C_{28} Diaromatic 8, 14-seco-pentacyclic triterpane	374
89	Unknown triaromatic triterpenoid (?)	340
90	C_{32} Trimethyl triaromatic isoprenoid	426
91	C_{33} Trimethyl triaromatic isoprenoid	440
92	Tetranor-oleana (ursa)-1, 3, 5(10),6,8,11,13-heptaene	342
93	Isomer of C_{33} trimethyl triaromatic isoprenoid	440
94	Isomer of unknown triaromatic tritepenoid	340
95	Pentamethyl diromatic triterpenoid	374
96	Isomer of C33 trimethyl triaromatic isoprenoid	237
97	24, 25, 26, 27-Tetra-oleana-1, 3, 5 (10), 6, 11, 13-heptaene	342
98	24, 25, 26, 27-Tetra-lupa-1, 3, 5(10), 6, 11, 13-heptaene	342
99	Tetramethyl diaromatic triterpenoid	360
100	Diaromatic pentacyclic terpane	374
101	Pentamethyl diaromatic triterpenoid	374
102	C_5-Pentacyclic diaromatic triterpenoid	390
103	Tetraaromatic hopane	324
104	Pentamethyl diaromatic pentacyclic terpane	374
105	Trimethyl tetraaromatic triterpenoid	342
106	Tetramethyl tetraaromatic triterpenoid	324
107	Isomer of C_{25} tetraaromatic oleanane	324

(continued)

Table 3 (continued)

S. No	Compounds	Molecular ion
108	C_{25} Tetraaromatic lupane	324
109	C_{25} Tetraaromatic oleanane	324
110	C_{31} Benzohopane	418
111	C_{32} 2, 3, 6-Trimethyl diaryl isoprenoids	434
112	24, 25, 26, 27, 28-Pentanor-ursa-1, 3, 5 (10), 6, 8, 11, 13, 15, 17-nonaene	324
113	Methylated tetraaromatic hopane	338
114	C_{33} 3, 4, 5-Trimethyl diaryl isoprenoid	448
115	Dimethyl picene	306
116	Dimethyl picene	306
117	C_{32} Benzohopane	432
118	C_{33} Benzohopane	446
119	C_{34} Benzohopane	460
120	Trimethyl picene	320
121	Trimethyl picene	320
122	C_{35} Benzohopane	474
123	C_{35} Benzohopane	474
D1–D4	C_{27} Diaromatic secobicadinane	362
D5–D8	C_{30} Diaromatic secobicadinane	402

C_{30} bicadinane suggests angiospermous organic matter input from dammar resin of the Dipterocarpaceae family, a tropical hardwood tree (Goswami et al. 2005).

The complete absence of diaromatic secobicadinanes in the AM samples but the presence of both C_{27} and C_{30} diaromatic secobicadianes in the JM samples (Table 1) signify a substantial contribution of tropical angiosperm to the JB samples. Moreover, the arrangement of long chain tricyclic terpanes and methylated hopanes, as well as the subordination of angiosperm markers in NB oils, represent algal/microbial input with limited contribution from angiosperms to the source of the AM samples.

The occurrence of triterpenoids, including hopane, C_{31} to $C_{34/35}$ benzohopane homologues along with 25-norarbora-5, 7, 9-triene, confirms the contribution of microbial organic matter. Moreover, the diagenetic/catagenetic origin of the organic matter with an influence of pyrogenic sources was indicated by the intermediate to high Fla/(Fla + Py) ratio in the source rock samples (Table 1).

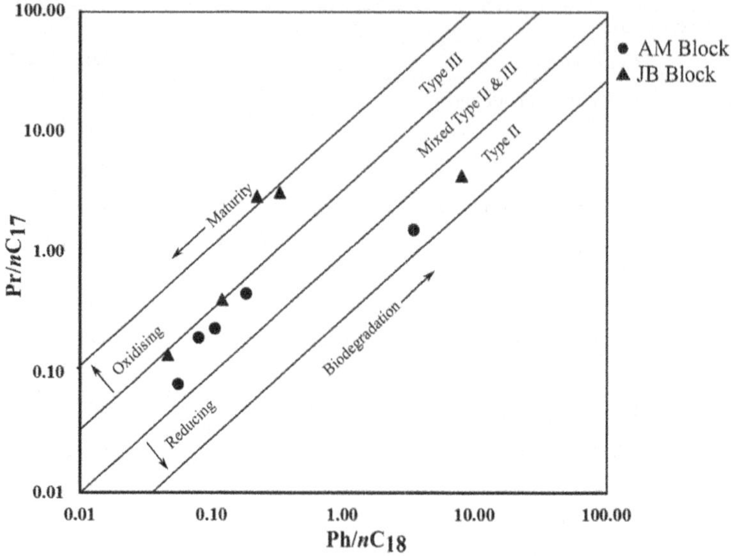

Fig. 8 Cross-plot of pristane/nC_{17} and phytane/nC_{18} of the shale samples from Cambay Basin

5.2 *Depositional Environment*

The depositional environment of the samples was examined based on some biomarker parameters. The lower Pr/nC_{17} and Ph/nC_{18} values indicate that the sedimentary environment was slightly reducing for the AM samples compared to the JB samples. This can be further substantiated by the Pr/Ph ratio. Pr/Ph ratio ranging from 1.06 to 3.08 in the AM samples designate dysoxic conditions, while for the JB samples, it varied between 1.44 and 12.26, indicating terrigenous input under an oxic to sub-oxic depositional environment. However, along with chlorophyll, many biomolecules may contribute to pristane and phytane, such as unsaturated isoprenoids in zooplankton, tocopherols, and archaeal ether lipids (Peters et al. 2005). So, caution must be taken while applying the Pr/Ph ratio in palaeoenvironmental interpretations. Thus, inferences drawn on the palaeoredox conditions from the Pr/Ph ratio must be collaborated with the other geological and geochemical datasets to reach sound conclusions.

The cross-plot Pr/Ph vs. DBT/P is incorporated in the study to comprehend the depositional environment of organic matter in the studied sample (Fig. 9). This plot illustrates that the JB samples were deposited in a fluvio-deltaic environment, whereas the AM samples fall in the region of the marine/lacustrine environment. The ternary plot of fluorene (F), dibenzofuran (DBT), and dibenzothiophene (DBT) is also employed to distinguish palaeodepositional environments (Li et al. 2013). The plot shows that the samples from the JB denote fluvial/deltaic palaeodepositional conditions, while the samples from the AM represent a transition region, indicating some intermittent marine/lacustrine-related sedimentary environment zones (Fig. 10).

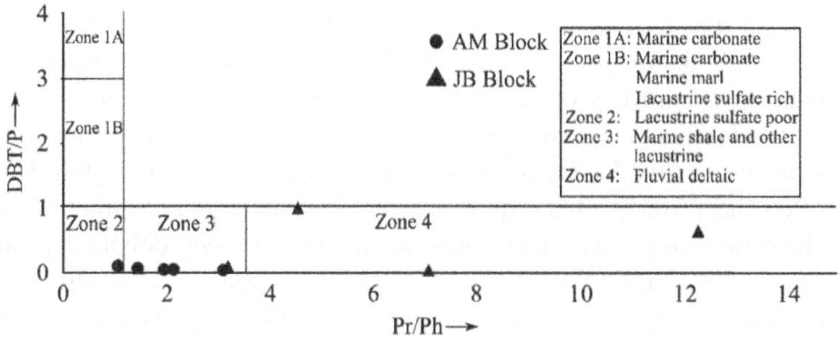

Fig. 9 Cross-plot between DBT/P and Pr/Ph with environmental interpretations (following Hughes et al. 1995)

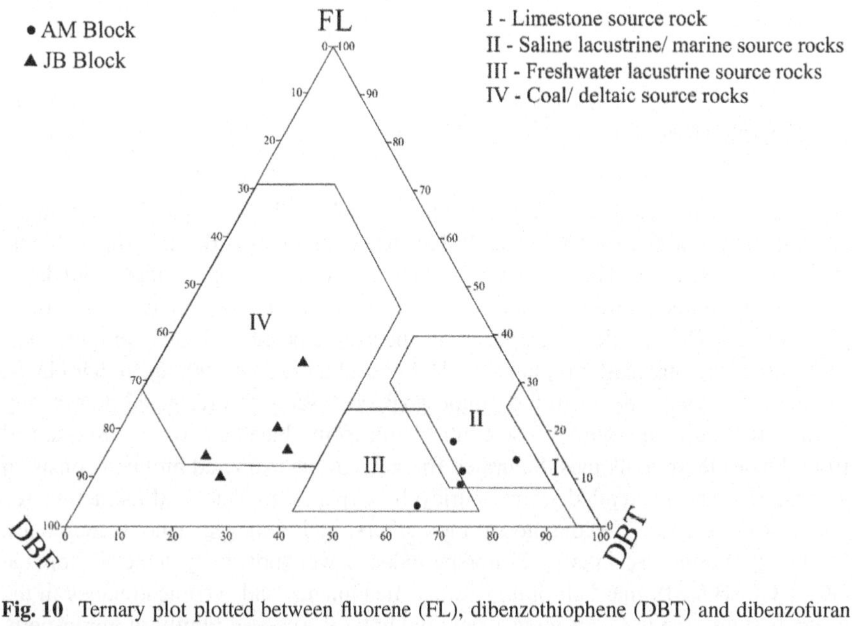

Fig. 10 Ternary plot plotted between fluorene (FL), dibenzothiophene (DBT) and dibenzofuran (DBF) (following Li et al. 2013)

The lacustrine environment is further substantiated by the preservation of C_{30} tetrapolyprenoid and C_{29} 28-nor-spergulane, which mark the unambiguous contribution of the lacustrine environment with low salinity (Huang et al. 2003; Holba et al. 2003). Additionally, the abundance of 3β-methylhopane homologous in the AM samples suggests the contribution of lacustrine organic matter (Farrimond et al. 2004; Eigenbrode et al. 2008). Furthermore, diaryl isoprenoids are derived from isorenieratene biosynthesised by the Chlorobiaceae family of green sulphur bacteria. The presence of these compounds in the AM represents the photic zone anoxia

(Summons and Powell 1987; Koopmans et al. 1996). Additionally, a higher abundance of benzohopanes in the AM samples reflects a lower Eh (suboxic-anoxic) depositional environment (Connan and Dessort 1987).

The Namada River system significantly contributed sediment supply to the southern blocks, i.e., the Jambusar-Broach and Namada-Tapti blocks of the Cambay Basin (Chowdhary 2005). This sedimentary influx containing allochthonous terrestrial angiosperm-rich organic matter was deposited in a fluvio-deltaic depositional environment. However, limited supply in restricted environmental conditions led to the long chain tricyclic terpane, diaryl isoprenoid, and benzohopane distributions with a subordinate contribution from angiosperms in the north and northeast blocks.

The discrepancy in the hydrocarbon biomarker distributions in the AM and JB samples clearly demonstrates that the geological setting and geometry of the Cambay Basin play important roles in source compositions and depositional conditions of organic matter in the source rocks.

6 Conclusions

Palaeogene shale samples from the Younger Cambay Shale, Older Cambay Shale, and Olpad Formations of Ahmedabad-Mehsana and Jambusar-Broach from Cambay Basin are investigated. This study reveals the variabilities of organic facies and depositional environments in both blocks in detail through in-depth analyses of molecular proxies. The profile of *n*-alkanes in the Ahmedabad-Mehsana samples indicates mixed organic matter input from algal/microbial and terrigenous higher plants. In contrast, organic matter in the Jambusar-Broach samples originated dominantly from terrestrial plants with a low contribution from algae/bacteria, as ascertained from the profile of *n*-alkanes. Saturated fractions of the extracted bitumen consisted of sesquiterpanes, extended series of tricyclic terpanes, hopane, and its methylated series, and steranes. The diagnostic biomarkers, including oleanane, bicadinanes, C_{30} tetrapolyprenoids, and C_{29} 28-norspergulanes, were positively detected from the GC × GC-TOFMS mass chromatograms. Bicadinanes and secobicadinanes in the samples suggest significant input from the Dipterocarpaceae family of angiosperm hardwood trees. Major aromatic triterpenoids recognised in the studies samples are related to oleanoid/ursanoid skeletons derived from amryin or lupeol, a constituent of angiosperms.

Besides, the low value of pristane/phytane, along with an abundance of extended series of tricyclic terpanes, 3β-methylated hopanes, diaryl isoprenoids, C_{30} tetrapolyprenoid, and C_{29} 28-norspergulane in the Ahmedabad-Mehsana samples suggest deposition of organic matter in the photic zone under suboxic to anoxic settings, possibly in a restricted marine or lacustrine sedimentary environment. This finding can be further corroborated by the environmental-specific cross-plots of biomarker parameters. Further, the occurrences and high abundances of oleanane, bicadinanes, secobicadinanes, and higher plant aromatic triterpenoids, along with higher pristane/phytane values in the Jambusar-Broach samples, clearly represent that the organic matter was deposited in an oxic fluvio-deltaic environment.

References

Alam M, Pearson M (1990) Bicadinanes in oils from the Surma Basin, Bangladesh. Org Geochem 15(4):461–464

Alexander R, Kagi R, Noble R (1983) Identification of the Bicyclic Sesquiterpenes Drimane and Eudesmane in Petroleum. J Chem Soc Chem Commun 5:226–228

Anderson K, Botto R (1993) The nature and fate of natural resins in the geosphere-III. Re-evaluation of the structure and composition of Highgate Copalite and Giessite. Org Geochem 20(7):1027–1038

Ames TR, Beton JL, Bowers A et al (1954) The chemistry of the triterpenes and related compounds. Part XXIII. The structure of taraxasterol, ψ-taraxasterol (heterolupeol), and lupenol-I. J Chem Soc 1905–1919.

Ariketi R, Bhui UK, Chandra S et al (2017) Evaluation of organic richness and thermal maturity of Cambay shale formation for shale gas exploration—a study from Ankleshwar Area, Cambay Basin, India. Geohorizons 17–27

Azevedo DA, Aquino Neto FR, Simoneit B et al (1992) Novel series of tricyclic aromatic terpanes characterized in Tasmanian tasmanite. Org Geochem 18(1):9–16

Banerjee A, Pahari S, Jha M et al (2002) The effective source rocks in the Cambay Basin, India. Am Asso Petrol Geol Bull 86(3):433–456

Banerjee A, Rao KLN (1993) Geochemical evaluation of part of the Cambay Basin, India. Am Asso Petrol Geol Bull 77(1):29–48

Bhandari LL, Choudhary LR (1975) Stratigraphic analysis of Kalol and Kadi formations. Am Asso Petrol Geol Bull 59(5):856–871

Biswas SK (1982) Rift basins in the western margin of India and their hydrocarbon prospects with special reference to Kutch basin. Am Asso Petrol Geol Bull 66(10):1497–1513

Bordoloi M, Shukla V, Nath S et al (1989) Naturally occurring cadinenes. Phytochemistry 28(8):2007–2037

Chandra K, Mishra CS, Samanta U et al (1994) Correlation of different maturity parameters in the Ahmedabad Mehsana block of the Cambay Basin. Org Geochem 21(3–4):313–321

Chattopadhyay A, Dutta S (2014) Higher plant biomarker signatures of Early Eocene sediments of North Eastern India. Mar Pet Geol 57:51–67

Chowdhary LR (2005) Petroleum geology of the Cambay Basin, Gujarat. India. Indian Petroleum Publishers, Dehradun, pp 74–111

Connan J, Dessort D (1987) Novel family of hexacyclic hopanoid alkanes (C_{32}–C_{35}) occurring in sediments and oil from anoxic paleoenvironments. Org Geochem 11(2):103–113

De Grande SMB, Aquino Neto FR, Mello MR (1993) Extended tricyclic terpanes in sediments and petroleums. Org Geochem 20(7):1039–1047

Eglinton G, Gonzales AG, Hamilton RJ et al (1962) Hydrocarbon constituents of the wax coatings of plant leaves: a taxonomic study. Phytochemistry 1(2):89–102

Eigenbrode JL, Freeman KH, Summons RE (2008) Methylhopane biomarker hydrocarbons in Hamersley Province sediments provide evidence for Neoarchean aerobiosis. Earth Planet Sci Lett 273(3–4):323–331

Ellis L, Singh RK, Alexander R et al (1996) Formation of isohexyl alkylaromatic hydrocarbons from aromatization-rearrangement of terpenoids in the sedimentary environment: a new class of biomarker. Geochim Cosmochim Acta 60(23):4747–4763

Farrimond P, Talbot HM, Watson DF et al (2004) Methylhopanoids: molecular indicators of ancient bacteria and a petroleum correlation tool. Geochim Cosmochim Acta 68(19):3873–3882

Ficken KJ, Li B, Swain DL et al (2000) An *n*-alkane proxy for the sedimentary input of submerged/floating freshwater aquatic macrophytes. Org Geochem 31(7–8):745–749

Goswami B, Bisht R, Bhatnagar A et al (2005) Geochemical characterization and source investigation of oils discovered in khoraghat-nambar structures of the assam-arakan basin, India. Organic Geochem 36(2):161–181

Grantham PJ, Posthuma J, Baak A (1983) Triterpanes in a number of Far Eastern crude oils. In: Bjorøy M et al (eds) Advances in organic geochemistry. Wiley, Chichester, pp 675–683

Hauke V, Graff R, Wehrung P et al (1992) Novel triterpene-derived hydrocarbons of the arborane/fernane series in sediments: Part II. Geochim Cosmochim Acta 56(19):3595–3602

Hazai I, Alexander G, Szekely T et al (1986) Investigations of hydrocarbons constituents of a young sub-bituminous coal by gas chromatography–mass spectrometry. J Chromatogr 367:117–133

Henderson W, Wollra V, Eglinton G (1968) Identification of steranes and triterpenoids from geological source by capillary gas liquid chromatography and mass spectrometry. In: Schenck P, Havenaar I (eds) Advances in organic geochemistry. Pergamon Press, Netherlands, pp 181–207

Holba AG, Dzou LI, Wood GD et al (2003) Application of tetracyclic polyprenoids as indicators of input from fresh-brackish water environments. Org Geochem 34(3):441–469

Huang B, Xiao X, Zhang M (2003) Geochemistry, grouping and origins of crude oil in the Western Pearl River Mouth Basin, offshore China Sea. Org Geochem 34(7):993–1008

Hughes WB, Holba AG, Dzou LIP (1995) The ratios of dibenzothiophene to phenanthrene and pristane to phytane as indicators of depositional environment and lithology of petroleum source rocks. Geochim Cosmochim Acta 59(17):3581–3598

Hunt JM (1995) Petroleum geochemistry and geology. W. H. Freeman and Company, New York, p 332

Jacob J, Disnar JR, Boussafir M et al (2007) Contrasted distributions of triterpene derivatives in the sediments of Lake Caçó reflect paleoenvironmental changes during the last 20,000 yrs in NE Brazil. Org Geochem 38(2):180–197

Jiang L, George SC (2019) Biomarker signatures of Upper Cretaceous Latrobe Group petroleum source rocks, Gippsland Basin, Australia: distribution and geological significance of aromatic hydrocarbons. Org Geochem 138:103905

Killops S, Raine J, Woolhouse A et al (1995) Chemostratigraphic evidence of higher-plant evolution in the Taranaki Basin, New-Zeland. Organic Geochem 23(5):429–445

Koopmans MP, Köster J, Van Kaam-Peters HME et al (1996) Diagenetic and catagenetic products of isorenieratene: molecular indicators for photic zone anoxia. Geochim Cosmochim Acta 60(22):4467–4496

Kumar S, Dutta S, Bhui UK (2021) Provenance of organic matter in an intracratonic rift basin: insights from biomarker distribution in Palaeogene crude oils of Cambay Basin, western India. Org Geochem 162:104329

Kundu J, Prakash A, Prasad B (1997) Structural framework and distribution of synrift depocentres in Cambay rift basin, western India. Am Asso Petrol Geol Bull 81(8):1392–1393

Kundu J, Wani MR, Thakur RK (1993) Structural style in South Cambay Rift and its control on Postrift Deltaic Sedimentation. In: Biswas SK, Pandey J, Dave A et al (eds) Proceeding second seminar on petroliferous basins of India. Indian Petroleum Publishers, Dehradun, pp 79–96

Kvenvolden K, Simoneit B (1990) Hydrothermally derived petroleum: examples from Guaymas Basin, Gulf of California, and Escanaba Trough, northeast Pacific Ocean. Am Assoc Pet Geol 74(3):223–237

Law CA (1999) Evaluating source rocks. In: Beaumont EA, Foster NH (eds) Exploring for oil and gas traps—treatise of Petroleum geology/handbook of petroleum geology. Tulsa, Okla, American Association of Petroleum Geologists, pp 1–44

Li M, Wang T, Zhong N et al (2013) Ternary diagram of fluorenes, dibenzothiophenes and dibenzofurans: indicating depositional environment of crude oil source rocks. Energy Explor Exploit 31(4):569–588

Lijmbach GWM (1975) On the origin of petroleum. In: Proceedings 9th World Petroleum Congress, vol 2. pp 357–369

Magoon JB, Dow WG (1994) The petroleum system-from source to trap. AAPG Mem 60:619
Moldowan JM, Sundararaman P, Schoell M (1986) Sensitivity of biomarker properties to depositional environment and/or source input in the Lower Toarcian of S.W. Germany. Org Geochem 10(4–6):915–926
Murray AP, Summons RE, Boreham CJ et al (1994) Biomarker and *n*-alkane isotope profiles for Tertiary oils: relationship to source rock depositional setting. Org Geochem 22(3–4):521–542
Neunlist S, Rohmer M (1985) Novel hopanoids from the methylotrophic bacteria Methylococcus capsulatus and Methylomonas methanica. (22S)-35-aminobacteriohopane-30,31,32,33,34-pentol and (22S)-35-amino-3 betamethylbacteriohopane- 30,31,32,33,34-pentol. Biochem J 231(3):635–639
Otto A, Wilde V (2001) Sesqui-, Di-, and Triterpenoids as Chemosystematic Markers in Extant Conifers. Rev Bot Rev 67(2):141–238
Pande A, Uniyal A, Chandra K (1994) Genetic correlation of biodegraded crude oils from Lower Assam, India Using Biomarker Compositions. Org Geochem 21(8–9):971–977
Pearson M, Alam M (1993) Bicadinanes and other terrestrial terpenoids in immature Oligocene sedimentary rocks and a related oil from the Surma Basin, N.E. Bangladesh. Org Geochem 20(5):539–554
Peters KE, Cassa MR (1994) Applied source rock geochemistry. In: Magoon LB, Dow WG (eds) The petroleum system—from source to trap. AAPG Memoir 60, 93–120
Peters KE, Walters CC, Moldowan JM (2005) The biomarker guide. Biomarkers and isotopes in petroleum exploration and earth history. Cambridge University Press, UK
Philp RP, Gilbert TD (1986) Biomarker distributions in Australian oil predominantly derived from terrigenous source material. Org Geochem 10(1–3):73–84
Philip RP, Gilbert TD, Friedrich J (1981) Bicyclic sequiterpenoids and diterpenoids in Australian crust oils. Geochim Cosmochim Acta 45:1173–1180
Raju ATR (1969) Geological evolution of Assam and Cambay tertiary basins of India. Am Asso Petrol Geol Bull 52(12):2422–2437
Raju ATR, Srinivasan S (1993) Cambay basin-petroleum habitat. In: Biswas SK (ed) Second seminar on petroliferous Basins, India, vol 2. Indian Petroleum Publication, pp 33–78
Romero-Sarmiento MF, Riboulleau A, Vecoli M et al (2010) Occurrence of retene in upper Silurian–lower Devonian sediments from North Africa: origin and implications. Org Geochem 41(3):302–306
Seifert WK, Moldowan JM (1979) The effect of biodegradation on steranes and terpanes in crude oils. Geochim Cosmochim Acta 43(1):111–126
Simoneit BRT (1986) Cyclic terpenoids of the geosphere. In: Johns RB (ed) Biological markers in the sedimentary record. Elsevier, Amsterdam, pp 43–99
Sivan P, Datta GC, Singh R (2008) Aromatic biomarkers as indicators of source, depositional environment, maturity and secondary migration in the oil of Cambay Basin, India. Org Geochem 39(11):1620–1630
Stout SA (1992) Aliphatic and aromatic triterpenoid hydrocarbons in a tertiary angiospermous lignite. Org Geochem 18(1):51–66
Summons RE, Powell TG (1987) Identification of arylisoprenoids in source rocks and crude oils: biological markers for the green sulphur bacteria. Geochim Cosmochim Acta 51(3):557–566
ten Haven HL, Rullkötter J (1988) The diagenetic fate of taraxer-14-ene and oleanene isomers. Geochim Cosmochim Acta 52(10):2543–2548
Tissot BP, Welte DH (1984) Petroleum formation and occurrence, 2nd edn. Springer Verlag, Berlin
van Aarssen BGK, de Leeuw JW (1992) High-molecular-mass substances in resinites as possible precursors of specific hydrocarbons in fossil fuels. Org Geochem 19(4–6):315–326
van Aarssen BGK, Kruk C, Hessels JKC et al (1990) Cis-cis-trans-Bicadinane, a novel member of an uncommon triterpane family isolated from crude oils. Tetrahedron Lett 31(32):4645–4648
Volkman J, Banks M, Denwer K et al (1989) Biomarker composition and depositional setting of Tasmanite oil shale from northern Tasmania, Australia. In: 14th International Meeting on Organic Geochemistry, Paris

Wang L, Wang N, Yao X et al (2003) Euphane and tirucallane triterpenes from the roots of Euphorbia kansui and their in vitro effects on the cell division of Xenopus. J Nat Prod 66(5):630–633

Woolhouse AD, Oung J-N, Philp RP et al (1992) Triterpanes and ring-a degraded triterpanes as biomarkers characteristic of tertiary oil derived from predominantly higher plant sources. Org Geochem 18(1):23–31

Yalcin MN, Welte DH, Misra KN et al (1988) 3-D computer aided basin modeling of Cambay basin, India—a case history of hydrocarbon generation. In: Kumar R, Dwivedi P, Banerjie V et al (eds) Petroleum geochemistry and exploration in Afro-Asian region. Rotterdam, Balkema, pp 417–450

Zundel M, Rohmer M (1985) Prokaryotic triterpenoids. 1. 3 beta-Methylhopanoids from Acetobacter species and Methylococcus capsulatus. Eur J Biochem 150(1):23–27

Application of Palynology in Petroleum and Coal Exploration

Bandana Samant, Abhilash Sen, and Neha Thakre

Abstract Palynology includes a wide range of organic walled microfossils, including spores, pollen, dinoflagellates, chitinozoa, fungal spores, acritarchs, etc. In this chapter, we discuss the use of palynological studies in the fields of petroleum and coal exploration. As palynomorphs are well known for being extremely resilient to the forces of nature (and also the effects of harsh chemical laboratory processing), in favorable conditions, they are abundantly well retained in litho-units and provide an important tool in stratigraphic correlation and paleo-environmental interpretation. Palynology is also used to infer thermal maturity, as well as to understand the production capabilities of basins and inferring the quality of coal deposits. In the past few decades, with the advent of cutting-edge processing and analyzing scientific tools, more and more uses of palynology in newer frontiers have emerged.

Keywords Petroleum · Kerogen · Thermal maturation · Paleoenvironment · Coal · Coal macerals · Coal bed methane

1 Introduction

Palynology currently has a place in a variety of diverse disciplines, including archaeology (both in terrestrial and underwater settings), prehistoric diet reconstruction, melissopalynology (study of pollen in honey), entomopalynology (study of pollen associated with insects) as well as cutting edge fields like medical palynology, and palynology for forensic studies (Batten, 2002). Though the scope for the application

Present Address:
B. Samant (✉)
Department of Geology, School of Environment and Earth Sciences, Central University of Punjab, Bathinda 151401, Punjab, India
e-mail: bandanabhu@gmail.com

B. Samant · A. Sen · N. Thakre
Postgraduate Department of Geology, RTM Nagpur University, Nagpur 440001, Maharashtra, India

B. Samant and D. Thakre (eds.), *Applications of Palynology in Stratigraphy and Climate Studies*, Society of Earth Scientists Series,
https://doi.org/10.1007/978-3-031-51877-5_12

of palynology is vast, a large portion of palynologists are employed for the 'traditional' study of palynology in geology for stratigraphic correlation and hydrocarbon exploration. The history of the application of palynology in geological studies spans many decades. Scophf and Potoniè were among the first to study microfossils and classify them for stratigraphic use (Potonié 1934, 1956; Schopf 1957, 1964). During and shortly after World War II there was a drastic increase in the study of palynology for the exploration of fossil fuels (Halbritter and Hesse, 2004). With the advent of modern technology like SEM, the ultrastructure of these microscopic units have been better understood. The correlation of the different types of palynomorphs in the sediments, with the type of kerogen that can be produced. Understanding the thermal maturity of the sediments through color changes in the pollens and spores, are a few of the quick, cost-effective techniques that have been developed by palynologists working in the industry to complement more rigorous geochemical studies. Seeing that fossil fuels (coal and petroleum) are the backbone on which our modern industry is developing. It can be said with some confidence that the need for palynologists for hydrocarbon exploration is not going to dwindle anytime soon (Fig. 1).

Fig. 1 Range of different palynomorphs (modified after Hopping 1967)

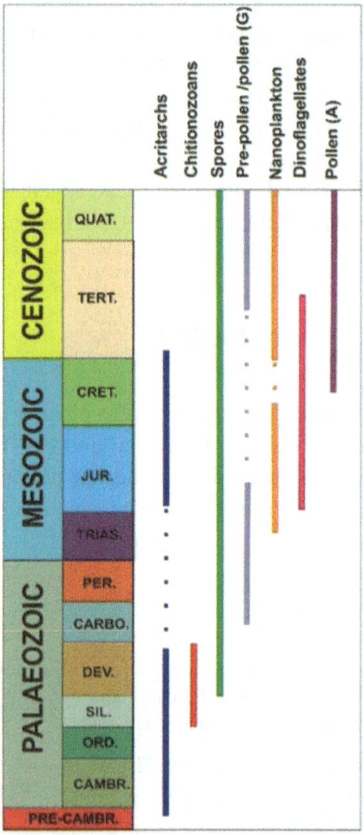

2 Application of Palynology in Hydrocarbon Exploration

As nations are struggling to combat global issues like climate change and overpopulation. The need for an adequate, if not surplus, quantity of energy resources is paramount. Amongst the energy resources currently in use, oil is probably the most popular and widely used. Currently, almost all of the 220 nations in the world use oil, though only roughly 42 countries produce appreciable quantities of the resource (Hall et al. 2003).

In this backdrop, the identification and exploration of oil reserves are of vital importance. The close partnership began on 20th August 1936, when Grimsdale, a paleontologist, working for the Royal Dutch (Shell) Group in Mexico, first proposed the use of fossil pollen as a marker for the correlation of Tertiary deposits in the area (Hopping 1967). Over the next few decades, as a more comprehensive database was generated (especially for the Cenozoic), the versatility and abundance of palynological data were appreciated, and the study of palynomorphs for stratigraphic correlation became popular.

3 Formation of Petroleum

Petroleum, may be used here to describe hydrocarbon resources, which mainly represent a group of compounds grading from gases (methane, ethane, propane)-liquids (oil)-plastic substances (asphalt), and solids (kerogen). The exact method of formation of petroleum is still widely debatable, and there are many theories indicating both organic and inorganic sources for the origin of petroleum (Selley 1998), but the most widely accepted theory is the formation of petroleum from the remains of organic matter (Fig. 2).

The total amount of carbon in the Earth crust has been estimated to be about 2.65×10^{20} g (Hunt 1977) of this, almost 82% of the carbon is in the form of carbonates in limestone and dolomites. And just 18% occurs as organic carbon in the form of coal oil and gas (Schidlowski et al. 1974). The process of formation of hydrocarbon resources starts with the conversion of inorganic carbon present in the atmosphere to organic carbon with the help of plants through photosynthesis (both on land and in the ocean). In this reaction, water and atmospheric carbon dioxide are converted to glucose and water. The animals that consume such plants convert this glucose into more complex carbon compounds. After the death of these organisms, the organic matter generally gets oxidized and destroyed, in certain cases, the organisms are preserved in an altered state as coal, oil or gas. Gas chromatography is used to correlate the source rock from which it has been derived (Bruce and Schmidt 1994) and to determine the type of organism and the depositional environment (Volkman 1988).

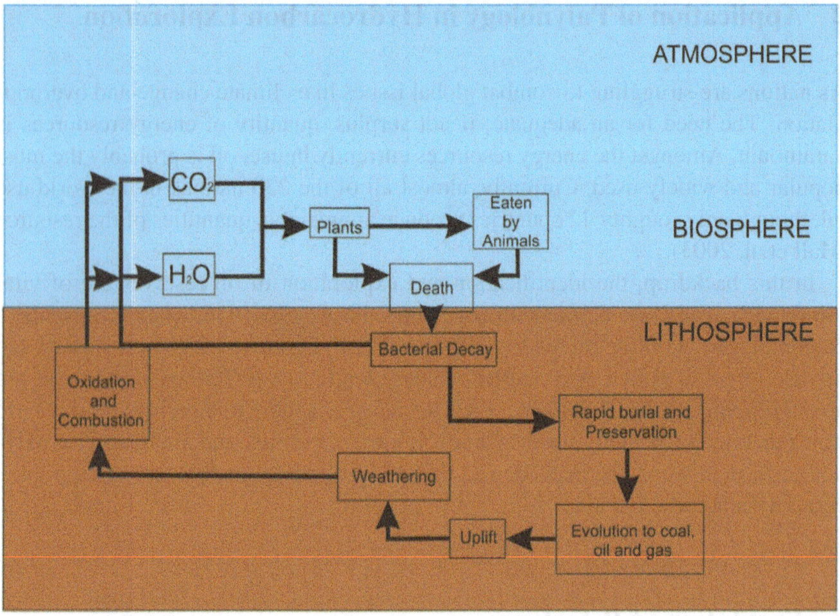

Fig. 2 Carbon cycle (modified after Selley 1998)

For the formation of hydrocarbon resources, certain criterion must be met. Firstly, there should be a significant amount of organic matter that is to be deposited. These areas are generally associated with regions of high biological productivity (both in terrestrial as well as oceanic basins). Secondly, the deposited organic matter should be preserved, this generally occurs when the basin is well stratified, that is there is a strong anoxic zone that prevents the oxidation of the deposited material. Also, there is an ideal temperature, Eh and pH condition that promotes the preservation of organic matter. After the deposition and preservation of the organic matter, comes the transformation of the organic matter into fuel resources. This process of conversion occurs in stages. After burial, the organic matter is subjected to increasing temperature and pressure conditions. There are three major phases that organic matter undergoes for the formation of petroleum, these are diagenesis, catagenesis and metagenesis (Tissot 1977). Diagenesis mainly involves the removal of oxygen from the organic matter, while leaving the hydrogen to carbon ratio unaltered, and forming kerogen. After this, the catagenesis stage takes over. With further increases in temperature and pressure, petroleum is released from the kerogen (first in the form of oil and then as a gas). At this stage, the hydrogen to carbon ratio decreases with no decrease in the oxygen to carbon contents. At very high temperatures and pressures (near the metamorphic barrier), metagenesis takes place. Then the formation of hydrocarbons (generally as methane) takes place by decreasing the hydrogen to carbon ratio untill only pure carbon remains (Selley 1998). During catagenesis, the kerogen matures and gives off oil and gas. Understanding the degree of maturation, of the kerogen in a sedimentary basin is essential for petroleum exploration. If the kerogen is immature no oil is formed. Conversely, when the kerogen is over mature neither oil nor gas remain.

4 Types of Kerogen

Kerogen is referred to as disseminated organic matter in sediments that is insoluble in normal petroleum solvents such as carbon bisulfide. Kerogen generally consists of carbon, hydrogen, and oxygen, with nitrogen and sulfur in minor quantities. Generally, there are three different types of kerogen based on the nature of the original organic matter (Tissot 1977; Dow 1977).

Type I kerogen is generally formed from algal organisms. These organisms have a higher hydrogen to oxygen ratio (1.2–1.7). In this type of kerogen, lipids are the dominant compound, with derivatives of oils, fats and waxes. The most common type of algae that makes up this type of hydrocarbon material is *Botryococcus*.

Type II kerogen, also known as liptinitic kerogen has an intermediate composition. It consists of algal detritus and materials derived from zooplankton and phytoplankton. This type of kerogen is rich in aliphatic compounds, having the hydrogen to carbon ratio greater than one.

Type III, kerogen, which is also known as humic kerogen, has a much lower hydrogen to carbon ratio (<0.87) compared to the other two types. Humic kerogen is low in aliphatic compounds but rich in aromatic compounds. It is predominantly made from lignin, which is present in large quantities in woody materials. Type III kerogen generally tends to form gas, and negligible quantities of oil (Seeley 1998; Bjørlykke 2015) (Fig. 3).

5 Thermal Maturation and Palynology

In hydrocarbon exploration, understanding the thermal histories of possible petroliferous basins is of utmost importance. This is because the degree of maturation of hydrocarbons (gas, liquid or solid) is directly related to the thermal conditions prevailing within the basin. There may be many factors controlling the temperature of a sedimentary basin. These are crustal thickness, geological setting, presence of igneous intrusions, thickness of overlying sedimentary formations, etc. Therefore, it becomes difficult to understand the relative changes in the thermal maturity of the sediments across a basin, (both along horizontal and vertical profile). In this regard, palynology becomes a useful and cost-effective method. The palynomorphs are sensitive to thermal variations and exhibit morphological changes (changes in form, color etc.) when exposed to different thermal ranges. Thus, studying these

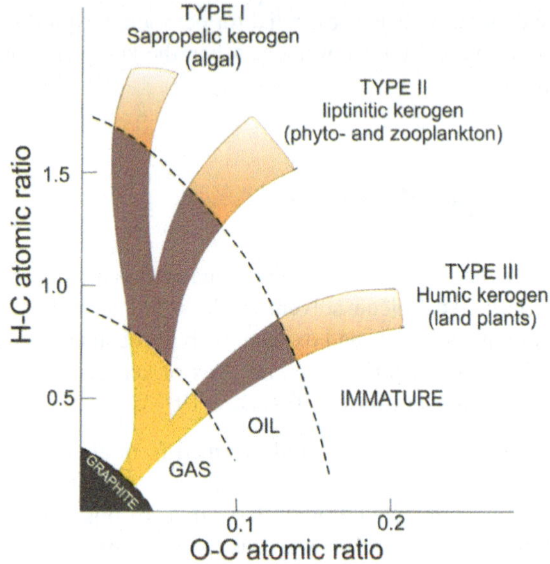

Fig. 3 Types of Kerogen (modified after Selley 1998)

palynomorphs gives us an idea about the temperature conditions of the sediments and the associated organic matter trapped in them. In this regard, Thermal Alteration Indices (Staplin 1969) can indicate the thermal maturity of the organic matter based on the change in their color (Fig. 4).

The change in color of the organic matter is controlled by the thickness of the spore-wall, variations in compositions, and degrees of oxidation/reduction of the wall (El Beialy et al. 2010). The Spore Coloration Index, introduced by Marshall (1990), grades sporomorph wall color from colorless to black. Similarly, the Thermal Alteration Index (TAI) by Olivier and Atud (2018), describes the colour of spores and pollen on a five point scale delineating oil and gas windows (Fig. 5).

6 Palynofacies

Palynomorphs are important indicators of local paleo-vegetation and, in consequence, become essential proxies for paleo-climate and paleo-depositional-environment assessment. In this regard, the study of paleo-palynology and palynofacies have developed over the past few years.

Palynofacies is generally the association and inter-relationship of the palynological matter found in the sediments/rocks which give an idea about the geological and biological conditions required for the formation of the palynological assemblages. Organic material studied in palynofacies analysis includes palynomorphs like spores, pollens, dinoflagellates, acritarchs, etc. and all other organic matter that is observed within the palynomorph size range, which is generally classified as palyno-debris.

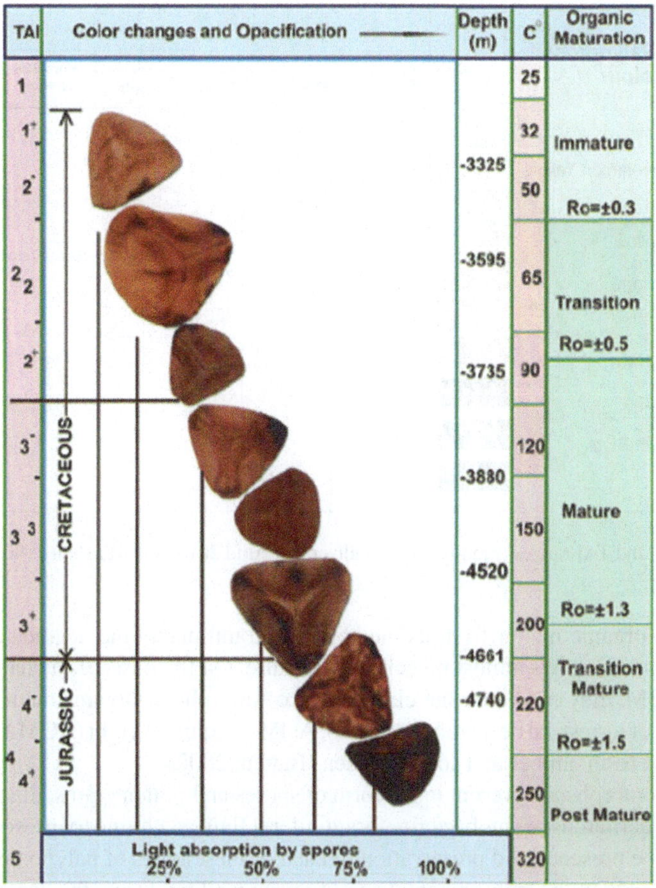

Fig. 4 Color alteration of palynomorphs (after Al-Ameri and Al-Obyadi 2011, published with permission from publisher, Springer)

The association of different types of palynomorphs and palyno-debris can be used as an indicator for the paleo-environmental conditions prevalent during that time.

Due to the diverse nature of palynodebris material, there have been many schemes for classifying it (Table 1). Among them, contributions made by Staplin (1969), Habib (1983), Batten (1983), and Pocock Stanley et al. (1987) are some of the more widely used classifications.

Most commonly, palynodebris may contain:

1. Structured Organic Matter (STOM): It consists of woody plant tissues, cuticles, and other solid detrital material that has retained most of its original structure. They may be phytoclasts and zooclasts, depending on their origin.
2. Unstructured Organic Matter (USTOM): These are generally the most abundant type of organic matter, showing an amorphous like form due to the separation

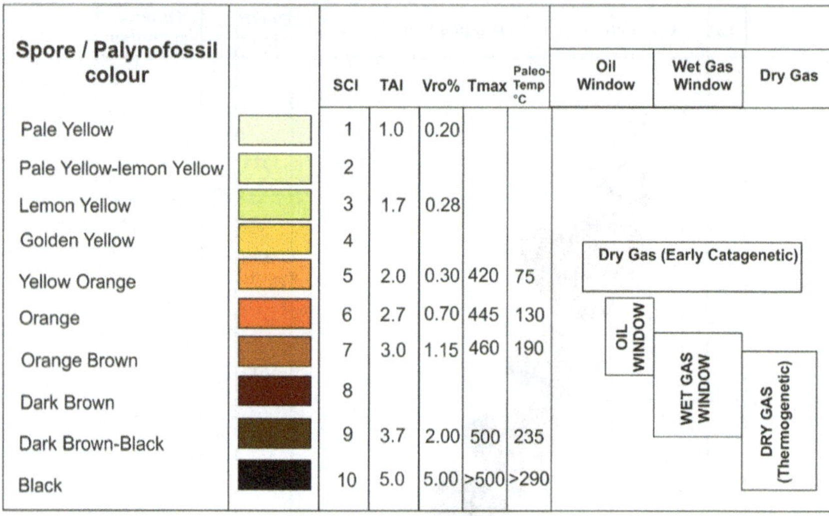

Fig. 5 SCI and TAI ranges (modified from Olliver and Atud 2018; and Marshall 1990)

of the organic matter from its membrane and ultimately the degradation of the material. It shows almost no cellular structure. On the basis of the nature of the USTOM, they can be further classified into Amorphous Organic matter (AOM), AOM of terrestrial derivation (AOMT), AOM of aquatic origin (AOMA), gelified matter, resin, amber and solid bitumen (Batten, 2002)

3. Palynomorphs are present in the form of spores and pollen grains, dinoflagellate cysts, acritarchs, cyanobacteria, foraminiferal linings, chitinozoans, scoleodonts etc. The presence and preservation of each of these types of palynomorphs give an idea about the environmental conditions prevailing during the deposition and preservation of the concerned sediments. (Batten 2002).

7 Application of Palynofacies in Paleoenvironmental Studies

Depending upon the nature of palynodebris, the source rock potential of sediments as well as their organic thermal maturation level can be understood. Palynomorphs, which are an important part of the palynofacies, give an idea about the nature of paleovegetation. Through the relative abundances of terrestrial and marine borne palynoflora, the depositional environment can be identified. Similarly, the thermal maturation of the sediments can be understood by analyzing the organic matter present in the sediments. Roncaglia (2004) established a negative correlation between the oxygen content in the water and the amorphous organic matter present in the sediments. That is, a higher percentage of amorphous organic matter would indicate that the water column in which the sediment was deposited had a relatively lower oxygen content. This is useful in understanding the preservation dynamics of a petroliferous sedimentary basin.

In the case of Indian petroliferous basins, there have been extensive biostratigraphic studies done on the basis of palynomorphs. Some of the important Cenozoic angiosperm pollen marker fossils reported from the major petroliferous basins of India with their respective age ranges are mentioned in Fig. 6a–c.

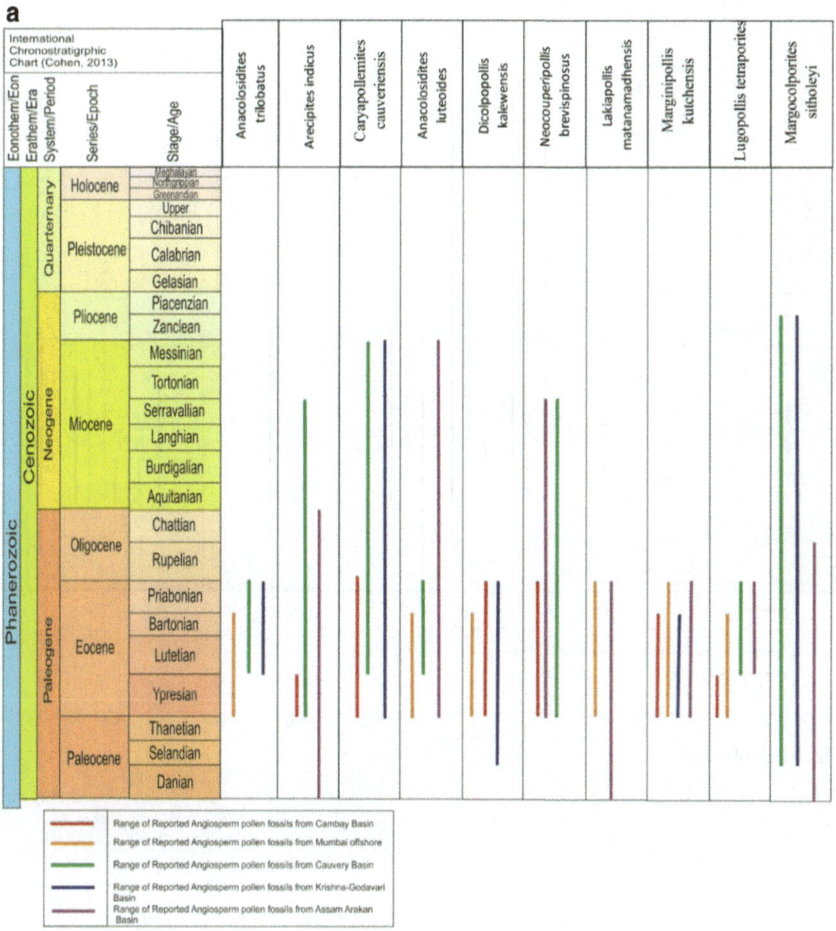

Fig. 6 **a** Important Cenozoic angiosperm pollen marker fossils reported from the major petroliferous basins of India modified after (Mehrotra et al. 2005). **b** Important Cenozoic Angiosperm Pollen marker fossils reported from the major petroliferous basins of India modified after (Mehrotra et al. 2005). **c** Important Cenozoic Angiosperm Pollen marker fossils reported from the major petroliferous basins of India modified after (Mehrotra et al. 2005)

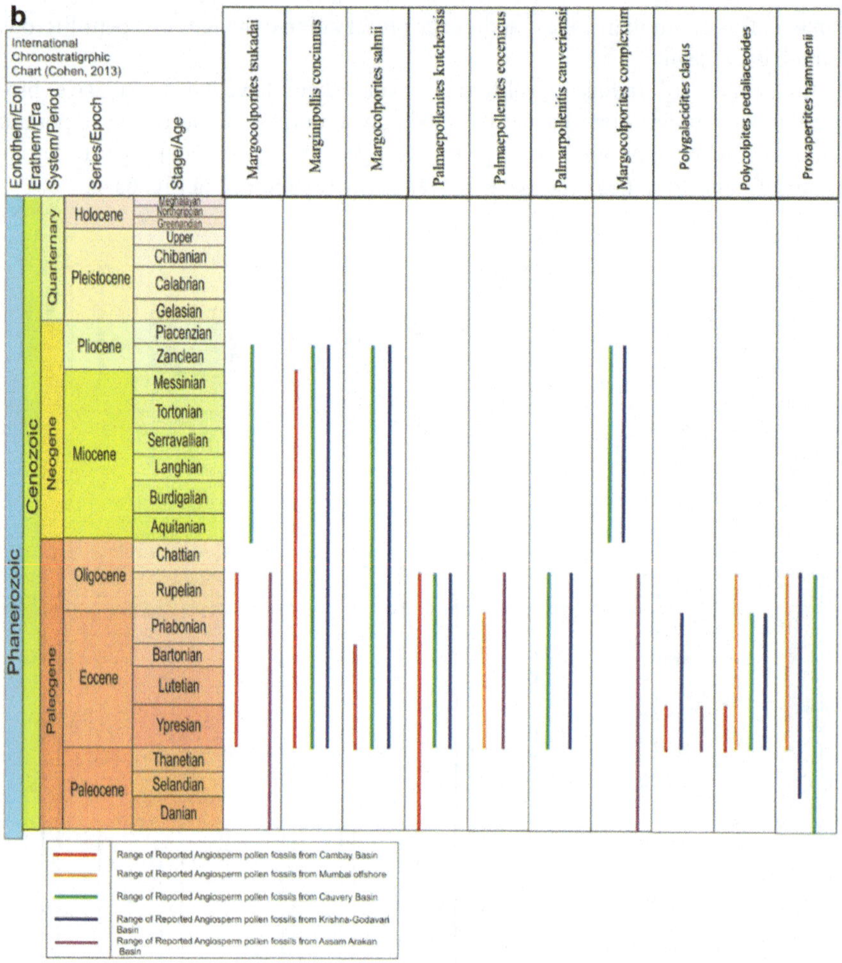

Fig. 6 (continued)

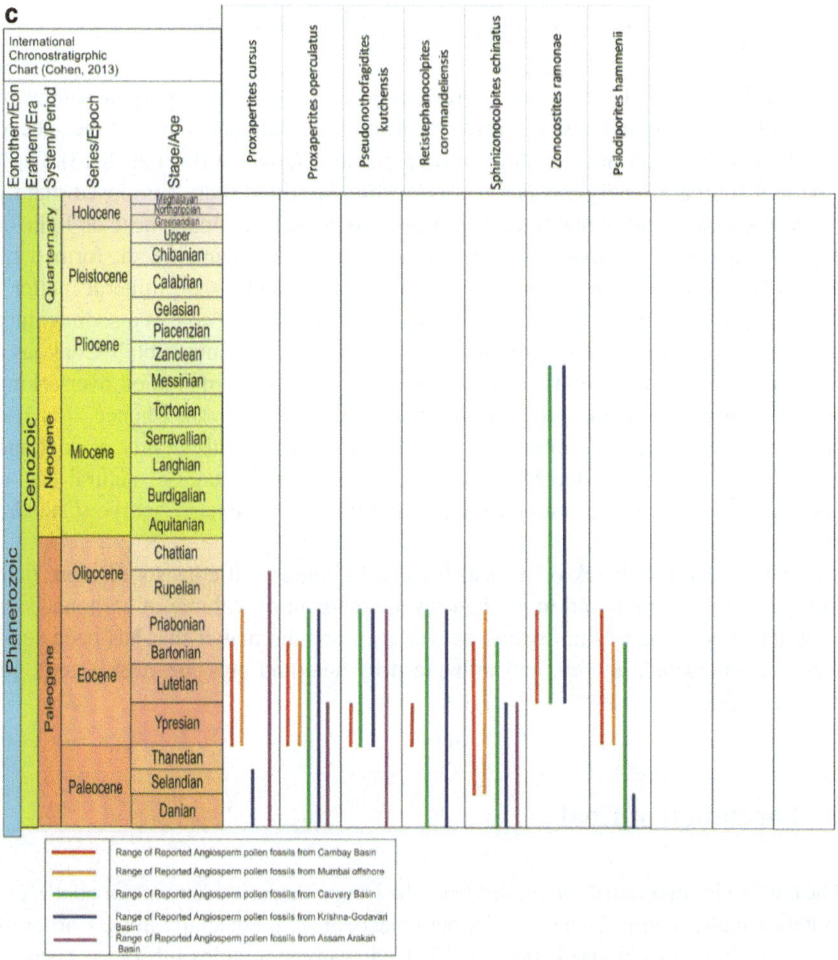

Fig. 6 (continued)

The extensive study of palynomorphs at high temporal and spatial resolution is a testimony to the importance of palynology as a biostratigraphic and paleothermometric tool in petroleum exploration studies. Microscopic studies of palynodebris serve as a cheap, quick and efficient proxy for hydrocarbon estimation. Generally, it is used to complement geochemical data and is applicable to a wide geographical and stratigraphic range.

8 Application of Palynology in Coal Exploration

Coal is the parent of the industrial revolution and the most important source of heat and electricity in the world. It has attracted and continues to attract the special attention of scientists and the entire world population (Jovanovski et al. 2023). Coal, a natural source of energy, is defined as "a solid stratified rock, composed mainly of hydrocarbons, and capable of being used as a fuel to supply heat or light or both" (Arber 1912). Alpern (1990) defined "coal a sedimentary rock formed by macerals and minerals and containing more organic matter than mineral matter". According to the United States Geological Survey (USGS), "coal is a sedimentary deposit composed predominantly of carbon that is readily combustible". Coal has a wide range of chemical and physical properties, and has been studied over a long period of time. This natural treasure has immense economic value; hence, it is also considered as a "black diamond". Today, a new avenue to the vitality of coal and coal deposits, i.e., Coal Bed Methane, an unconventional source of natural gas, is considered an alternative source for exploring the world's energy source (Chandra et al. 2014).

Coal is generally black or brownish-black in colour and consists of more than fifty percent by weight and more than seventy percent by volume of carbonaceous material. It is formed from organic as well as inorganic matter that has been sedimented, compacted, altered, and influenced by heat and pressure over a geologic time (USGS).

9 Formation of Coal

The earliest known coal deposits are from the Devonian period, and major coal accumulation has taken place since the Upper Palaeozoic era, especially in the Carboniferous and Permian periods (Thomas 2013). Coal is a very complex substance, because a vast range of parameters determine its quality. These parameters are (1) the plants, plant remains, and other organic matter in the peat swamp (2) biochemical processes and the degree of preservation of organic remains (3) the location and condition of the swamp (4) the inorganic component which is the mineral compositions which get accumulated with the plant material and (5) coalification (Schweinfurth 2003).

Coal is formed by the accumulation of organic remains in swamps, ponds, deltas, or river floodplains called "peat swamps" or "bogs" in wet or humid climate with, a stable to slowly sinking land surface, and moderate to weak eroding forces such as rivers and ocean waves (Schweinfurth 2003). Another very important factor that controls coalification is that the accumulation must exceed the oxidation or biodegradation of organic matter. Over a period of time, burial and compaction by overlying sediments (derived from rivers and seas) plus the effects of temperature (from the interior of the earth or vicinity to volcanic sources) cause peaty material to transform into a very complicated, hard, chiefly organic, and well stratified sedimentary rock, called coal.

The chief attributes of a coal are its thickness, lateral extent, rank, maceral composition, and quality. Except for the rank, which is controlled by burial and post tectonic history, the other properties are defined by factors governing the mire where the peat originally formed. These factors are: type of mire, type(s) of flora, growth rate, degree of humification, base-level variation, and input rate of clastic sediment (McCabe and Parrish 1992; Thomas 2013).

There are two main theories related to the accumulation of vegetal matter and formation of coal , (a) the 'In situ' theory and (b) the 'Drift' theory. 'In situ' theory explains that marshes and old forests got buried at the place of their origin. A large proportion of coal is generated from the vegetation that has formed peat, close to where its growth has taken place. Such coals are underlain by seat earths or rootlet beds, and are known as autochthonous coals. These types of coal seams are fairly constant in composition over wide areas. The ash content is lower, and detrital material is almost absent. On the other hand, 'Drift' theory considers that vegetable matter is transported from the place of its origin by running water, glaciers and deposited in deep lakes, estuaries and river valleys where it is later covered with sediments. These types of coals are known as allochthonous coals, for example, large rafts of peat or trees drifting on lakes or estuaries. Allochthonous coals do not have an underlying rootlet bed, but rest directly on the bed below. Because of their drifted nature of formation, the coal seams show a varied composition, especially a high amount of detritus material such as sand and clay. Here the ash content is also high (Chandra et al. 2014). The factors that influence the characteristics of coal during the formation of peat include the type of deposition, the peat-forming vegetation, the supply of nutrients, acidity, bacterial activity, temperature, and redox potential.

$$\text{inflow} + \text{precipitation} = \text{outflow} + \text{evapotranspiration} + \text{retention}$$

This equation must be balanced for mire and peat to accumulate. The primary conditions for peat accumulation are therefore, a balance between the production of vegetation and organic decay. Both are a function of climatic factors, vegetation and the decomposition of vegetation, and such decomposition within the peat profile is known as humification. In the peat profile, the upper part is subjected to fluctuations in the water table, and here the humification is most active. The two most important conditions for the preservation of organic matter are rapid burial or anoxic conditions (McCabe and Parrish 1992; Thomas 2013), the latter being present in the waterlogged part of the peat profile. Additionally, an organic-rich system will become anoxic faster than an organic-poor one as the decay process consumes oxygen. This process is influenced by higher temperatures, decay rates get accelerated in hot climates. The acidity of groundwater also determines the rate of humification, i.e., microbial activity is suppressed by high acidity (Thomas 2013).

The transformation of organic matter into coal is a very complex process. It occurs in distinct stages, namely the biochemical and geochemical (dynamo chemical) stages. This process is generally referred as the coalification process (Chandra et al. 2014).

9.1 Biochemical Stage

It is the initial conversion of plant debris into the precursor of coal micro-constituents. This biochemical transformation is controlled by factors like hydrogen-ion (pH) concentration and the redox potential (EH) of the medium. The rapid sedimentation of organic matter prevents atmospheric oxidation and the decomposition and degradation of vegetable matter by microorganisms. The accumulated organic material subsides under its own weight. Here, anoxic conditions prevail with favorable anaerobic bacteria, which take part in decomposing and degrading the vegetable matter at a relatively lower rate (Chandra et al. 2014).

Protoplasm, proteins, and starches and, to a lesser degree, cellulose, decompose faster in anaerobic conditions. Since lignin is highly resistant, it partially decomposes. The waxy and strong protective layers of plants such as, cuticles, spores, pollen and resins are highly resistant to both aerobic and anaerobic bacteria. Therefore, these are mostly preserved unaltered. The quality of coal is mainly dependent on the nature and degree of decomposition and degradation (Chandra et al. 2014). Based on temperature and pressure conditions, wood is converted into coal at various stages in a series:

$$\text{Wood} \rightarrow \text{peat} \rightarrow \text{lignite} \rightarrow \text{bituminous coal} \rightarrow \text{anthracite}$$

To transform vegetable matter into coal, the overlying pressure and temperature factors play a crucial role. As the depth increases, the maturity of the coal seam increases. The time factor also plays a significant role in coalification. The coalification process, takes roughly one million years (Chandra et al. 2014).

9.2 Geochemical Stage

The progression from organic matter into coal is also called the geochemical stage. This results in an increase in the rank or maturity of the coal. Two plant components, namely, cellulose and lignin that mainly build-up the plant structure and make up the entire wood composition, play a significant role in the formation of coal. During coalification there is a decrease in the oxygen percentage, which is due to a loss of water and carbon dioxide. Subsequently, there is increase in carbon content, which is why the process is called carbonization (Chandra et al. 2014). The process of carbonization consists of three sub-processes namely, dehydration, decarboxylation, and demethanation (Jovanovski et al. 2023).

In dehydration, water is released from the maturing coal via the following reactions (Hatcher et al. 1992):

$$2\,R - OH \rightarrow R - O - R + H_2O$$

$$2\,R - CH_2 - O - CH_2 - R \rightarrow R - CH = CH - R + H_2O$$

Decarboxylation takes out carbon dioxide from the maturing coal, which occurs through a reaction:

$$RCOOH \rightarrow RH + CO_2$$

Where as, de-methanation proceeds through the next two reactions (R represents the remainder of a cellulose or lignin molecule to which the reacting groups are attached)

$$2R - CH_3 \rightarrow R - CH_2 - R + CH_4$$

$$R - CH_2 - CH_2 - CH_2 - R \rightarrow R - CH = CH - R + CH_4$$

At an initial stage of coalification, the processes of dehydration and decarboxylation occur, while demethanation comes into the picture much later. At this stage, the coal has already reached the bituminous rank.

The coal samples can be studied as hand specimen, or with the aid of a petrological microscope. The petrographic characterization of coal involves the qualitative and quantitative estimation of macroscopic and microscopic constituents. When observed under the microscope, coal is seen to be composed of particles and layers of different types of carbonaceous material. These distinct entities show the coalified remains of various plant parts or plant derived substances that were present at the initial stage of coalification. On the basis of their physical characteristics, hardness, and microscopic characteristics, they can be differentiated from each other, which can be useful in coal utilization.

The plant parts, consisting of tree trunks, roots, branches, leaves, grasses, algae, spores, and other plant parts, collectively form organic matter. These plant parts and residues get petrified and result in coalification, which consists of a fundamental unit called a maceral. It is a micropetrographic constituent of coal. The inorganic mineral matter, when infused into peat during sedimentation and compaction, permineralized the organic remains present in peat. This helps to retain the original structures present in organic matter; which ultimately can be used as an aid to identify the original plant material present during the process of coalification (Schweinfurth 2003).

10 Coal Macerals

Macerals are classified into three main groups: (1) vitrinite, (2) liptinite, and (3) inertinite (McCabe, 1984). These groups are further divided into various subgroups (Table 2). Coal can be made up largely of one type of maceral or, more usually, associations of macerals. These associations, when studied under the microscope, are called microlithotypes. The maceral-mineral association at the microloithtype scale is known as "Carbominerites". Generally, the macerals of one particular group share similarities in their origin, mode of conservation, colour, and reflectance, but have differences in their morphology at maceral and submaceral levels. In addition, the macerals of a particular group show a similar nature in their chemical composition, such as vitrinite rich in oxygen, liptinite in hydrogen and inertinite in carbon (Chandra et al. 2014) (Fig. 7).

The maceral derives its types from different parts of vegetable material. These different types of macerals, have distinct characterization based on the specific plant parts. For example, roots, bark, stems, and other plant tissues result in a vitrinite maceral whereas spores and leaves give rise to liptinites, and fungal matter converts to inertinite. The study of respective maceral types to identify their original organic matter helps to estimate the coal's quality. The reflectance varies according to the rank of the coal. These interlinks are very complex because the original vegetable material has been converted into a petrified form beyond recognition.

10.1 Vitrinite

This group is also known as the huminite group (Thomas 2013). It is gray to white in colour and its reflectance lies between the macerals of the liptinite and inertinite

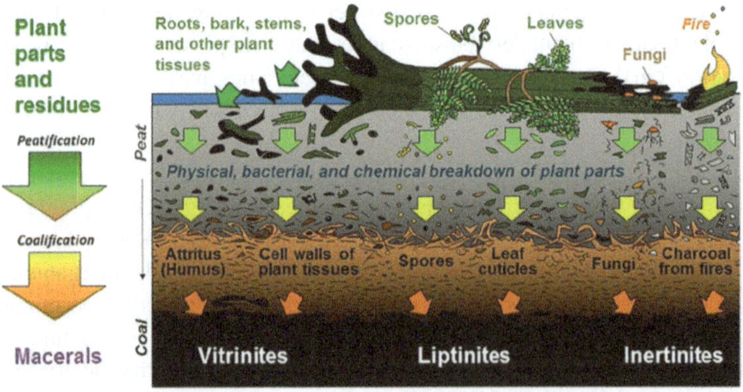

Fig. 7 Transformation of plant and fungal tissue and burned material from peat into different macerals (After Jovanovski et al. 2023; published with permission from publisher, Springer)

groups. The macerals of this group are further named as telinite, collotelinite, gelinite, corpogelinite, vitrodetrinite and collodetrinite. The classification of the macerals of vitrinite is after I.C.C.P-1994. These macerals are differentiated on the basis of the presence or absence of structures of organic matter and their size (Chandra et al. 2014).

The vitrinite maceral group is more common than other maceral group and results from the coalification of plant materials like woody tissue, bark, leaves etc. It is sensitive to temperature and becomes tougher, more compact and vitreous as it is subjected to higher temperatures, either with the effect of subsurface depth, or by proximity to an external heat source, like a volcano. The vitrinite reflectance index is very useful in determining the level of heat and maturity to which coal and other organic matter, such as petroleum, have been subjected (Schweinfurth 2003).

Vitrinite Reflectance

It is a measure of the percentage of incident light reflected from the surface of polished vitrinite particles in sedimentary rocks. It is a standard method for calculating the relative amount of coalification (i.e., rank and maturation of coal) that a coal (or other organic substance) has undergone.

10.2 Exinite

The Exinite maceral group results from waxy or oily plant material such as spores, algae, and resin (Schweinfurth 2003); this group is subdivided into several subgroups based on the original plant material—i.e., sporinite, alginite and resinite. Another name for this group is Liptinite. Macerals from this group bear a distinct morphology with a dark gray to black colour in low grade coal. Exinite macerals are more enriched in hydrogen than either vitrinite or fusinite. Exinite -rich coals generate larger amounts and higher grades of liquid fuel when subjected to destructive distillation than vitrinite- or fusinite-rich coals (Schweinfurth 2003).

This group is further classified into two types, namely, primary and secondary. Distinct morphology makes the primary liptinite macerals different from the secondary ones; the former are derived from the hydrogen rich constituents of plants such as spores, pollens, resins, cuticles, algae, suberin, and others. The secondary liptinite macerals result from coalification hence, they are considered as an expelled hydrocarbon. These can be seen in the cleats, fissures, and cavities of the vitrinite and inertinite groups of macerals (Chandra et al. 2014). The Liptinite group of macerals might not be easily recognized, because of their high maturity and lack of fluorescence. The sporinite macerals consists of exines of pollen and spores. The cutinite maceral has a distinguished, long thread like structure with a saw-toothed end at one end. The colour of cutinite ranges from black to dark gray but it is lighter than sporinite when observed from the same coal sample. Resinite macerals are generally found as a cell filling and as oval, circular or rod shaped, secluded bodies of varying sizes. Alginite maceral looks darker than all the other macerals in this group. Suberinite maceral results from the corkified cell wall of bark tissues, roots, stems, and fruits. Leptodetrinite maceral is composed of the detritus of sporinite, cutinite, resinite, alginate, and suberinite maceral (Chandra et al. 2014).

10.3 Inertinite

This group of macerals is generally referred to as "inert" because they do not react to heat during the entire process of petrification and carbonisation (Chandra et al. 2014). Inertinite comprises a group of common macerals that are produced from partially oxidized, or burned plant cell walls. Fusinite is the main representative of this group. Sometimes the upper layers of a peat swamp dry out, and some surface peat either slowly oxidizes or catches fire, or some plants get converted into charcoal, and then fusinite forms. Generally, coal has a maceral composition mostly ranging from vitrinite to fusinite, depending on the original organic matter and the level of preservation (Schweinfurth 2003). Fusinite maceral shows good preservation of a cell structure of the wood, or sclerenchyma. Semifusinite maceral exhibits woody or parenchymatous cell structures, which are relatively rarer than fusinite maceral. Macrinite, on the other hand, is a non-granular groundmass that is amorphous in nature. The Secretinite maceral is an oxidation product of resins (Chandra et al. 2014). Vitrinite dominated coal is vitreous black, clean, fractured conchoidally, and often shows iridescence (called "peacock coal") on fresh surfaces. Fusinite dominated coal is similar to charcoal and is dull, black, friable, and dusty (Schweinfurth 2003 USGS) (Figs. 8 and 9).

11 Palaeoclimate of Gondwana Sediments Using Palynology

Floral remains in the sediments can be treated as an excellent tool to study paleoclimate. As plants cannot move from one place to another in extreme climatic conditions; they simply respond to this either by adapting new morphological characters or perishing due to extremities. Hence, palynomorphs can be considered as good paleoclimate proxies. Palynology has been extensively used to study the paleoclimate of Gondwana sediments.

Lower Gondwana: Anil and Shaila Chandra (1988), on the basis of plant fossils, suggested that during Lower Gondwana, the climate gradually transformed from extremely cold in the Talchir stage to moderately cold in the Karhabari, Barakar, Barren Measure (Kulti), and Raniganj stages. As the temperature moderated, there

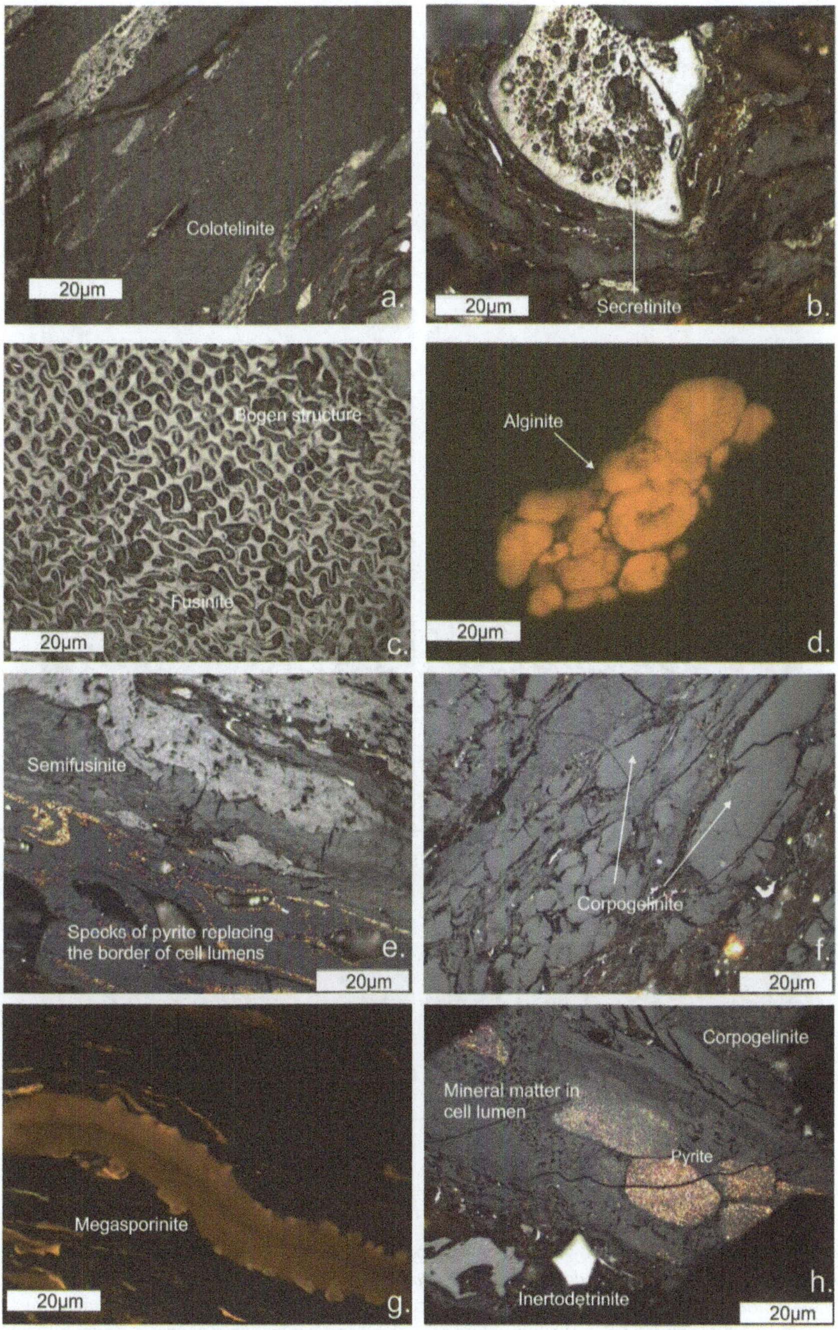

Fig. 8 Photomicrographs showing different types of macerals. **a** Colotelinite; **b** Secretinite (*white arrow*); **c** Bogen structure; **d** Alginite (*white arrow*); **e** Semifusinite as well as Specks of pyrite replacing the border of cell lumens; **f** Corpogelinite (*white arrow*); **g** Megasporinite; **h** Inertodetrinite as well as Corpogelinite. (Photographs, courtesy Prof. A.S. Naik, Banaras Hindu University, Varanasi, India)

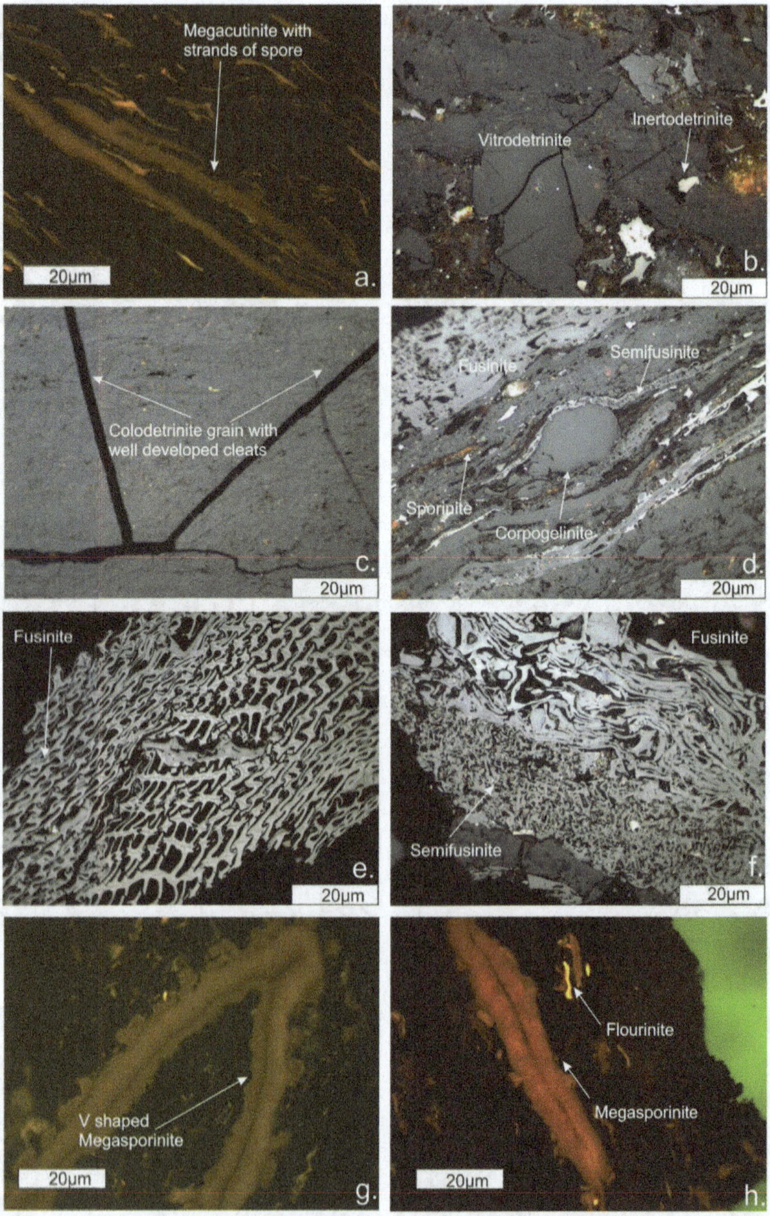

Fig. 9 Photomicrographs showing different types of macerals. **a** Megacutinite with strands of spore; **b** Vitrodetrinite, Inertodetrinite (*white arrow*); **c** Colodetrinite grains with well developed cleats (*white arrow*); **d** Sporinite, Fusinite, Corpogelinite, Semifusinite (*all marked by white arrows*); **e** Fusinite (*white arrow*); **f** Semifusinit (*white arrow*), Fusinite; **g** V shaped Megasporinite; **h** Flourinite and Megasporinite (*marked with white arrows*). (Photographs, courtesy Prof. A. S. Naik, Banaras Hindu University, Varanasi, India)

was an increase in humidity and rainfall, which caused the luxuriant growth of pteridophytic and gymnospermous plants. Afterwards, there was a dry and semi-arid climate, which is indicated by the decline in the flora during the Kamthi Formation.

Upper Gondwana: It is dominated by *Ptilophyllum* flora (Lele 1967; Shah et al. 1971), consisting of a rich assemblage of Equisetales, Filicales, Pteridosperms, cycadophytes, Coniferales and Ginkgoales. The large number of cycadophytes and the presence of depterid ferns i.e., Marattiaceae and Osmundaceae in the floral community represent a hot, moist subtropical to tropical climate. Contrary to this, the prevalence of conifers, ginkgophytes and few ferns like Schizaeaceae, Gleicheniaceae and polypodiaceae indicate a temperate climate. Flora shows a gradual change in climatic conditions during Upper Gondwana (Lele 1976).

In addition to megaflora, the morphological characteristics of the palynomorphs are also good indicators of climatic conditions. Tiwari and Tripathi (1987) suggested the temperature and humidity conditions of the Gondawana sediments based on pollen morphology (Table 3).

Table 1 Classification of palynodebris (modified after Masran and Pocock 1981)

Classification of palynodebris		
Category	Sub category	
Structured terrestrial	Leaf Root Stem	
Spores and pollen	Monolete and trilete spores Gymnospermous pollen Angiospermous pollen	
Biodegraded terrestrial	Stem and root Sheets of cells with walls Sheets of cells lacking walls Semi-fusinite (medium Eh)	
Inertinite	Semi-fusinite (high Eh)	
	Organic pyrite	
	Fusinite	Degradofusinite
		Pyrofusinite
Fungi	Mycelia	
	Single celled spores	
Resins		
Amorphous	Yellow amber (medium Eh)	Granular-bacterial Structure-less Spherulitic

(continued)

Table 1 (continued)

Classification of palynodebris		
	Grey amber (low Eh)	Granular-bacterial Structure-less Spherulitic
Structured aqueous	Algal	Colonial Unicellular
	Dinoflagellates	
	Acritarchs	
	Others	

Table 2 Macerals and group macerals recognized in hard coals (after McCabe, 1984)

Maceral group	Maceral	Morphology	Origin
Vitrinite (huminite)	Tellinite	Cellular structure	Cells walls of trunks, branches, roots, leaves
	Collinite	Structure less	Reprecipitation of dissolved organic matter in a gel form
	Vitrodetrinite	Fragments of vitrinite	Very early degradation of plant and humic peat particles
Exinite (liptinite)	Sporinite	Fossil form	Mega-microspores
	Cutinite	Bands which may have appendages	Cuticles—the outer layer of leaves, shoots and thin stems
	Resinite	Cell filling layers or dispersed	Plant resins, waxes and other secretions
	Alginite	Fossil form	Algae
	Liptodetrinite	Fragments of Exinite	Degradation residues
Inertinite	Fusinite	Empty or mineral filled cellular structure; cell structure usually well preserved	Oxidized plant material—mostly charcoal from burning of vegetation
	Semifusinite	Cellular structure	Partly oxidized plant material
	Macrinite	Amorphous 'cement'	Oxidized gel material
	Inertodetrinite	Small patches of fusinite, semi-fusinite or macrinite	Redeposited inertinites
	Micrinite	Granular, rounded Grains ~1 μm in diameter	Degradation of macerals during coalification
	Sclerotinite	Fossil form	Mainly fungal remains

Table 3 Interpreting temperature and humidity from Gondwana flora using morphological characters (After Tiwari and Tripathi 1987)

Sr. No	Morphological characters	Temperature	Humidity
1	Thin central body, diploxylonoid, leathery, saccus, taeniae	Warm/ moderately cool	Medium to low
2	Thin central body, haploxylonoid, non-leathery saccus, striations, monosaccate	Warm	Very high
3	Dense central body, diploxylonoid, leathery saccus, striations	Warm	Medium
4	Thin central body, haploxylonoid, non-leathery, saccus	Cool	High

12 Age and Coal Seam Demarcation Using Palynology

Palynology can be applied to solve geologic or stratigraphic problems that involve the delineation of particular strata or stratigraphic columns using marker palynomorphs. These marker palynomorphs can also be useful in coal seam demarcation. These marker palynomorphs can be used to demarcate the coal seam strata. Fourteen Permian spore-pollen assemblage zones have been proposed in the Gondwana sediments of peninsular India (Jha 2006). The ages of these palynozones primarily depend on their stratigraphic position and the invertebrate fauna recorded from the intervening isolated marine beds in central India (Prasad et al. 1995).

Smith and Butterworth (1967) suggested the different phases of seam splitting and their use to link seams to their different splits. In coal seams, each palynoassemblage, as a combined entity, tends to show the combined vegetation of different parts of the basin, whereas individual components of the palynoassemblage may differ both quantitatively and qualitatively, depending upon the sedimentary processes and local concentrations of the parent vegetation (Richardson and McGregor 1986).

Correlation is carried out by sampling from either a channel cut through the entire seam or separate points representing particular parts of the seam. Channel sampling is preferable, as this method can be applied coherently at outcrop, to a core, and most importantly, to drill cuttings. Both qualitative and statistical methods of correlation rely on two assumptions: (1) Palynomorphs were distributed uniformly across a coal mire at the time of accumulation such that, at any locality, a statistical average of all types of miospores would be found. (2) There is sufficient difference in the composition of the flora of successive coal mires to produce miospore assemblages that are recognizably different (Raistrick 1937; McLean and Murray 1996).

If it is assumed that each coal seam consists of a distinct palynomorph assemblage from the same geographic location, the question of correlation then involves essentially evaluating sampling error in the statistical sense (Gray and Guennel 1961). Tomlinson (1957) showed that such errors can be minimized, if a sample of an assemblage is documented. Common practice is to carry out correlation based on the abundance of samples identified at a particular level (Smith and Butterworth 1967), or of genera and species combined (Guennel 1958).

13 Coal Bed Methane (CBM) and Palynology

The methods for subsurface correlation of coal bearing strata used for hydrocarbon exploration can also be applied for Coal Bed Methane, because coal seams are rich in fossiliferous sediments hence, biostratigraphic correlation of reservoirs can be done directly. Mapping and correlation using the palynology of inter-seam sediments with the help of seam correlation is helpful in exploring methane reserves. The coal seams, which are affected by structural features like faults and washouts, can be correlated using palynology. This type of correlation gives an outline for coal thickness and helps to estimate CBM reserves (McLean and Murray 1996).

14 Summary

Presently, our society is gradually moving towards renewable sources of energy. However, for these sources to contribute significantly to global energy demand, a lot of research and development are still required. For the foreseeable future, it is fossil fuels (petroleum and coal) that will be the backbone for the development of any nation. In this context, palynology plays a significant role in the fields of petroleum and coal exploration. As plants cannot relocate themselves in search of food or in extreme climatic conditions, they can only respond to this by adapting to new conditions or they die out. This peculiar nature of plants makes them very special for deciphering the paleoclimate from paleosediments.

The palynomorphs, are sensitive to thermal variations and exhibit morphological changes (changes in form, color etc.) when exposed to different temperature ranges. Thus, studying these palynomorphs gives us an idea about the temperature conditions of the sediments and the associated organic matter trapped in them. Coal, similarly, can be studied by its organic entity called macerals, which are left over components of plants during coalification. Macerals are a good indicator of coal rank, its maturity, and subsequently its quality. Palynology helps to identify and study these macerals, which provide a good foundation for coal exploration.

Palynology, because of its quick and cost-effective nature, is an important tool in the geological exploration of most fossil fuels. Recent developments in palynofacies and palynodebris analysis have shown how intricately organic matter in sediments plays a role in the formation of kerogen and ultimately oil/gas production. With newer techniques constantly developing, it will be interesting to see what the future holds for palynology.

Acknowledgements BS and AS are thankful to the Science and Engineering Research Board, New Delhi, (No. CRG/2020/001339) and NT is thankful to UGC, NET- JRF (No. 411/(CSIR-UGC NET JUNE 2019) for financial assistance. We are also thankful to the Head, Postgraduate Department of Geology, RTM Nagpur University for working facility. We thank to research student Mr. Anup Dhobale, for helping with the figures of this chapter.

References

Al-Ameri TK, Al-Obaydi RY (2011) Cretaceous petroleum system of the Khasib and Tannuma oil reservoir. East Baghdad oil field, Iraq. Arab J Geosci 4(5):915–932

Alpern B (1990) Letter to I.C.C.P. on the definition of coal. International Committee for Coal and Organic Petrology News, No. 3

Arber EAN (1912) The natural history of coal. Cambridge University Press. 103

Batten DJ (1983) Identification of amorphous sedimentary organic matter by transmitted light microscopy. Geol Soc Lond Spec Publ 12(1):275–287

Batten DJ (2002) Palynofacies. In: Jansonius J, McGregor DC (ed) Palynology principles and applications, vol 1. American Association of Stratigraphic Palynologists Foundation, pp 1011–1065

Bjørlykke K (2015) Introduction to petroleum geology. In: Bjørlykke K (ed) Petroleum geoscience. Springer, Berlin, Heidelberg, pp 1–29

Bruce LG, Schmidt GW (1994) Hydrocarbon fingerprinting for application in forensic geology: review with case studies. AAPG Bull 78(11):1692–1710

Chandra D, Singh RM, Singh MP (2014) Text book of coal. Varanasi Tara book agency, India

Chandra S, Chandra A (1988) Vegetational changes and their climatic implications in coal-bearing Gondwana. J Paleosci Paleobot 36:74–86

Dow WG (1977) Kerogen studies and geological interpretation. J Geochem Explor 7:79–100

El Beialy SY, El Atfy HS, Zavada MS et al (2010) Palynological, palynofacies, paleoenvironmental and organic geochemical studies on the Upper Cretaceous succession of the GPTSW-7 well, North Western Desert, Egypt. Mar Pet Geol 27(2):370–385

Gray HH, Guennel GK (1961) Elementary statistics applied to palynologic identification of coal beds. Micropaleontology 7:101–106

Guennel GK (1958) Miospore analysis of the Pottsville coals, vol 13. Department of Conservation and Geological Survey of Indiana Bulletin, pp 1–101

Habib D (1983) Sedimentation rate dependent distribution of organic matter in the North Atlantic Jurassic Cretaceous. In: Sheridan RE, Gradstein FM, et al (eds) Initial reports of deep-sea drilling project, vol 76. pp 781–792

Halbritter H, Hesse M (2004) Principal modes of in foldings in tricolp(or)ate Angiosperm pollen. Grana 43:1–14

Hall C. Tharakan, P, Hallock J, Cleveland C, Jefferson, M (2003) Hydrocarbons and the evolution of human culture. Nature 426(6964):318–322

Hatcher PG, Faulon JL, Wenzel KA, Cody GD (1992) A structural model for lignin-derived vitrinite from high-volatile bituminous coal (coalified wood). Energy Fuels 6:813–820

Hopping CA (1967) Palynology and the oil industry. Rev Palaeobot Palynol 2(1–4):23–48. https://doi.org/10.1016/0034-6667(67)90136-4

Hunt JM (1977) Distribution of carbon as hydrocarbons and asphaltic compounds in sedimentary rocks: geologic notes. AAPG Bull 61(1):100–104

Jha N (2006) Permian palynology from India and Africa—a phytogeographical paradigm. Palaeontol Soc India 51(1):43–55

Jovanovski G, Boev B, Makreski P (2023) Chemistry and geology of coal: nature, composition, coking, gasification, liquefaction, production of chemicals, formation, peatification, coalification, coal types, and ranks. Springer Nature, Switzerland, pp 1–16

Lele KM (1976) Paleoclimatic implications of Gondwana Floras. Geophytology 207–229

Marshall AE, Smith AHV (1965) Assemblages of miospores from some Upper Carboniferous coals and their associated sediments in the Yorkshire coalfield. Palaeontology 7:656–673

Marshall JEA (1990) Determination of thermal maturity. In: Briggs DEG, Crowther P (eds) Palaeobiology—a synthesis. Blackwell Scientific Publications, Oxford, UK, pp 511–515

Masran THC, Pocock SAJ (1981) The classification of plant derived organic matter in sedimentary rocks. In: Brooks J (ed) Organic maturation studies and fossil fuel exploration, vol 76. Academic press, London, pp 145–175

McCabe PJ (1984) Depositional environments of coal and coal-bearing strata. In: Rahmani RA, Flores RM (eds) Special publication sedimentology of coal and coal-bearing sequences. International Association of Sedimentologists, 7th edn. Blackwell Scientific Publications, Oxford, pp 13–42

McCabe PJ, Parrish JT (1992) Tectonic and climatic controls on the distribution and quality of cretaceous coals, in controls on the distribution and quality of cretaceous coals. Geol Soc Am Spec Pap 267:1–15

Mclean D, Murray I (1996) Subsurface correlation of Carboniferous coal seams and inter-seam sediments using palynology: application to exploration for coalbed methane. In: Gayer R, Harris I (eds) Coalbed methane and coal geology. Geological society special Publication No 109, pp 315–324

Mehrotra NC, Venkatachala BS, Kapoor PN (2005) Palynology in hydrocarbon exploration: the Indian Scenario. The Geological Society of India. Memoir 61:61–99

Njoh Olivier A, Atud T (2018) Spores and pollen colour changes and thermal maturation of mid-cretaceous lacustrine organic-rich sediments in the Mamfe basin, SW Cameroon. J Geogr Geol 10(1)

Pocock Stanley AJ, Vasanthy G, Venkatachala BS (1987) Introduction to the study of particulate organic materials and ecological perspectives. J Palynol 23–24:167–188

Potonié R (1934) I. Zur Morphologie der fossilen Pollen und Sporen. Arb Inst Paläobotanik Petrographie Brennsteine 4:5–24

Potonié R (1956) Synopsis der Gattungen der Sporae dispersae, I. Teil: Sporites. Beih Geol Jahrb 23:1–103

Prasad B, Jain AK, Mathur YK (1995) A standard palynozonation scheme for the Cretaceous and pre-Cretaceous subsurface sediments of Krishna-Godavari Basin, India. Geosci J 16(2):155–233

Raistrick A (1937) The correlation of coal-seams by microspore content. Colliery Enginners 14:299–302

Richardson JB, McGregor DC (1986) Silurian and Devonian spore zones of the old red sandstone continent and adjacent regions. Bull Geol Surv Can 364:1–79

Roncaglia L (2004) Palynofacies analysis and organic-walled dinoflagellate cysts as indicators of palaeo-hydrographic changes: an example from Holocene sediments in Skálafjord, Faroe Islands. Marine Micropal 50:21–42

Schopf JM (1957) Spores and related plant microfossils – Paleozoic. In: Ladd HS (ed) Treatise on marine ecology and paleoecology, vol 2, Paleoecology: Geological Scociety of America Memoir 67/2, pp 703–707

Schopf JM (1964) Practical problems and principles in study of plant microfossils. In: Cross, AT (ed) Palynology in oil exploration – A symposium: Society of Economic Paleontologists and Mineralogists Special Publication 11, p. 29–57

Schidlowski M, Eichmann R, Junge CE (1974) Evolution des iridichen Sauerstof-Budgets und Entwicklung de Erdatmosphare. Umschau 22:703–707

Schweinfurth SP (2003) Coal—a complex natural resource, An overview of factors affecting coal quality and use in the United States, (With a contribution on coal quality and public health by Robert B. Finkelman) U.S. Geological Survey Circular 1143 U.S

Selley RC (1998) Elements of petroleum geology. Department of Geology, Imperial College, London United Kingdom, pp 37–145
Shah SC, Singh G, Gururaja MN (1971) Observation on the post-triassic gondwana sequence of India. Paleobotanist 20(1–3):221–237
Smith AHV, Butterworth MA (1967) Miospores in the coal seams of the carboniferous of Great Britain. Spec Pap Palaeontol 1:1–324
Staplin FL (1969) Sedimentary organic matter, organic metamorphism, and oil and gas occurrence. Bull Can Pet Geol 17(1):47–66
Thomas L (2013) Coal geology. 2nd edn. Dargo Associates Ltd, Wiley Publication, Wiley-Blackwell
Tiwari RS, Tripathi A (1987) Palynological zones and their climatic inferences in coal bearing Gondwana of Peninsular India. Paleobotanist 36:87–101
Tissot BP (1977) The application of the results of organic chemical studies in oil and gas exploration. In: Hobson GD (ed) Dev Petrol Geol 1:53–82
Tomlinson RC (1957) Coal Measures miocrospore analyses; a statistical investigation into sampling procedures and some other factors. Geol Surv G B Bull 12:18–26
USGS. What is Coal. https://www.usgs.gov/faqs/what-coal. Accessed 08 April 2023 https://epcamr.org/home/content/reference-materials/coal-typesformation-and-methods-of-mining/
Volkman JK (1988) Biological marker compounds as indicators of the depositional environments of petroleum source rocks. Geol Soc Lond Spec Publ 40(1):103–122

[illegible]

The manufacturer's authorised representative in the EU is Springer Nature Customer Service Centre GmbH, Europaplatz 3, 69115 Heidelberg, Germany. If you have any concerns regarding our products, please contact ProductSafety@springernature.com

Printed and bound by CPI Group (UK) Ltd, Croydon, CR0 4YY
28/09/2026
02224086-0012